UPDATES IN VOLCANOLOGY - NEW ADVANCES IN UNDERSTANDING VOLCANIC SYSTEMS

Edited by **Károly Németh**

Updates in Volcanology - New Advances in Understanding Volcanic Systems

http://dx.doi.org/10.5772/3390
Edited by Károly Németh

Contributors

Umberta Tinivella, Michela Giustiniani, Inna Derbeko, Karoly Nemeth, Gabor Kereszturi, Koji Umeda, Jiaqi Liu

Published by InTech

Janeza Trdine 9, 51000 Rijeka, Croatia

Publishing Process Manager Iva Lipovic
Typesetting InTech DTP Team
Cover InTech Design Team

Additional hard copies can be obtained from orders@intechopen.com

Updates in Volcanology - New Advances in Understanding Volcanic Systems, Edited by Károly Németh
p. cm.
ISBN 978-953-51-0915-0

For EU product safety concerns:
IN TECH d.o.o., Prolaz Marije Krucifikse Kozulić 3, 51000 Rijeka, Croatia,
info@intechopen.com or visit our website at intechopen.com.

Meet the editor

Dr Károly Németh works as a Senior Research Officer at the Massey University, New Zealand. In 2012/2013 he has been on research leave as an Associate Professor at the King Abdulaziz University in Jeddah, Kingdom of Saudi Arabia. His research focuses on monogenetic volcanoes, volcanic field evolution and phreatomagmatism. He has worked in Argentina, Chile, Hungary, Idaho, Japan, Korea, Libya, Mexico, New Zealand, Samoa, Saudi Arabia, Slovakia, Solomon Islands and Vanuatu. He is currently working on to understand the eruption styles and hazards of the monogenetic volcanoes of Auckland in New Zealand. Dr. Németh has also spent time working with local communities in the Pacific to understand their view on volcanic hazards. He is currently an Editorial Board Member of the Journal of Volcanology and Geothermal Research, Bulletin of Volcanology, Central European Journal of Geosciences and The Open Geology Journal.

Contents

Preface IX

Section 1 From Small to Large Volume Volcanoes 1

Chapter 1 **Monogenetic Basaltic Volcanoes: Genetic Classification, Growth, Geomorphology and Degradation 3**
Gábor Kereszturi and Károly Németh

Chapter 2 **The Role of the Andesitic Volcanism in the Understanding of Late Mesozoic Tectonic Events of Bureya-Jziamysi Superterrain, Russian Far East 91**
I.M. Derbeko

Section 2 Volcanic Hazards 117

Chapter 3 **Application of the Bayesian Approach to Incorporate Helium Isotope Ratios in Long-Term Probabilistic Volcanic Hazard Assessments in Tohoku, Japan 119**
Andrew James Martin, Koji Umeda and Tsuneari Ishimaru

Section 3 Volcanic Resources and Geoconservation 149

Chapter 4 **Volcanic Rock-Hosted Natural Hydrocarbon Resources: A Review 151**
Jiaqi Liu, Pujun Wang, Yan Zhang, Weihua Bian, Yulong Huang, Huafeng Tang and Xiaoyu Chen

Chapter 5 **Volcanic Natural Resources and Volcanic Landscape Protection: An Overview 181**
Jiaqi Liu, Jiali Liu, Xiaoyu Chen and Wenfeng Guo

Section 4 Non-Magmatic Volcanism 223

Chapter 6 **An Overview of Mud Volcanoes Associated to Gas Hydrate System 225**
Umberta Tinivella and Michela Giustiniani

Preface

This new book is a successor of the first volcanology book published by InTech Open Access Publishing in 2011, titled "Updates in Volcanology – A comprehensive Approach to the Volcanological Problems" edited by Professor Francesco Stoppa.

On the basis of the success of the first book, a new initiative generated this book we now offer to the reader. Volcanology is a dynamically developing science, reflecting the need to improve our understanding of volcanic systems in order to develop better strategies to cope with the volcanic hazards the growing population of the Earth faces at present and in the future. The natural curiosity to understand the driving forces that govern the diversity of volcanism on Earth and other planets is the basis of many recent developments in volcanic research. These include understanding the nature and variety of explosive volcanism, volcanic conduit dynamics, magma ascent, physical properties of various laterally moving pyroclastic gravity currents and the interaction between internal versus external driving forces determining the nature and style of volcanic eruptions. In addition, a significant development can be seen in the understanding of the interplay between volcanism and the evolution of sedimentary basins, as well as in identifying the basic rules of volcano instability, collapse and erosion on both short and long term time scales. These developments in volcanology resulted in a huge surge of scientific research outputs in peer-reviewed journals and books presented by great variety of publishers. InTech Open Access Publishing recently offered its publication avenues to make public the quickly advancing research in the field of volcanology.

With "Updates in Volcanology – New Advances in Understanding Volcanic Systems", InTech Open Access Publishing again offers high quality research output options for leading experts in volcanology, and hopefully will keep this tradition open for the community in the future. The present book offers six fairly substantial chapters over a broad span of research fields, ranging from small volume volcanoes to understanding non-magmatic volcanism. The six chapters are grouped in four sections, and ordered to guide the readers through physical volcanology to more exotic fields of volcanism such as mud volcanoes.

In the first section, titled "From small to large volume volcanoes", the reader can find two chapters. The first chapter, presented by Kereszturi and Németh, is an extensive and very detailed review on our knowledge of monogenetic volcanism. The chapter guides the reader through our current knowledge on the formation of various types of monogenetic volcanoes, such as lava spatter cones, scoria cones, maars, tuff rings and tuff cones. Presenting a state-of-the-art review about various eruption mechanisms and the resulting volcanic landforms, followed by a comprehensive summary of basic elements of the erosion of these types of volcanoes, the chapter also provides some unique ways to deal with erosion processes that could be applied to many volcanic fields. The chapter also provides a huge set of examples from various volcanic fields on the Earth, based on direct field-based research and observations of the Authors. The same section contains a chapter about andesitic volcanism from a less known area in Siberia, Russia by Deberko. This chapter takes the reader back to the Late Mesozoic to the Russian Far East and explores the role of andesitic volcanism in understanding the tectonic events of the Bureya-Jziamysi superterrain. The chapter describes a comprehensive volcano-chemical method to identify major andesite volcanic phases that could be linked to various stages of terrain evolution. The chapter is very data-rich and certainly will provide valuable information not only for volcanologists but also for researchers directly interested in understanding the volcano-tectonic evolution of the Russian Far East. The numerous geochemical and geochronological data will certainly serve as a fundamental resource for future geological exploration work as well.

The second section of the book is titled "Volcanic hazards" and contains a single, but extensive, chapter presented by Martin et al. Volcanic hazard studies increasingly utilize geostatistical methods to understand the time and space variations of volcanic events in both dispersed- and focused-vent volcanic systems. The chapter provides a detailed study on the application of the Bayesian approach to incorporate helium isotopes ratios into long-term probabilistic volcanic hazard assessments for Tohoku, Japan. Japan is the home of many active volcanoes, including long-lived composite volcanoes, as well as dispersed-vent-dominated volcanic fields. Composite volcanoes are commonly associated with dispersed-vent systems and the distinction between them is often not straightforward. In this chapter, the authors provide a very up-to-date method to combine the volcanic hazard posed by volcanic fields and the associated central volcanoes. There is a wealth of geological data on the Tohuku area in Japan and past volcanic events at various scales are very well documented; therefore, the region is very well suited to geostatistical analysis in order to understand the spatio-temporal evolution of the volcanism in the area. This chapter not only documents primary research outputs, but will also be valuable to anyone interested in volcanic hazards and various geostatistical methods.

The third section of the book is titled "Volcanic resources and geoconservation", reminding the reader of two of the main consequences of volcanism that benefit human society. Jiaqi Liu et al. provide a very detailed summary to highlight the importance of volcaniclastic host rocks (and sediments) in the development of major hydrocarbon reservoirs. This chapter provides an initial review of volcanic rock-hosted natural hydrocarbon resources in general, before presenting some examples from relatively less known areas, such as the Songliao basin in China. This chapter explains basic rules on how volcanism may impact the formation of oil and gas reserves and how volcanism can act as a catalyst for organic matter extraction and evolution. Following the exciting physico-chemical background provided on the link between volcanism and hydrocarbon formation at the start of the chapter, the second part of the chapter describes how volcaniclastic basin filling successions could act as hydrocarbon resources and is relevant to the general volcanic community. The second chapter in this section has also been provided by Jiaqi Liu et al. who give a detailed summary on the link between volcanic resources and volcanic landscapes with particular attention to volcanic landscape geoconservation. This chapter will be a valuable read for research students and others working on protected areas of volcanic landscapes. The dynamic evolution of volcanic landscape management and conservation programs worldwide shows that a professional approach to geoconservation and geoeducation is an important part of research associated with volcanic regions. This chapter also provides information from some relatively less well known areas with which the global volcanic community could become linked in the future.

The fourth section of the book, titled "Non-magmatic volcanism", reflects on a peculiar type of volcanism that has undergone significant recent developments. "Non-magmatic" refers to the process that feeds this type of volcanism, signalling the lack of deep source magma involvement in the formation of the resulting landform, which is similar, in morphological terms, to volcanoes generated by magmatism. This section contains a single very detailed review article on mud volcanism associated with gas-hydrates, presented by Tinivella and Giustiniani. "Mud volcanoes and mud volcanism are some of nature's most anonymous, mysterious and undiscussed geological features, ….." the Authors state at the beginning of their work is a view that many readers will share after reading this chapter, which highlights the uniqueness of this type of volcanism. The review style article certainly would be a great help for anyone unfamiliar with the subject, and the chapter will be particularly useful for research students wanting to get a good and clear introduction to this type of volcanism. The chapter also provides a great wealth of examples that can help the reader to understand where to look for mud volcanoes on Earth.

Overall, Updates in Volcanology – New Advances in Understanding Volcanic Systems hopefully will serve as another good snapshot of the diversity of recent volcanic research. As the Editor of this book, it is my wish that the two "Updates in Volcanology" books published thus far will be just the start of a long-lasting book series serving to document diverse and high quality research outputs in the field of volcanology.

Dr Károly Németh
Volcanic Risk Solutions, CS-INR, Massey University,
Palmerston North, New Zealand
Geological Hazards Research Unit, King Abdulaziz University,
Jeddah, Kingdom of Saudi Arabia

Section 1

From Small to Large Volume Volcanoes

Monogenetic Basaltic Volcanoes: Genetic Classification, Growth, Geomorphology and Degradation

Gábor Kereszturi and Károly Németh

Additional information is available at the end of the chapter

http://dx.doi.org/10.5772/51387

1. Introduction

Plate motion and associated tectonics explain the location of magmatic systems along plate boundaries [1], however, they cannot give satisfactory explanations of the origin of intraplate volcanism. Intraplate magmatism such as that which created the Hawaiian Islands (Figure 1, hereafter for the location of geographical places the reader is referred to Figure 1) far from plate boundaries is conventionally explained as a result of a large, deep-sourced, mantle-plume [2-4]. Less volumetric magmatic-systems also occur far from plate margins in typical intraplate settings with no evidence of a mantle-plume [5-7]. Intraplate volcanic systems are characterized by small-volume volcanoes with dispersed magmatic plumbing systems that erupt predominantly basaltic magmas [8-10] derived usually from the mantle with just sufficient residence time in the crust to allow minor fractional crystallization or wall-rock assimilation to occur [e.g. 11]. However, there are some examples for monogenetic eruptions that have been fed by crustal contaminated or stalled magma from possible shallower depths [12-19]. The volumetric dimensions of such magmatic systems are often comparable with other, potentially smaller, focused magmatic systems feeding polygenetic volcanoes [20-21]. These volcanic fields occur in every known tectonic setting [1, 10, 22-28] and also on other planetary bodies such as Mars [29-33]. Due to the abundance of monogenetic volcanic fields in every tectonic environment, this form of volcanism represents a localized, unpredictable volcanic hazard to the increasing human populations of cities located close to these volcanic fields such as Auckland in New Zealand [34-35] or Mexico City in Mexico [36-37].

Importantly, research on monogenetic volcanoes and volcanic fields is focused on their "source to surface" nature, i.e. once the melt is extracted from the source it tends to ascend to the surface [11, 16-17, 38]. The rapid melt generation and short eruptive history of volcanoes fed by these magmas mean they can be used as 'probes' of various processes, particularly to detect short- and long-term changes occurring during emplacement of a single vent and/or a volcanic field. They also provide evidence of the evolution of magmatic systems that fed numerous individual small-volume volcanoes over time spans of millions of years [39-44]. This research has led to an understanding of the processes of melt extraction [17, 45-46], interactions in the lithospheric mantle [47-49], ascent within the lower to middle crust [16, 50] and in the shallow crust region [10, 51-54]. Other studies have elucidated plumbing and feeder systems of monogenetic volcanoes [8-9, 55-57], eruption mechanisms [58-61] and associated volcanic hazards [34, 62-67] as well as surface processes [68-71] and long-term landscape evolution [72-74].

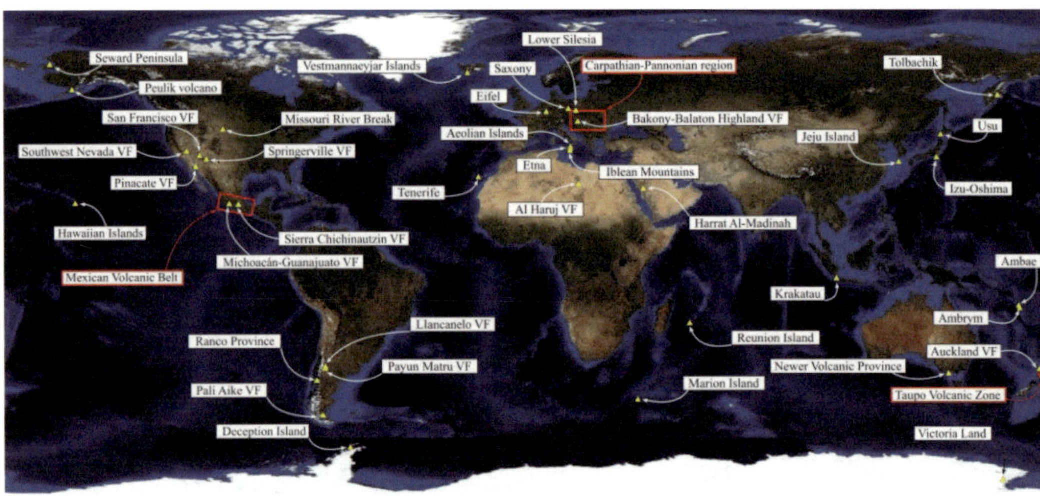

Figure 1. Overview map of the location of the volcanic field and zones mentioned in the text. The detailed location of specific volcanic edifices mentioned in the text can be downloaded as a Google Earth extension (.KMZ file format) from http://www.intechopen.com/.

Eruption of magma on the surface can be interpreted as the result of the dominance of magma pressure over lithostatic pressure [50, 75-76]. On the other hand, freezing of magma en route to the surface are commonly due to insufficient magma buoyancy, where the lithostatic pressure is larger than the magma pressure, or insufficient channelling/focusing of the magma [50, 76-78]. Once these small-volume magmas (0.001 to 0.1 km^3) intrude into the shallow-crust, they are vulnerable to external influences such as interactions with groundwater at shallow depth [79-82]. In many cases, the eruption style is not just determined by internal magma properties, but also by the external environmental conditions to which it has been exposed. Consequently, the eruption style becomes an actual balance between magmatic and environmental factors at a given time slice of the eruption. However, a combination of eruption styles is responsible for the formation of monogenetic volcanoes with wide range of morphologies, e.g. from conical-shaped to crater-shaped volcanoes. The morphology that results from the eruption is often connected to the dominant eruptive mechanisms, and therefore, it is an important criterion in volcano classifications. Diverse sources of information regarding eruption mechanism, edifice growth and hazards of monogenetic volcanism can be extracted during various stages of the degradation when the internal architecture of a volcano is exposed. Additionally, the rate and style of degradation may also help to understand the erosion and sedimentary processes acting on the flanks of a monogenetic volcano. The duration of the construction is of the orders of days to decades [83-84]. In contrast, complete degradation is several orders of magnitude slower process, from ka to Ma [68, 71, 73]. Every stage of degradation of a monogenetic volcano could uncover important information about external and internal processes operating at the time of the formation of the volcanic edifice. This information is usually extracted through stratigraphic, sedimentary, geomorphic and quantitative geometric data from erosion landforms. In this chapter, an overview is presented about the dominant eruption mechanism associated with subaerial monogenetic volcanism with the aim of understanding the syn- and post-eruptive geomorphic and morphometric development of monogenetic volcanoes from regional to local scales.

2. Monogenetic magmatic systems

Melt production from the source region in the mantle is triggered by global tectonic processes such as converging plate margins, e.g. Taupo Volcanic Zone in New Zealand [85-88] and in the Carpathian-Pannonian region in Central Europe [89-93] or diverging plate margins, for example sea-floor spreading along mid-oceanic ridges [94-95]. Melting also occurs in sensu stricto "convection plumes" or "hot spots" [2, 4, 96], which could alternatively result from small-scale, mantle wedge-driven convection cells [97]. This is often a passive effect of topographic differences between thick, cratonic and thin, oceanic lithosphere, as suspected by numerical modelling studies [97-101].

Typical ascent of the magma feeding eruptions through a monogenetic volcano starts in the source region by magma extraction from melt-rich bands. These melt-rich bands are commonly situated in a low angle (about 15–25°) to the plane of principal shear direction introduced by deformation of partially molten aggregates [95, 102-103]. The degree of efficiency of melt extraction is dependent on the interconnectivity, surface tension and capillary effect of the solid grain-like media in the mantle, which are commonly characterized by the dihedral angle between solid grains [104-105]. When deformation-induced strain takes place in a partially molten media, it increases the porosity between grains and triggers small-scale focusing and migration of the melt [104]. With the continuation of local shear in the mantle, the total volume of melt increases and enhances the magma pressure and buoyancy until it reaches the critical volume for ascent depending on favourable tectonic stress setting, depth of melt extraction and overlying rock (sediment) properties [16, 42]. The initiation of magma (crystals + melt) ascent starts as porous flow in deformable media and later transforms into channel flow (or a dyke) if the physical properties such as porosity/permeability of the host rock are high enough in elastic or brittle rocks in the crust [50, 75, 106-107]. The critical volume of melt essential for dyke injections is in the range of a few tens of m^3 [76], a volume which is several orders of magnitude less than magma batches feeding eruptions on the surface, usually ≥0.0001 km^3 [39, 108]. An increase in melt propagation distance is possible if small, pocket-fed initial dykes interact with each other [50, 76], which is strongly dependent on the direction of maximum (σ_1) and least principal stresses (σ_3), both in local and regional scales [109] and the vertical and horizontal separation of dykes [50, 76, 110]. These dykes move in the crust as self-propagating fractures controlled by the density contrast between the melt and the host rock from the over-pressured source zone [50]. The dykes could remain connected with the source region or propagate as a pocket of melt in the crust [111-112]. The geometry of such dykes is usually perpendicular to the least principal stress directions [108, 111]. The lateral migration of the magma en route is minimal in comparison with its vertical migration. This implies the vent location at the surface is a good approximation to the location of melt extraction at depth, i.e. the magma footprint [42, 54, 108]. The important implication of this behaviour is that interactions between magma and pre-existing structures are expected within the magma footprint area [54, 108]. Correlation between pre-existing faults and dykes are often recognized in volcanic fields [10, 53, 108, 113-115]. The likelihood of channelization of magma by a pre-existing fracture such as a fault, is preferable in the case of high-angle faults, i.e. 70–80°, and shallow depths [53] when the magma pressure is less than the tectonic strain taken up by faulting [42, 53].

These monogenetic eruptions have a wide variation in eruptive volumes. Volumetrically, two end-members types of volcanoes have been recognized [5, 109, 116]. Large-volume (≥1 km^3 or polygenetic) volcanoes are formed by multiple ascent of magmas that use more or less the same conduit system over a long period of time usually ka to Ma and have complex phases of construction and destruction [86, 117-119]. The spatial concentration of melt ascents, and temporally the longevity of such systems are usually caused by the formation of magma storage systems at various levels of the crust beneath the volcanic edifices [120-122]. In this magma chamber stalled magma can evolve by differentiation and crystallization in ka time scales [123]. On the other hand, a small-volume (≤1 km^3 or monogenetic) volcano is referred to as "[it] erupts only once" [e.g. 116]. The relationship between large and small volume magmatic systems and their volcanoes is poorly understood [1, 5, 109, 124-127]. Nevertheless, there is a wide volumetric spectrum between small and large (monogenetic and polygenetic) volcanoes and these two end-members naturally offer the potential for transition types of volcanoes to exist. An ascent event is not always associated with a single batch of magma, but commonly involves multiple tapping events (i.e. multiple magma batches), creating a diverse geochemical evolution over even a single eruption [9, 11, 16-17, 45, 128]. Multiple melt batches involved in a single event may be derived from the mantle directly or from some stalling magma ponds around high density contrast zones in the lithosphere such as the upper-mantle/crust boundary [9, 128] and/or around the ductile/brittle boundary zone in the crust [16].

A volcanic eruption on the surface is considered to be a result of a successful coupling mechanism between internal processes, such as melt extraction rate and dyke interaction en-route to the surface [50, 76, 110], and external processes, such as local and regional stress fields in the crust [42, 109]. Therefore, the spatial and temporal location of a volcanic event represents the configuration of the magmatic system at the time of the eruption. However, mantle-derived, usually primitive magmas feeding monogenetic magmatic systems are uncommon and rarely erupt individually. They tend to concentrate in space forming groups of individual volcanoes or clusters [7, 24, 129-130], and in time constitute volcanic cycles [39, 42, 131-132]. The spatial component of volcanism is dependent on the susceptibility of magma to be captured by pre-existing structures such as faults [10, 53-54], and the regional stress field at the time of the melt ascent [7, 43, 109, 133]. Temporal controls are also significantly influenced by internal and external forces. The monogenetic magmatic systems can be classified into two groups [131, 134]. The volume-predictable systems [134-135] are internally-controlled, i.e. it is magmatically-controlled [42]. In this system, an eruption on the surface is a direct result of successful separation of melt from a heterogeneous mantle, which is independent from the tectonics. Therefore, the total volume of magma erupting at the surface is usually a function of magma production rates of the system and repose time since the previous eruption [42]. These magmatic systems are usually characterized by high magma flux, promoting frequent dyke injections and high magmatic contribution to local extensional strain accumulation. These could trigger earthquakes, faulting and surface deformations, such as ruptures, associated with the high rates of magma intrusions [111, 136] similar to the intrusion at tensional rift zones [e.g. 137-138]. Magma ascent is often dominated by the regional-scale direction of stress rather than the location of pre-existing faults and topography [111]. In contrast, the time-predictable magmatic system [131, 139] is a passive by-product of tectonic shear-triggered melt extraction [42, 95, 103, 131]. Without tectonic forces, the melt would not be able to be extracted from partially molten aggregates [42]. Consequently, this magma generation process is externally- or tectonically-controlled [42]. The overall magma supply of these volcanic fields is generally low. Magmatic pressure generated by the magma injections are commonly suppressed by lithostatic pressure, resulting in a greater chance of interaction between magma and pre-existing structures in the shallow crust [53, 111, 140]. Dyke capturing commonly takes place if the orientation of the dyke plane is not parallel with the direction of maximum principal stress, causing vent alignments and fissure orientation to not always be perpendicular with the least principal stress direction [42, 54].

Restriction of magma ascent to a small area usually results in monogenetic volcanoes forming volcanic fields in a well-defined geographic area. These eruptions normally take place from hours to decades resulting in the accumulation of small-volume eruptive products on the surface predominantly from basaltic magmas. However, a monogenetic volcanic field could experience monogenetic eruptions over time scales of Ma [5, 39, 141-142] and the lifespan is characterized by waxing and waning stages of volcanism and cyclic behaviour [39, 108]. In a single monogenetic volcanic field, tens to thousands of individual volcanoes may occur [143] with predominantly low SiO_2 content eruptive products ranging from ca. 40 wt% up to 60 wt% [16, 40, 128, 144-146]. However, monogenetic volcanism does not depend on the chemical composition because there are similar small-volume monogenetic volcanoes that have been erupted from predominantly silica-rich melt such as Tepexitl tuff ring, Serdán-Oriental Basin, Mexican Volcanic Belt, Mexico [147] or the Puketarata tuff ring, Taupo Volcanic Zone, New Zealand [148].

3. Construction of monogenetic volcanoes

The ascent of magma from source to surface usually involves thousands of interactions between external and internal processes, thus the pre-eruptive phase works like an open system. Once single or multiple batch(es) of magma start their ascent to the surface, there is continuous degassing and interactions with the environment at various levels en route. On the surface, the ascending magma ascent can feed a volcanic eruption that can be explosive or effusive. Important characteristics of the volcanic explosion are determined at shallow depth (≤1–2 km) by the balance between external and internal factors such as chemical composition or availability of external water. The volcanic eruptions are usually characterised by discrete eruptive and sedimentary processes that are important entities of the formation and emplacement of a monogenetic vent itself.

3.1. Internal versus external-driven eruptive styles

The current classification of volcanic eruptions is based mainly on characteristics such as magma composition, magma/water mass ratio, volcanic edifice size and geometry, tephra dispersal, dominant grain-size of pyroclasts and (usually eye-witnessed) column height [e.g. 149]. If the ascending melt or batches of melts reach the near-surface or surface region, it will either behave explosively or intrusively/effusively. Explosive magma fragmentation is triggered either by the dissolved magmatic volatile-content [150] or by the thermal energy to kinetic energy conversion and expansion during magma/water interactions [151-152], producing distinctive eruption styles. These eruption styles can be classified on the basis of the dominance of internal or external processes.

Internally-driven eruptions are promoted by dissolved volatiles within the melt that exsolve into a gas-phase during decompression of magma [153-155]. The volatiles are mainly H_2O with minor CO_2, the latter exsolving at higher pressure and therefore greater depths than H_2O [e.g. 156]. Expansion of these exsolved gases to form bubbles in the magma suddenly lowers the density of the rising fluid, causing rapid upward magma acceleration and eventually fragmentation along bubble margins [150, 155, 157-159]. The growth of gas bubbles by diffusion and decompression in the melt occurs during magma rise, until the volume faction exceeds 70–80% of the melt, at which point magma fragmentation occurs [160-161]. Magmas with low SiO_2 contents, such as basalts and undersaturated magmas have low viscosity, allowing bubbles to expand easily in comparison to andesitic and rhyolitic magmas. Thus these low-silica magmas generate mild to moderate explosive types of eruptions such as Hawaiian [e.g. 162], Strombolian [e.g. 153], violent Strombolian [e.g. 163] and in very rare instances sub-Plinian types [e.g. 164, 165]. There is a conceptual difference between Hawaiian and Strombolian-style eruptions because in the former case magmatic gases rise together with the melt [154], whereas in Strombolian-style eruptions an essentially stagnant magma has gas slugs that rise and bubble through it – generating large gas slug bursts and foam-collapse at the boundary of the conduit [153, 166]. According to the rise speed-dependent model, bubbles form during magma ascent [150], while in the case of the foam collapse model, bubbles up to 2 m in diameter are generated deeper, in the upper part of a shallow magma chamber, based on acoustic measurements at the persistently active Stromboli volcano in the Aeolian Islands, Italy [153].

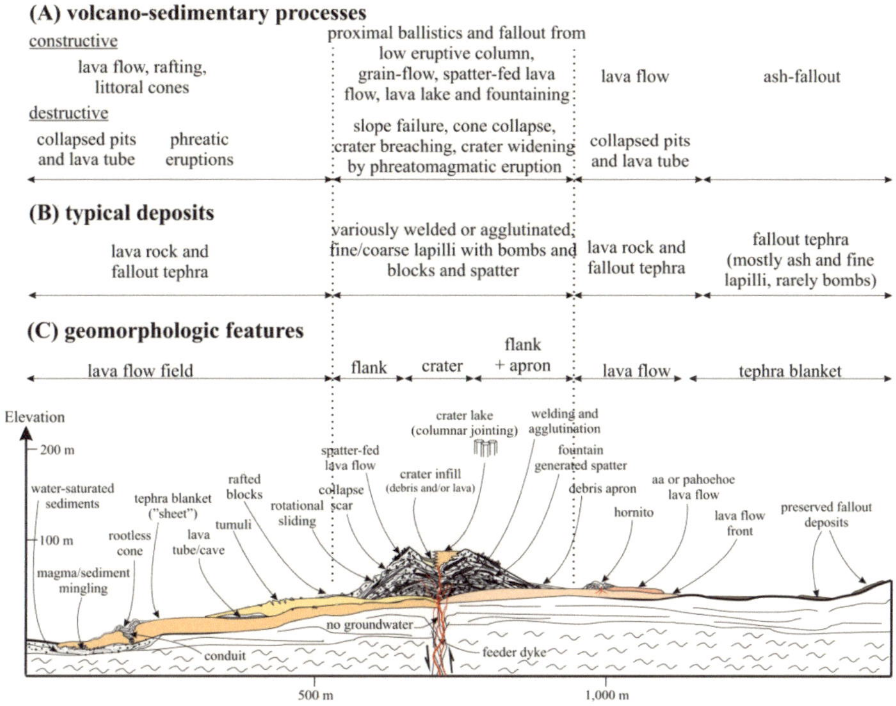

Figure 2. Schematic cross-section through a spatter cone showing the typical volcano-sedimentary processes and geomorphologic features.

A Hawaiian eruption results from one of the lowest energy magma fragmentation that are driven mostly by the dissolved gas content of the melt, which produces lava fountaining along fissures or focussed fountains up to 500 m in height [150, 162, 167-168]. The lava fountaining activity ejects highly deformable lava 'rags' at about 200–300 m/s exit velocity with an exit angle that typically ranges between 30–45° from vertical [169-170]. The nature and the distribution of the deposits associated with lava fountaining depend on the magma flux and the magma volatile content [162, 171-172]. Magmatic discharge rates during lava fountain activity range typically between 10 and 10^5 kg/s [162, 166]. The duration of typical lava fountaining activity may last only days or up to decades. An example for the former is Kilauea Iki, which erupted in 1959 [167, 170], while an example for the latter is Pu'u 'O'o-Kupaianaha, which began to erupt in 1983 [173]. Both are located on the Kilauea volcano in the Big Island of Hawaii, USA. Pyroclasts generated by lava fountaining are coarsely fragmented clots of magma which do not travel far from and above the vent [170-171]. They commonly land close to the vent and weld (i.e. mechanical compaction of fluid pyroclasts due to overburden pressure), agglutinate (i.e. flattening and deformation of fluid pyroclasts) or coalesce (i.e. homogenously mixed melt formed by individual fluidal clots) due to the high emplacement temperature of fragmented lava lumps on the depositional surface and/or the fast burial of lava fragments, which can retain heat effectively for a long time [171-172, 174]. The degree of welding and agglutinating of lava spatter is dependent on the [170-172, 174-175]:

1. accumulation rate and thickness of the deposit,
2. duration of eruption,
3. lava fountain height,
4. initial temperature determined by the magma composition,
5. heat loss rate, as well as
6. the grain size.

As a result of the limited energy involved in this type of magma fragmentation, the coarsely fragmented lava clots are transported ballistically, while the fines are transported by a low eruption column, as in the case of Plinian eruptions [162, 176]. The fragments tend to accumulate in proximal position, forming a cone-shaped pile, a spatter cone (Figure 2), which is built up by alternation of lava spatter and lava fountain-fed flows <100 m in diameter and a few tens of meters in height [170-172, 177-179].

Based on the grain size and the limited areal dispersion of tephra associated with typical Strombolian-style eruptions, it is considered a result of a mild magma fragmentation [149, 155]. However, larger volumes of tephra are produced than Hawaiian-style eruptions [159, 180]. Tephra production is derived from relatively low, non-sustained eruption columns [111, 153, 158, 181]. Individual explosions last <1 min and eject 0.01 to 100 m^3 of pyroclasts to <200 m in height with an exit velocity of particles of 3–100 m/s [180]. The magma discharge rate of 10^3 to 10^5 kg/s is based on historical examples of volcanoes erupted from water-rich, subduction-related magma [156]. The near surface fragmentation mechanism and limited energy released in a single eruption results in coarse lapilli-to-block-sized pyroclasts, predominantly between 1 and 10 cm in diameter, accumulating in close proximity to the vent [84, 182-183]. The exit velocity and angle of ballistic trajectories of particles of 20–25° determines the maximum height of the edifice and produces a limited size range of clasts in these volcanic edifices [184-185]. The repetition of eruptions produces individual, moderately-to-highly vesicular pyroclasts, called scoria or cinder, that do not agglutinate in most situations, but tend to avalanche downward forming talus deposits on the flanks of the growing cone [185-187]. Due to the mildly explosive nature of the eruptions, and the relatively stable pyroclast exit angles and velocity, a well-defined, conical-shaped volcano is constructed and is commonly referred to as a scoria or cinder cone (Figure 3). These cones have a typical basal diameter of 0.3 to 2.5 km, and they are up to 200 m high [153, 179, 182, 185, 188-189].

A more energetic magma fragmentation than is normally associated with Strombolian-activity cause violent Strombolian eruptions [163, 190]. In the 'normal' Strombolian-style eruptions, the magma is separated by gas pockets, which rise periodically in the magma through the conduit forming a coalescence of gas pockets, or a slug flow regime [153]. When the gas segregation increases, the eruptions become more explosive due to episodic rupture of liquid films of large bubbles, causing alternation of the flow regime from slug flow to churn flow, which is a typical characteristic of the violent Strombolian activity [163]. Based on numerical simulations, the increases in the gas flux, which creates the "churn flow", is caused by factors such as an increased length of conduit, the change in magma flux from 10^4 to 10^5 kg/s, the gas content, and/or the ascent speed variations that allow magma to vesiculate variably within the conduit [156, 163, 191]. Larger energy release during more explosive eruptions produces a higher degree of fragmentation, and hence finer-grained, ash-lapilli dominated beds [191], as well as higher eruption columns (<10 km) that disperse tephra efficiently over longer distances [83, 163].

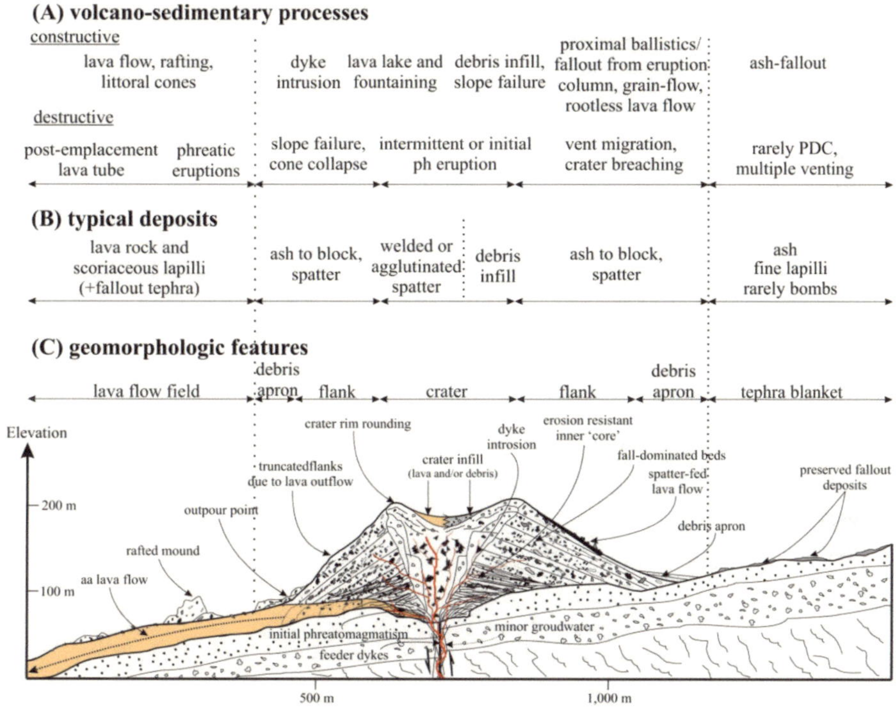

Figure 3. Schematic cross-section through a typical scoria cone showing the typical volcano-sedimentary processes and geomorphologic features. Abbreviations: PDC – pyroclastic density current, ph – phreatomagmatic eruption.

Externally-driven fragmentation occurs when the melt interacts with external water leading to phreatomagmatic or Surtseyan-style eruptions [152, 192-195]. These explosive interactions take place when magma is in contact with porous- or fracture-controlled groundwater aquifers or surface water [151, 194, 196-202]. In special cases when explosive interactions take place between lava and lake, sea water or water-saturated sediments, littoral cone [203-204] and rootless cone [205-208] are generated. Processes and eruption mechanisms associated with these eruptions are not discussed in the present chapter. The evidences of the role of water in the formation of tuff rings and maar have been proofed by many studies [e.g. 152,

201, 209]. However, there are similar eruptive processes and eruption styles have been described from eruptions of silica-undersaturated magmas (e.g. foidite, melilitite and carbonatite) in environments, where the role of external water on the eruptive style is limited [e.g. 210-212].

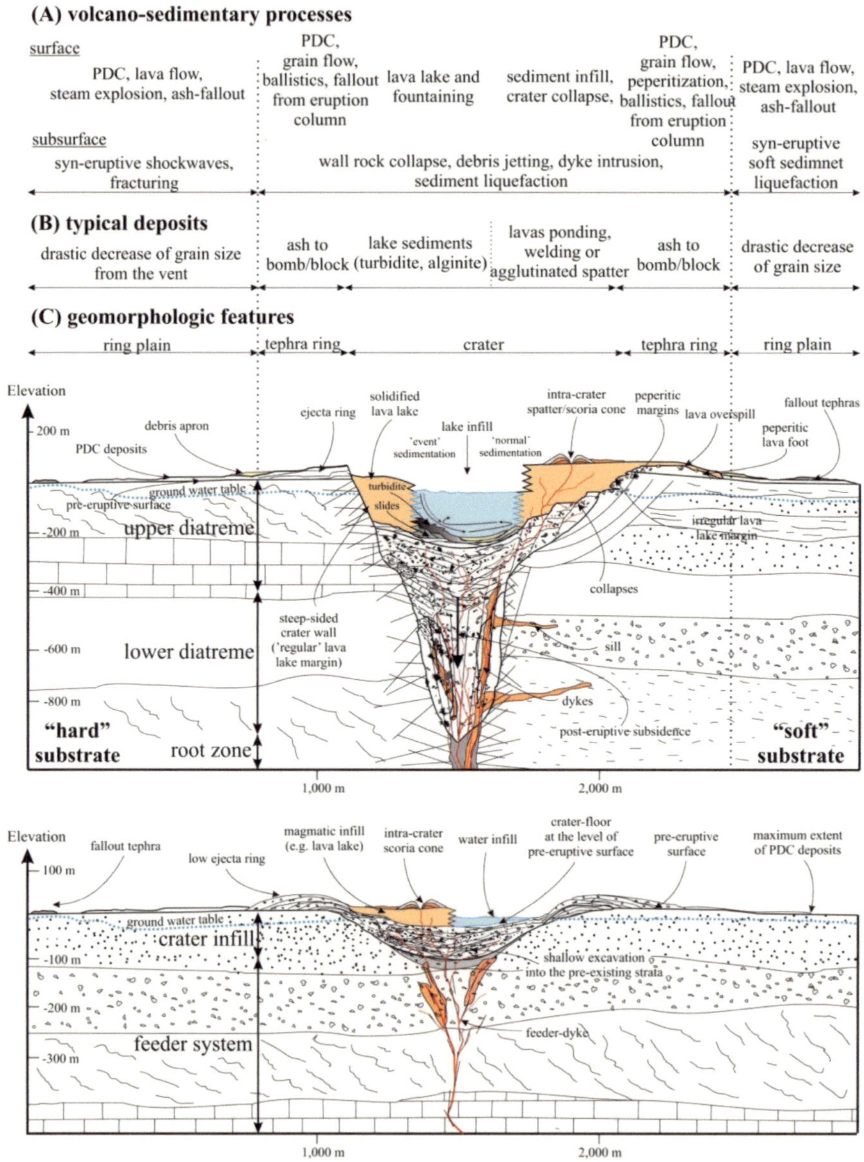

Figure 4. Schematic cross-sections through a maar-diatreme (top figure) and a tuff ring (bottom figure) showing the typical volcano-sedimentary processes and geomorphologic features. Note that the left-hand side represents the characteristics of a maar-diatreme volcano formed in a hard-substrate environment, while the right-hand side is the soft rock environment. Abbreviations: PDC – pyroclastic density current.

Phreatomagmatic eruptions (rarely called Taalian eruptions) are defined by some as being in subaerial environments [194]. These eruptions may produce a series of volcanic craters which vary in size between 0.1 km and 2.5 km in diameter [213]. The largest ones are very likely to be generated by multiple eruptions forming amalgamated craters such as Lake Coragulac maar, Newer Volcanics Province, south-eastern Australia [214] and/or formed in specific environment such as Devil Mountain maar, Seward Peninsula, Alaska [215]. The fragmentation itself is triggered by a molten fuel-coolant interaction (MFCI) processes requiring conversion of magmatic heat to mechanical energy [151, 193-194, 216-220]. The MFCI proceeds as follows [151, 220]:

1. coarse premixing of magma and water producing a vapour film between fuel and coolant,
2. collapse of the vapour film, generating fragmentation of magma and producing shock waves,
3. rapid expansion of superheated steam to generate thermohydraulic explosions, as well as
4. post-eruption (re)fragmentation of molten particles.

In some cases the MFCI process is self-driven and, after the initial interactions, the fragmentation does not involve any other processes [194, 196, 217, 221-222]. The series of eruptions may excavate a crater that cuts into the pre-existing topography, forming a hole-in-the-ground structure called a maar (Figure 4) [152]. If the explosion locus stays at shallow depths, the resulting volcano is tuff ring, which has a crater floor normally near the pre-eruptive surface (Figure 4) [223]. Both eruptions result in a surface accumulation of tephra by fallout and pyroclastic density currents, mostly base surges, forming a usually circular ejecta ring around the crater [51, 58, 195, 200, 202, 224-228]. These eruptions produce pyroclastic deposits that have a diversity of juvenile pyroclasts (e.g. various shape, grain-size, vesicularity and microlite content) and variety of accidental lithic clasts derived from the underlying strata [52, 79, 229-231]. Pyroclastic successions of phreatomagmatic volcanoes can form coarse grained, chaotic breccias related to vent construction, conduit wall collapse or migration, as well as well-stratified, lapilli and ash-dominated beds with various degrees of sorting and large ballistically ejected, fluidal-shaped juvenile bombs or angular to heavily milled accidental lithic blocks [41, 59, 81, 223, 232-237]. Due to density current transportation of pyroclasts, the accumulating deposits are stratified and are commonly cross- or dune-bedded [81, 229-230, 238-240]. The craters of most of these phreatomagmatic volcanoes are filled by either post-maar eruptive products such as solidified lava lakes commonly showing columnar jointing and/or intra-crater scoria/spatter cones [81, 241-244], or non-volcanic sediments, such as lacustrine alginate, volcaniclastic turbidite deposits [195, 245-250].

Surtseyan-style eruptions occur when the external water is 'technically' unlimited during the course of the eruption when eruptions occur through a lake or the sea [222, 251-254]. In contrast with phreatomagmatic eruptions, Surtseyan-style eruptions require a sustained bulk mixing of melt and coolant, which generates more abrupt and periodic eruptions [194, 196]. During Surtseyan-style eruptions, water is flashed to steam which tears apart large fragments of the rising magma tip [222, 252, 254-256]. This process is far less efficient than self-sustained typical MFCI and causes a near continuous ejection of tephra [194]. This tephra feeds subaqueous pyroclastic density currents, which build up a subaqueous volcanic pile that may emerge to become an island in the course of the eruption [222, 253-254, 257-259], as was the case during the well-documented eruption of Surtsey tuff cone, Vestmannaeyjar Islands, Iceland in 1963–1967 AD [260-261]. After emergence, a conical volcano can cap the edifice and build a typical steep-sided tuff cone (Figure 5). The tuff cone gradually grows by rapidly expelled and frequent (every few seconds) tephra-laden jets that eject muddy, water-rich debris, which may initiate mass flows later on in the inner-crater wall and on the outer, steepening flank of the growing cone [253, 260, 262-265]. These shallow explosions eventually produce a cone form, although it often has irregular geometry with a breached or filled crater by late-stage lava flows or asymmetric crater rim [223, 234, 259, 264]. The diameters of craters of these tuff cones are comparable to the tuff rings and maars, but the elevation of the crater rims are higher, reaching up to 300 m [223]. Monogenetic volcanoes that formed by Surtseyan-eruptions typically have no diatreme below their crater, however, some recent research suggested that diatremes may exist beneath a few tuff cones, such as Saefell tuff cone, south Iceland [266] or Costa Giardini diatreme, Iblean Mountains, Sicily [267].

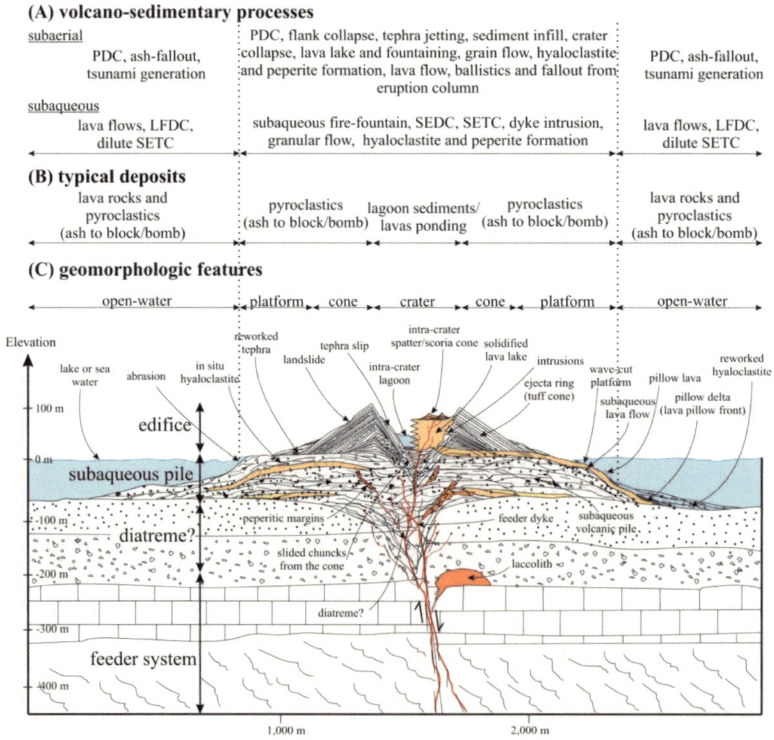

Figure 5. Schematic cross-section through a tuff cone showing the typical volcano-sedimentary processes and geomorphologic features. Abbreviations: PDC – pyroclastic density current, SEDC – subaqueous eruption-fed density current, SETC – subaqueous eruption-fed turbidity current, LFDC – lava flow-fed density current

3.2. Spectrum of basaltic monogenetic volcanoes

As documented above, five types of monogenetic volcanoes are conventionally recognized [177, 179, 223]:

1. lava spatter cones,
2. scoria or cinder cones,
3. maars or maar-diatremes,
4. tuff rings and
5. tuff cones.

This classification is primarily based on the morphological aspects and dominant eruption styles of these volcanoes. Furthermore, there is a strong suggestion that a given eruption style results in a given type of volcanic edifice, e.g. Strombolian-style eruptions create scoria cones [e.g. 111, 182]. The conventional classification also fails to account for the widely recognized diversity or transitions in eruption styles that may form 'hybrid' edifices, e.g. intra-maar scoria cones with lava flows or scoria cones truncation by late stage phreatomagmatism [229, 244, 268-271]. The variability in the way a monogenetic volcano could be constructed also means that the conventional classification hides important details of complexity that may be important from volcanic hazard perspective (e.g. a volcano built up by initial phreatomagmatic eruptions and later less dangerous Strombolian eruptions). The diversity of pyroclastic successions relates to fluctuation of eruption styles that may be triggered by changing conduit conditions, such as geometry, compositional change, and variations in both magma and/or ground water supply [41, 52, 150, 163, 272-273]. Due to the abundance of intermediate volcanoes, a classification scheme is needed, where the entire eruptive history can be parameterized numerically.

In the present study, the construction of a small-volume volcano is based on two physical properties (Figure 6):

1. eruption style and associated sedimentary environment during an eruption and
2. number of eruption phases.

A given eruption style is a complex interplay between internal and external controlling parameters at the time of magma fragmentation. The internally-driven eruption styles are, for example, controlled by the ascent speed, composition, crystallization, magma degassing, number of magma batches involved, rate of cooling, dyke and conduit wall interactions, depth of gas segregation and volatile content such as H_2O, CO_2 or S [9, 11, 17, 111, 128, 150, 154-156, 163, 188, 191, 274-276]. These processes give rise to eruption styles in basaltic magmas that are equivalent to the Hawaiian, Strombolian and violent Strombolian eruption styles. However, due to the small-volume of the ascending melt, the controls on magma fragmentation are dominated by external parameters, including conduit geometry, substrate geology, vent stability/migration, climatic settings, and the physical characteristics of the underlying aquifers [39, 82, 234, 277-279]. Another important parameter in the construction of a monogenetic volcanic edifice is the number of eruptive phases contributing to its eruption history (Figure 6). The complexity of a monogenetic landform increases with increasing number or combination of eruptive phases. These can be described as "single", "compound" and "complex" volcanic edifices or landforms [280-281]. In this classification, the volcano is the outcome of combinations of eruption styles repeated by *m* phases. For example, a one-phase volcano requires only one dominant eruption mechanism during its construction. Due the single eruption style, the resulting volcano is considered to be a simple landform with possibly simple morphology. However, monogenetic volcanoes tend to involve two or multiple phases (Figure 6). Their construction requires two or more eruption styles and the result is a compound or complex landforms respectively, e.g. maar-like scoria cones truncated by late stage phreatomagmatic eruptions [e.g. 82, 270, 282] or a tuff cone with late-stage intra-crater scoria cone(s) [e.g. 265, 283]. These phases may occur at many scales from a single explosion (e.g. a few m^3) to an eruptive unit comprising products of multiple explosions from the same eruption style.

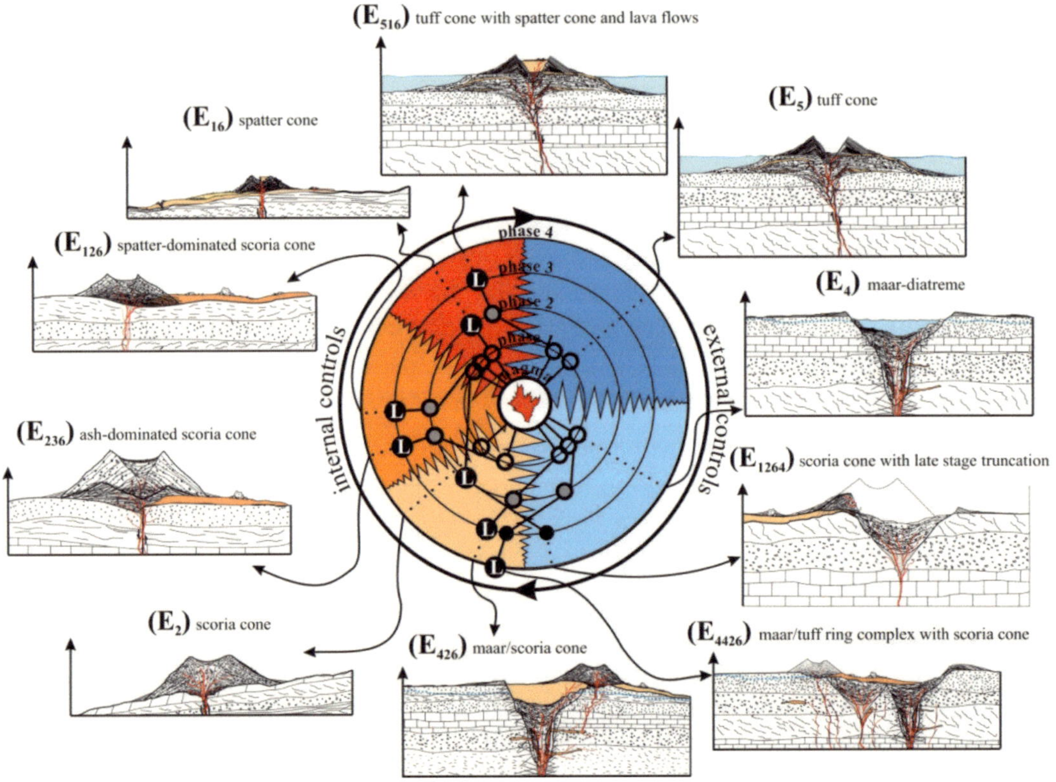

Figure 6. Eruption history (E) defined by a spectrum of eruptive processes determined by internal and external parameters at a given time. The initial magma (in the centre of the graph in red) is fragmented by the help of internal and external processes which determine the magma fragmentation mechanism and eruption style (phase 1). If a change (e.g. sudden or gradual exhaustion of groundwater, shift in vent position or arrival of new magma batch) occurs, it will trigger a new phase (phase 2, 3, 4,..., n); moving away from the pole of the diagram) until the eruption ceases. Note that black circles with white "L" mean lava effusion. If the eruption magma is dominantly basaltic in composition, the colours correspond to Surtseyan (dark blue), phreatomagmatic (light blue), Strombolian (light orange), violent Strombolian (dark orange) and Hawaiian (red) eruption styles.

To put this into a quantitative context, this genetic diversity can be expressed as set of matrices, similar to Bishop [280]. In Bishop [280], the quantitative taxonomy is represented by matrices of volcanic landforms that were based on surface morphologic complexity and eruption sequences. The role of geomorphic signatures is reduced due to the fact that in the case of an eruption centre built up by multiple styles of eruptions, not all eruption styles contribute to the geomorphology. For example, a scoria cone constructed by tephra from a Strombolian-style eruption, might be destroyed by a late stage phreatomagmatic eruption, as documented from Pinacate volcanic field in Sonora, Mexico [82, 284] and Al Haruj in Libya [282]. In these cases, the final geomorphologies resemble to maar craters, but the formation such volcanoes are more complex than a classical, simple maar volcano. In the proposed classification scheme, the smallest genetic entity (i.e. eruption style and their order) was considered to define the eruption history of a monogenetic volcano quantitatively. Considering only the typical, primitive basaltic composition range (SiO_2 ≤52% w.t.), internally-driven eruption styles are the Hawaiian, Strombolian and violent Strombolian eruption styles [111]. At the other end of the spectrum, externally-driven eruption styles are the phreatomagmatic and Surtseyan-types [81, 285]. In addition, the effusive activity can also be involved in this genetic classification. The abovementioned eruption/effusion styles can build up a volcano in the following combination: 6×1, 6×6, 6×6^2 or 6×m (or n×m) matrices, depending on the number of volcanic phases involved in the course of the eruption. This means that an eruption history (E) of a simple volcano (E_{simple}) could be written as:

$$E_{simple} = [1 \quad 2 \quad 3 \quad 4 \quad 5 \quad 6] \tag{1}$$

where the elements 1, 2, 3, 4 and 5 corresponds to explosive eruptions such as Hawaiian, Strombolian, violent Strombolian, Taalian (or phreatomagmatic) and Surtseyan-type eruptions, respectively, while the 6 is the effusive eruption. For a more complex eruption history involving two ($E_{compound}$) and multiple ($E_{complex}$) eruption styles can be written as:

$$E_{compound} = \begin{bmatrix} 11 & 12 & 13 & 14 & 15 & 16 \\ 21 & 22 & 23 & 24 & 25 & 26 \\ 31 & 32 & 33 & 34 & 35 & 36 \\ 41 & 42 & 43 & 44 & 45 & 46 \\ 51 & 52 & 53 & 54 & 55 & 56 \\ 61 & 62 & 63 & 64 & 65 & 66 \end{bmatrix} \tag{2}$$

For instance, a monogenetic volcano with an eruption history of fire-fountain activity associated with a Hawaiian-type eruption and effusive activity could be described as having a compound eruption history (or E_{16} in Figure 6). While an example of a monogenetic volcano with a complex eruption history could be a volcanic edifice with a wide, 'maar-crater-like' morphology, but built up from variously welded or agglutinated scoriaceous pyroclastic rock units (e.g. E_{1264} in Figure 6), similar to Crater Elegante in Pinacate volcanic field, Sonora, Mexico [284]. In some cases, gaps, paucity of eruptions or opening of a new vent site after vent migration between eruptive phases is observed/expected based on reconstructed stratigraphy [e.g. 52, 279] and geochemistry [e.g. 9, 286]. In this classification system, the recently recognized polymagmatic or polycyclic behaviour of monogenetic volcanoes, e.g. an eruption fed by more than one batch of magma with distinct geochemical signatures [17, 23, 287], can be integrated. For example, the volcano could be E_{44} if the controls on eruption style remained the same or E_{42} if that chemical change is associated with changes in eruption style. The number of rows and columns in these matrices could be increased until all types of eruption style are described numerically, thus an n×m matrix is created. Increasing the number of volcanic phases will increase the range of volcanoes that could possibly be created. Of course, the likelihood of various eruptive combinations described by these matrices is not the same because there are 'unlikely' (e.g. E_{665}) and 'common' eruptive scenarios (e.g. E_{412}).

$$E_{complex} = \begin{bmatrix} 111 & 112 & 113 & 114 & 115 & 116 \\ . & & & & & . \\ . & & & & & . \\ . & & & & & . \\ 661 & 662 & 663 & 664 & 665 & 666 \end{bmatrix} \quad (3)$$

In summary, a volcano from monogenetic to polygenetic can be described as a matrix with elements corresponding to discrete volcanic phases occurring through its evolution. The major advantage of these matrices is that their size is infinite (n×m), thus an infinite number of combinations of eruption styles could be described (Figure 6). In this system each volcano has a unique eruptive history, in other words, each volcano is a unique combination of *n* number of eruption styles through *m* number of volcanic phases (Figure 6). This matrix-based classification scheme helps to solve terminological problems and to describe volcanic landforms numerically. For example, the diversity of scoria cones from spatter-dominated to ash-dominated end-members [68, 288] cannot be easily expressed within the previous classification scheme. This completely quantitative coding of volcanic eruption styles into matrices could be used for numerical modelling or volcanic hazard models, e.g. spatial intensity of a given eruption style.

4. Geomorphology of monogenetic volcanoes

4.1. Historical perspective

The combination of eruption styles (listed above) and related sedimentary processes are often considered to be the major controlling conditions on a monogenetic volcano's geomorphic evolution [84]. Thus, the quantitative topographic parameterization of volcanoes is an important source of information that helps to reveal details about their growth, eruptive processes and associated volcanic hazards and its applicable to both conical [119, 190, 289-292] and non-conical volcanoes [199, 293]. These methods are commonly applied to both polygenetic [119, 289, 294-297] and monogenetic volcanic landforms [68-69, 298-299]. Morphometric measurements on monogenetic volcanoes began with the pioneering work of Colton [300], who noticed a systematic change in the morphology of volcanic edifices over time due to erosional processes such as surface wash and gullying. A surge of research in volcanic morphometry, focused mostly on scoria cones, occurred from the 1970s to 1990s, when the majority of morphometric formulae were established and tested [70-71, 84, 179, 185, 199, 213, 223, 293, 301-304]. This intense period of research was initiated by National Aeronautics and Space Administration (NASA) in the 1960s and 1970s due to an increasing interest in extraterrestrial surfaces that could be expected to be encountered during landings on extraterrestrial bodies such as the Moon or Mars [e.g. 33]. Additional interests were to understand magma ascent, the lithospheric settings of extraterrestrial bodies, the evolution of volcanic eruptions, the geometry of volcanoes in different atmospheric conditions, surface processes and seeking H_2O in extraterrestrial bodies [30, 33, 305-308]. Given the lack of field data from extraterrestrial bodies, many parameters that were able to be measured remotely, such as edifice height (H_{co}), basal (W_{co}) and crater diameters (W_{cr}) were introduced. There were measured manually from images captured by Mariner and Viking orbiter missions [e.g. 33] and Luna or Apollo missions for the Moon [e.g 309] in order to compare these data with the geometry of volcanic landforms on the Earth [e.g. 179, 293]. Dimensions, such as crater diameter, were measured directly from these images, while the elevation of the volcanic edifices was estimated from photoclinometry (i.e. from shadow dimensions of the studied landform) [33, 309-310]. Because elevation measurements were indirect, the horizontal dimensions such as W_{co} and W_{cr} were preferred in the first morphometric parameterization studies [179]. The increased need for Earth analogues led to intense and systematic study of terrestrial small-volume volcanoes [179, 185, 189, 293]. The terrestrial input sources, such as topographic/geologic maps and field measurements, were more accurate than the extraterrestrial input resources; however, they were still below the accuracy required (i.e. the contour line intervals of ≥20 m were not dense enough to capture the topography of a monogenetic volcano having an average size of ≤1500–2000 m horizontally and of ≤100–150 m vertically). The extensive research on monogenetic volcanoes identified general trends regarding edifice growth, eruption mechanism and subsequent degradation [71, 84, 185, 293]. In addition, morphometric signatures were recognized that associated a certain type of monogenetic volcanic landform with the discrete eruption style that formed it. The morphometric signatures of Earth examples were then widely used to describe and identify monogenetic volcanoes on extraterrestrial bodies such as the Moon and Mars [179]. Basic morphometric parameters were calculated and geometrically averaged to get morphometric signatures for four types of terrestrial, monogenetic volcanoes [179], including spatter cones (W_{co} = 0.08 km, W_{cr}/W_{co} = 0.36 km and H_{co}/W_{co} = 0.22 km), scoria cones (W_{co} = 0.8 km, W_{cr}/W_{co} = 0.4 km and H_{co}/W_{co} = 0.18 km), as well as maars and tuff rings (W_{co} = 1.38 km, W_{cr}/W_{co} = 0.6 km and H_{co}/W_{co} = 0.02 km). These morphometric signatures are still used in landform recognition [e.g. 31, 311].

In terrestrial settings, the morphometric studies of monogenetic volcanoes in volcanic fields and on polygenetic volcanoes have targeted

1. the characterization of the long-term and short-term evolution of magmatic systems [39, 42, 131, 134],
2. understanding of eruption mechanisms and processes [84, 182, 185, 312-314],
3. expression of tectonic influences on edifice growth [298, 315-316],
4. dating of conical landforms such as scoria or cinder cones [189, 304, 317-320],
5. examination of erosion processes [69-70, 290, 299, 321-327] and landscape evolution [39, 68, 73, 328],
6. reconstruction of the original size, geometry and facies architecture of polygenetic [329-332] and monogenetic edifices [291, 333-336], as well as
7. detection of climate and climate change influences on degradation [71, 324-325, 337].

Morphology quantified via morphometric parameters could be a useful tool to address some of these questions in volcanology, geology and geomorphology. The morphology of a volcanic edifice contains useful information from every stage of its evolution, including eruptive processes, edifice growth and degradation phases. However, the geomorphic information extracted through morphometric parameters often show bi- or even multi-modality, i.e. the morphometry is a mixture of primary and secondary attributes [e.g. 338]. The following section explores the dominant volcanological processes that influence the geomorphology of a monogenetic volcano.

4.2. Syn-eruptive process-control on morphology

The eruption styles shaping the volcanic edifices may undergo many changes during the eruption history of a monogenetic volcano (Figure 6). A given volcano's morphology and the grain size distribution of its eruptive products are generally viewed as the primary indicator of the eruption style that forms a well-definable volcanic edifice (i.e. "Strombolian-type scoria cones"). This oversimplification of monogenetic volcanoes, together with the widely used definition that "they erupts only once" [116], suggest a simplicity in terms of magma generation, eruption mechanism and sedimentary architecture. This supposedly simple and homogenous inner architecture of each classical volcanic edifice, such as spatter cones, scoria cones, tuff rings and maars, led to the identification of a "morphometric signature". The morphometric signature of monogenetic volcanoes was used in the terrestrial environment, e.g. to ascribe a relationship between morphometry and "geodynamic setting" [337], as well as extraterrestrial environments, e.g. for volcanic edifice recognition [31, 179, 339-341]. Certain types of volcanoes could be discriminated from each other based on their morphometric signature, but some general assumptions need to be made. For example,

1. the morphometric signature concept is entirely based on the assumption that a volcanic landform directly relates to a certain well-defined eruption style,
2. thus the pyroclast diversity within the edifice is minimal (i.e. homogenous), as well as

3. the resultant volcanic landform is emplaced in a closed-system with no transitions between eruption styles, especially from externally to internally-driven eruption styles and vice versa.

Consequently, the edifice studied was believed to have a relatively simple eruption history, which is a classical definition of a "monogenetic volcano". As demonstrated above, monogenetic volcanoes develop in an open-system. This section explores the volcanological/ geological constrains of geomorphic processes responsible for the final volcanic edifice and the morphometric development of two end-member types of monogenetic volcanoes such as crater-type (4.2.1.) and cone-type edifices (4.2.2).

4.2.1. Crater-type monogenetic volcanoes

Crater-type monogenetic volcanoes such as tuff rings and maar volcanoes (Figure 4), are characterized by a wide crater with the floor above or below the syn-eruptive surface, respectively [81, 152, 223, 234]. Their primary morphometric signature parameters are major/ minor crater diameter and depth, crater elongation and breaching direction, volume of ejecta ring, and crater or slope angle of the crater wall [313, 342-345]. Of these morphometric parameters, the crater diameters were used widely for interpreting crater growth during the formation of a phreatomagmatic volcano. For the genetic integration of crater growth and, consequently, the interpretation of crater diameter values of terrestrial, dominantly phreatomagmatic volcanoes, there are fundamentally two end-member models.

The first model is the incremental growth model (Figure 7A). In this model, the crater's formation is related to many small-volume eruptions and subsequent mass wasting, shaping the crater and underlying diatreme [81, 151, 199-200, 209, 221, 285, 346-347]. Growth initiates when the magma first interacts with external water, possibly groundwater along the margin of the dyke intrusions, triggering molten-fuel-coolant interactions (MFCI) [192-193, 220, 348]. These initial interactions excavate a crater on the surface, while the explosion loci along the dyke gradually deepen the conduit beneath the volcano towards the water source, resulting in a widening crater diameter [199]. This excavation mechanism initiates some gravitational instability of the conduit walls, triggering slumping and wall rock wasting, contributing to the growing crater [81, 195, 199, 223, 229-230, 349]. This classical model suggests that

1. crater evolution is related to diatreme growth underneath, and

2. the crater's growth is primarily a function of the deep-seated eruption at the root zone.

However, it is more likely that the pyroclastic succession created at the rim of the crater preserves only a certain stage of the evolution of whole volcanic edifice. For instance, the possibility of juvenile and lithic fragments being erupted and deposited within the ejecta ring from a deep explosion (i.e. at the depth of a typical diatreme, about 2 km) is highly unlikely. Rather than being dominated by the deep-seated eruptions, explosions can occur at variable depths within the diatreme [347]. The individual phreatomagmatic eruptions from various levels of the volcanic conduit create debris jets (solids + liquid + magmatic gases and steam),

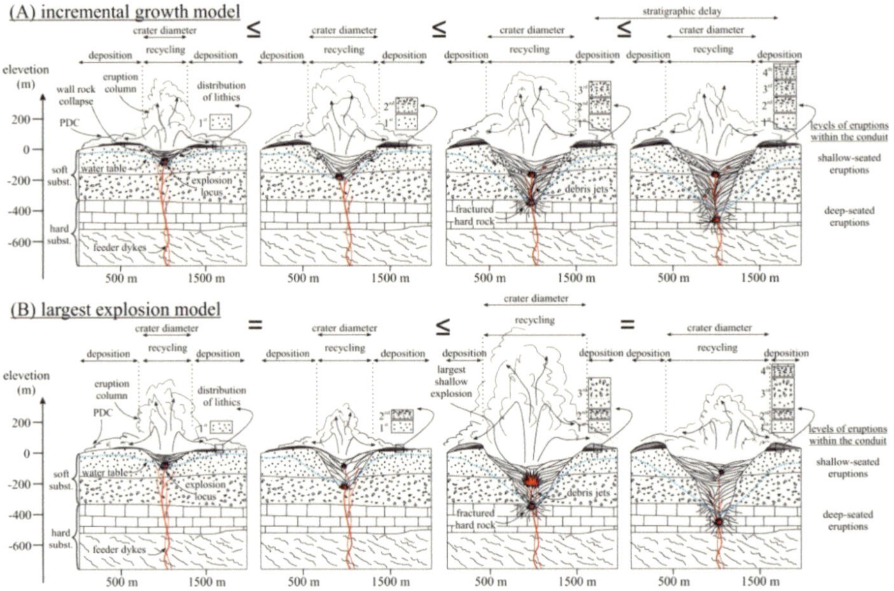

Figure 7. Crater growth envisaged by the incremental growth (A) and largest explosion models (B). In the top figure, the crater grows by each explosion and the subsequent mass wasting processes (e.g. slumping of wall rock). This results in the gradual growth of the crater and the underlying diatreme (if any). On the other hand, in the bottom figure the crater can reach its final size in the middle of the eruption history by the largest near-surface explosion and the further explosions make no contribution to its size, geometry and morphology. Abbreviations: PDC – pyroclastic density current.

which are responsible for the transportation of tephra [285, 347, 350]. Every small-volume explosion causes upward transportation of fragmented sediment in the debris jet, giving rise to small and continuous subsidence/deepening of the crater floor [198-199]. This is in agreement with stratigraphic evidence from eroded diatremes, such as Coombs Hills, Victoria Land, Antarctica [285, 350] and Black Butte diatreme, Missouri River Breaks, Montana [351].

The second model is where the crater geomorphology is dominated by largest explosion event during the eruption sequence. Thus, the crater size directly represents the 'peak' (or maximum) energy released during the largest possible shallow explosion [202, 313, 343-344, 352]. This model of crater growth (Figure 7B) for phreatomagmatic volcanoes is proposed on the basis of analogues from phreatic eruption, such as Uso craters, Hokkaido, Japan in 2000 [352], and experiments on chemical and nuclear explosions [344]. In this model, the crater diameter (D) is a function of the total amount of ejected tephra (V_{ejecta}) [313]:

$$D = 0.97V_{ejecta}^{0.36} \tag{4}$$

which can further be converted into explosion energy (E) as:

$$E = 4.45 \times 10^{6} D^{3.05} \tag{5}$$

which approximates the largest energy released during the eruptions. This relationship between crater size and ejected volume was based on historical examples of phreatomagmatic eruptions [313]. These historical eruptions are, however, associated with usually polygenetic volcanoes, such as Taupo or Krakatau, and not with classical monogenetic volcanoes, except the Ukinrek maars, near Peulik volcano, Alaska. In this model, the largest phreatomagmatic explosion governs the final morphology of the crater, so the crater size correlates with the peak energy of the maar-forming eruption directly [343]. Scaled experiments showed that there is a correlation between the energy and the crater depth and diameter [353] if the explosions take place on the surface [344]. Most of the explosions modelled in Goto et al. [344] were single explosions, only a few cratering experiments involved multiple explosions at the same point [344], which is more realistic for monogenetic eruptions. In these multiple explosions, the crater did not grow by subsequent smaller explosions, possibly because the blast pressure was lower than the rock strength when it reached the previously formed crater rim [344]. As noted by Goto et al. [344], such experimental explosions on cratering are not applicable to underground eruptions; therefore, they do not express the energy released by deep-seated eruptions generating three-phase (solid, gas and fluid) debris jets during diatreme formation [e.g. 350]. Morphologically, these deep-seated eruptions have a minor effect on crater morphology and diameter, and their deposits rarely appear within the ejecta ring around the crater.

Theoretically, both emplacement models are possible because both mechanisms can contribute significantly to the morphology of the resulting landform. The incremental growth model is based on statigraphy, eye-witnessed historical eruptions and experiments [81, 152, 198-199, 346-347], while the largest explosion model is based on analogues of chemical or nuclear explosion experiments, phreatic eruptions and impact cratering [313, 344, 353-354]. Based on eye-witnessed eruptions and geological records, the crater diameter as a morphometric signature for maar-diatreme and tuff ring volcanoes is the result of complex interplay between the eruptions and the substrate. The dominant processes, such as many, small-volume explosions with various energies migrating within the conduit system vertically and horizontally, as well as gradual mass wasting depending on the physical properties of rock strength, are what control the final crater diameter. On the other hand, the substrate beneath the volcano also plays an important role in defining crater morphology, as highlighted for terrestrial volcanic craters [209, 229, 355], as well as extraterrestrial impact craters [356]. In different substrates, different types of processes are responsible for the mass wasting. For example, an unconsolidated substrate tends to be less stable due to explosion shock waves that may liquefy water-rich sediments, and induce grain flow and slumping, enlarging the crater [229]. On the other hand, in a hard rock environment, the explosions and associated shock waves tend to fracture the country rock, depending on its strength, leading to rock falls and sliding of large chucks from the crater rim [229]. The crater walls in these two contrasting environments show different slope angles [229, 345]. These differences in the behaviour of the substrate in volcanic explosions may cause some morphological variations in the ejecta distribution and the final morphology of the crater (Figure 8).

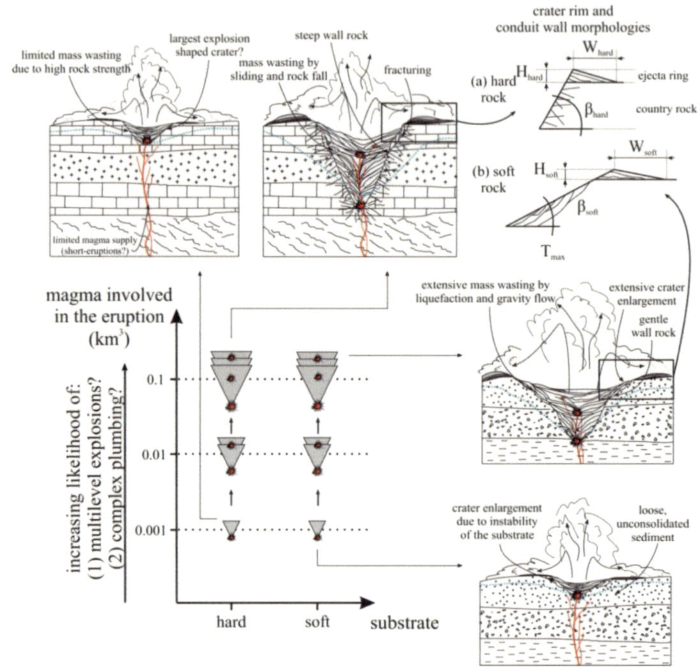

Figure 8. Morphology of volcanic crater and diatreme generated by phreatomagmatic eruptions in soft and hard substrate environments. The parameters (β – wall rock angle, H – ejecta ring height over the surrounding topography, W – flank width of the ejecta ring) for crater morphologies have the following possible relationships: $\beta_{hard} > \beta_{soft}$; while the $H_{hard} > H_{soft}$ (?) and $W_{hard} > W_{soft}$ (?).

The crater diameter is an important morphometric parameter in volcanic landform recognition; however, the final value of the crater diameter is the result of a complex series of processes, usually involving syn-eruptive, mass wasting processes of the crater walls, e.g. the 1977 formation of the Ukinrek maars in Alaska [357-358]. This makes the direct interpretation of crater diameter values more complicated than predicted by simple chemical and nuclear cratering experiments [313, 344]. Thus, the incremental growth (multiple eruptions + mass wasting) model seems to be a better explanation of the growth of a crater during phreatomagmatic eruptions [198-200, 267, 346, 359] and kimberlite volcanism [360-361]. Thus, the morphometric data of a fresh maar or tuff ring volcano contain cumulative information about the eruption energy, the location and depth of (shallow) explosion loci, as well as the stability of the country rock and associated mass wasting. The largest eruption dominated model may only be suitable to express energy relationships without the effects of mass wasting from the crater walls. This probably exists in only a few limited sites. For example, these eruptions should take place from a small-volume of magma, limiting the duration of volcanic activity and reducing the possibility of development of a diatreme underneath (Figure 8). Additionally, these eruptions should be in a consolidated hard rock environment, with high rock strength and stability. The following model for the interpretation of crater diameter and morphology data can be applied for phreatomagmatic volcanoes (Figure 8). This model integrates both conceptual models for crater and edifice growth, but the majority of craters experience complex development instead of the dominance of the largest explosion event. The likelihood of the largest explosion event dominated morphology is limited to simple, short-lived eruptions due to limited magma supply or vent migration in a hard rock environment (Figure 8). If the magma supply lasts, the development of a diatreme underneath starts that has a further effect on the size, geometry and morphology of the crater of the resultant volcano.

The crater diameter is often a function of basal edifice diameter (W_{co}) or height (H_{co}), creating ratios which are commonly used in landform recognition in extraterrestrial environments [177, 179, 340]. W_{co} is often difficult to measure because of the subjectivity in boundary delimitation [293]. This high uncertainty in delimitation of the crater boundary is a result of the gradual thinning of tephra with distance from the crater, with usually a lack of a distinct break in slope between the ejecta ring's flanks and the surrounding tephra sheet, e.g. the Ukinrek maars, Alaska [357-358]. Any break in slope could also be smoothed away by post-eruptive erosional processes. The crater height estimates vary greatly for maar-diatreme and tuff ring volcanoes, but they are usually ≤50 m [199, 223, 293]. This small elevation difference from the surrounding landscape gives rise to some accuracy issues, particularly regarding the establishment of the edifice height.

To demonstrate the limitations (e.g. input data accuracy, data type, and genetic oversimplification) of morphometric signature parameters on phreatomagmatic volcanoes, two examples (Pukaki and Crater Hill) were selected from the Quaternary Auckland volcanic field in New Zealand. Both volcanoes above were used to establish the average morphometric signature of an Earth analogue phreatomagmatic volcano [e.g. 293]. The early morphometric parameters were measured from topographic maps having coarse contour line intervals (e.g. 20–30 m), which cannot capture the details of the topography accurately. Some cross-checks were made on the basic morphometric parameters established from topographic maps and Digital Elevation Models (DEMs) derived from airborne Light Detection And Ranging (LiDAR) survey. The results showed that the differences in each parameter could be as high as ±40%. In addition, both the Pukaki and the Crater Hill volcanoes from Auckland were listed as "tuff rings" [293], due to the oversimplified view of monogenetic volcanism in the 1970s and 80s. Their eruption history, including volume, facies architecture and morphology, are completely different. The present crater floor of Pukaki volcano is well under the syn-eruptive surface, thus it is a maar volcano sensu stricto following Lorenz [199]. This was formed by a magma-water interaction driven phreatomagmatic eruption from a small volume of magma of 0.01 km^3 estimated from a DEM and corrected to Dense Rock Equivalent (DRE) volume [362-363]. The present facies architecture of this volcano seems quite simple (e.g. like an E_4 volcano in Figure 6). On the other hand, Crater Hill has a larger eruptive volume of 0.03 km^3 [362-363] and experienced multiple stages of phreatomagmatism (at least 3) and multiple stages of magmatic eruptions (at least 5) with many transitional layers between them, forming an initial tuff ring and an intra-crater scoria cone [80]. Later eruption formed an additional scoria cone and associated lava flow that filled the crater with lava up to 120 m in thickness [80, 364]. Consequently, Crater Hill is an architecturally complex volcano with complex eruption history, i.e. at least an E_{4226}. The important implication of the examples above and the usual complex pattern and processed involved on the establishment of the final geomorphology (e.g. incremental crater growth) are that morphometric signature (if it exists) can be only used for phreatomagmatic volcanoes if the eruption history of the volcano is reconstructed. In other words, parameters to express morphometric signature cannot be compared between volcanoes with different eruption histories (i.e. they are characterized by different phases with different eruption styles). Comparison without knowledge of the detailed eruption history could be misleading. Furthermore, the morphometric signature properties used in extraterrestrial volcano recognition for crater-type volcanoes should be reviewed, using a volcanological constraint on the reference volcano selection.

4.2.2. Cone-type monogenetic volcanoes

Cone-type monogenetic volcanoes, such as spatter- (Figure 2), scoria (or cinder; Figure 3) and tuff cones (Figure 5), are typically built up by proximal accumulation of tephra from low to medium (0.1–10 km in height) eruption columns and associated turbulent jets, as well as block/bombs that follow ballistic trajectories [188, 223, 288, 365-367]. Deposition from localized pyroclastic density currents is possible, mostly in the case of tuff cones [223, 265, 283] and rarely in the case of scoria cones from violent Strombolian eruptions [188]. The primary morphology of cone-type monogenetic volcanoes could be expressed by various morphometric parameters, including height (H_{co}), basal (W_{co}) and crater diameter (W_{cr}) and their ratios (H_{co}/W_{co} or W_{cr}/W_{co}), inner and outer slope angle or elongation. On a fresh edifice, where no post-eruptive surface modification has taken place, these morphometric parameters are related to the primary attributes of eruption dynamics and syn-eruptive sedimentary processes. However, there are potentially two valid models to explain their dominant construction mechanisms, including a ballistic emplacement with drag forces and fallout from turbulent, momentum-driven jets at the gas-thrust region [84, 182, 185, 368-369]. In both models, the angle of repose requires loose, dry media. This criterion is rarely fulfilled in the case of a tuff cone [e.g. 223, 234] and littoral cones that form during explosive interactions between lava and water [e.g. 203, 204]. In these cases the ejected fragments have high water-contents that block the free avalanching of particles upon landing [223, 252, 259, 283, 286]. This is inconsistent with other magmatic cone-type volcanoes; the growth processes of tuff and littoral cones are not discussed in further detail here.

The ballistics model with and without drag for scoria cone growth was proposed as a result of eye-witness accounts of eruptions of the NE crater at Mt. Etna in Sicily, Italy [185]. This model is based on the assumption that the majority of the ejecta of a volcanic cone is coarse lapilli and block/bombs (≥8–10 cm in diameter), thus they follow a (near) ballistic trajectory after exiting the vent (Figure 9A). Consequently, the particle transport is momentum-driven, as documented for the bomb/block fraction during bursting of large bubbles in the upper conduit during Strombolian style explosive eruptions [180, 183, 370]. The particle velocity of such bomb/blocks was up to 70–80 m/s for a sensu stricto Strombolian style eruption measured from photoballistic data [371-372]. However, recent studies found that the typical exit velocities are about 100–120 m/s [180, 183] and they could reach as high as 400 m/s [373]. These studies also showed that the typical particle diameter is cm-scale or less instead of dm-scale [180, 183], which cannot be derived purely from impact breakage of clasts upon landing [182, 368]. Especially during paroxysmal activity at Stromboli [374] or more energetic violent Strombolian activity [83, 163], the dominance of fine particles in the depositional records contradicts the ballistics emplacement model for the cones.

To solve the debate about cone growth, the jet fallout model was proposed [182], based on the fact that there is a considerably high proportion of fines in cone-building pyroclastic deposits [60, 182, 288]. This proposed behaviour is similar to the proximal sedimentation from convective plumes, forming cones with similar geometry to scoria cones, e.g. the 1886 eruption of Tarawera, Taupo Volcanic Zone, New Zealand [375] or the 1986 eruption of Izu-Oshima volcano, Japan [368]. The fines content does not fulfil the criteria of pure ballistic trajectory, thus turbulent, momentum-driven eruption jets (Figure 9A) should be part of the cone growth mechanisms [182, 368, 376]. As documented above, scoria cones demonstrate a wider range of granulometric characteristics [182] than previously thought [185]. These slight or abrupt changes of grain size within an edifice imply that the term "scoria cone" is not as narrow and well-defined as proposed in earlier studies [71, 185, 189, 301]. Consequently, there should be a spectrum of characteristics within the "scoria cones" indicating the existence of spatter-, lapilli- and ash-dominated varieties [68, 288]. Such switching from classical lapilli-dominated to ash- or block-dominated cone architectures reflects syn-eruptive reorganization of conduit-scale processes, including

1. multiple particle recycling and re-fragmentation during conduit cleaning [180, 377] or
2. changes in magma ascent velocity (i.e. increase or decrease in the efficiency of gas segregation) that in turn effect the viscosity of the magma [150, 163, 378].

The latter case may or may not cause a change in eruption style (e.g. a shift from normal Strombolian-style to violent Strombolian-style eruption) that could effectively lead to changes in the grain size distribution of the ejecta by possible skewing towards finer fractions (≤1–2 cm). This switching has significant consequences on pyroclast transport as well. The higher efficiency of magma fragmentation and production of finer pyroclasts (e.g. ash) causes more effective pyroclast-to-gas heat transfer in the gas-thrust region. A buoyant eruption column is created when the time of heat transfer is shorter than the residence time of fragments in the lowermost gas-thrust region [111, 182, 379]. Thus, particle transport shifts from momentum- to buoyancy-driven modes [182-183]. Based on numerical simulations, these changes in the way pyroclasts are transported are consistent with modelled sedimentation trends from jet fallout as a function of vent distance [182]. Once the eruption has produced medium ($Md\phi \leq 10$–20 mm) and coarse fragments ($Md\phi \leq 50$–100 mm), pyroclasts show an exponential decrease in sedimentation rates away from the vent [182]. This is in agreement with the trend predicted by the ballistic emplacement model [185]. However, once the overall fragment size is dominated by fines ($Md\phi \leq 2$–3 mm), the maximum sedimentation rate departs further towards the crater rim [182]. The threshold particle launching velocity is about ≥50 m/s [182]. The fragment diameter of $Md\phi \leq 2$–3 mm is consistent with the calculated threshold for formation of buoyant, eruptive columns during violent Strombolian activity [111], but is significantly finer than fragments (Mdϕ ≤10–12 mm) generated during some paroxysmal events recorded at Stromboli in Italy [374].

In the case of scoria cone growth the final morphology is not only dependent on the mode of pyroclast transportation via air, but also on significant post-emplacement redeposition. If particles are sufficiently molten and hot, their post-emplacement sedimentation processes usually involve some degree of welding/agglutination and rootless lava flows may form [171-172]. In this case, high irregularity in the flank morphology is expected due to the variously coalescent large lava clots and spatter [171-172, 380]. If the particles are brittle and cool, as well as having enough kinetic energy to keep moving, they tend to avalanche on the inclined syn-eruptive depositional surface [84, 141, 172, 185, 234, 288]. The avalanching grain flows often give rise to inversely-graded horizons or segregated lenses within the overall homogenous, clast-supported successions of the accumulating pyroclast pile, while the hot particles cause spatter-horizons in the statigraphy (Figure 10). In the earlier case, the criterion to sustain efficient grain flow processes on the initial flank of a pyroclastic construct is that the particles have to be granular media (i.e. loose and sufficiently chilled). The properties such as grain size, shape and surface roughness determine the angle of repose, which is a material constant [381]. These are all together responsible for the formation of usually smooth cone flank morphologies.

Classically, scoria cones are referred to as being formed by Strombolian-style eruptions, in spite of the fact that Stromboli volcano, Aeolian Islands, is not a scoria cone. In reality, scoria cones are formed by "scoria cone-forming" eruptions. Thus, the term scoria cone includes every sort of small-volume volcano with a conical shape and basaltic to andesitic composition. Additionally, during scoria cone growth, three major styles of internally-driven eruption types can be distinguished, Hawaiian, Strombolian and violent Strombolian, and an additional externally-driven eruption style, such as phreatomagmatism-dominated, is also expected (e.g. Figure 6). From the eruption styles listed above, at least the first three could individually form a "scoria cone", or similar looking volcano, which is rarely or never taken into account during interpretation of geomorphic data of a monogenetic volcano. The cone growth mechanism is, here, considered to be a complex interplay between many contrasting modes of sedimentation of primary pyroclastic materials, including transport via air (by turbulent jets and as ballistics) and subsequent redistribution by particle avalanching (Figure 9A). It is also important to note that cone growth is not only a constructive process; there could also be destructive phases. These processes (e.g. flank failure or crater breaching) alter the morphology in a short period of time. Consequently, the edifice growth is not a straightforward process (e.g. a simple piling up of pyroclastic fragments close to the vent), but rather a combination of constructive and destructive phases at various scales. The spatial and temporal contexts of such constructive and destructive processes are important factors from a morphometric stand point. In this chapter, two modes of cone growth mechanisms are recognized (Figures 9B and C) cones formed by:

1. a distinct and stable eruption style (e.g. E_{simple}) and by
2. various magma eruption styles with transitions between them (e.g. $E_{compound}$ or $E_{complex}$), during the eruption histories.

The simple cone growth model is applicable to cone growth from a single and stable eruption style, e.g. Strombolian, Hawaiian or violent Strombolian styles only (Figure 9B). Theoretically, if an edifice is formed by a repetition of one of these well-defined and stable eruption style without fluctuation of in efficiency or any changes, the crater excavation and diameter, as well as the mode of pyroclast transport, should vary in a narrow range, i.e. fragmentation mechanism 'constant' (Figure 9B). The first explosions when the gas bubbles manage to escape from the magma leading to the explosive fragmentation of the melt, usually take place on the pre-eruptive surface or at a few tens of meters deep [57, 84, 185, 382-383]. After the first eruption, there is a time involved to either excavate the crater or pile up ejecta around the vent, which is in turn dependent on the eruption style and the tephra accumulation rate. Once the threshold crater rim height is reached, the height and its position are attached to properties of certain eruption dynamics (Figure 9B). In other words, the eruption style and related pyroclast transport distribute tephra to limited vertical and horizontal directions. Due to the steadiness of eruption style, particle fallout from the near-vent, dilute jets at the gas-thrust region and ballistics have an 'average' vertical distance that they travel. This 'average' will determine the width and relative offset of the crater rim above the crater floor, which grows rapidly during the initial establishment of the crater morphology (first cartoon in Figure 9B) and then slows down (second and third cartoons in Figure 9B). Of course the location, morphology of the crater rim and floor are not just dependent on the efficiency of the magma fragmentation, but the subsequent wall rock failure, as documented by Gutmann [369], similar to the development of maar-diatreme volcanoes (Figure 7). Due to the stability of the conduit and a single eruption style, the role of such failures in the control of morphology is minimal in comparison with complex modes of edifice growth (see later).

This growth model is applicable to simple eruptions, with steady eruption styles and possibly steady magma discharge rates, such as the violent-Strombolian eruptions during the Great Tolbachik fissure eruptions in Kamchatka, Russia [84, 382] or the Strombolian-style eruption during the growth of the NE crater at Mt. Etna [185]. During the Tolbachik fissure eruptions, the rim-to-rim crater diameter of Cone 1 grew rapidly from 56 m to 127 m during the first 5 days, and later slowed and stayed in a narrow range around 230–280 m during the rest of the eruptions [84, 382-385]. This is similar to certain stages of growth of the NE crater at Mt. Etna [84, 185]. This means the crater widens initially until a threshold width is reached, which corresponds to the maximum strength of the pyroclastic pile and the limits of the eruption style (Figure 9B). This trend seems to be consistent with an exponential growth of crater width over time until the occurrence of lava flows, as documented at Cone 1 and 2 of the Tolbachik fissure eruptions [e.g. 383]. The maximum range of the crater width appeared to be reached once the lava outflowed from the foot of the cones. This can be interpreted as an actual decrease in magma flux fuelling the explosion, and therefore the pyroclast supply for flank growth. Assuming that the magma is torn apart into small particles and are launched with sufficiently high initial velocity to the air, e.g. Strombolian eruptions, the particles have enough time to cool down, thus upon landing they initiate avalanching due to their kinetic energy. These processes will smooth the syn-eruptive surface to the angle of repose of the ejected pyroclasts if the pyroclast-supply is high enough to cover the entire flank of the growing edifice. If the angle of repose of the tephra, θ, depending on granulometric characteristics, and the height of the crater rim, H, are known at every stage of the eruption ($i = 1,2,\ldots, n$), the flank width (W_i) would indicate at a certain stage of growth (Figure 9B):

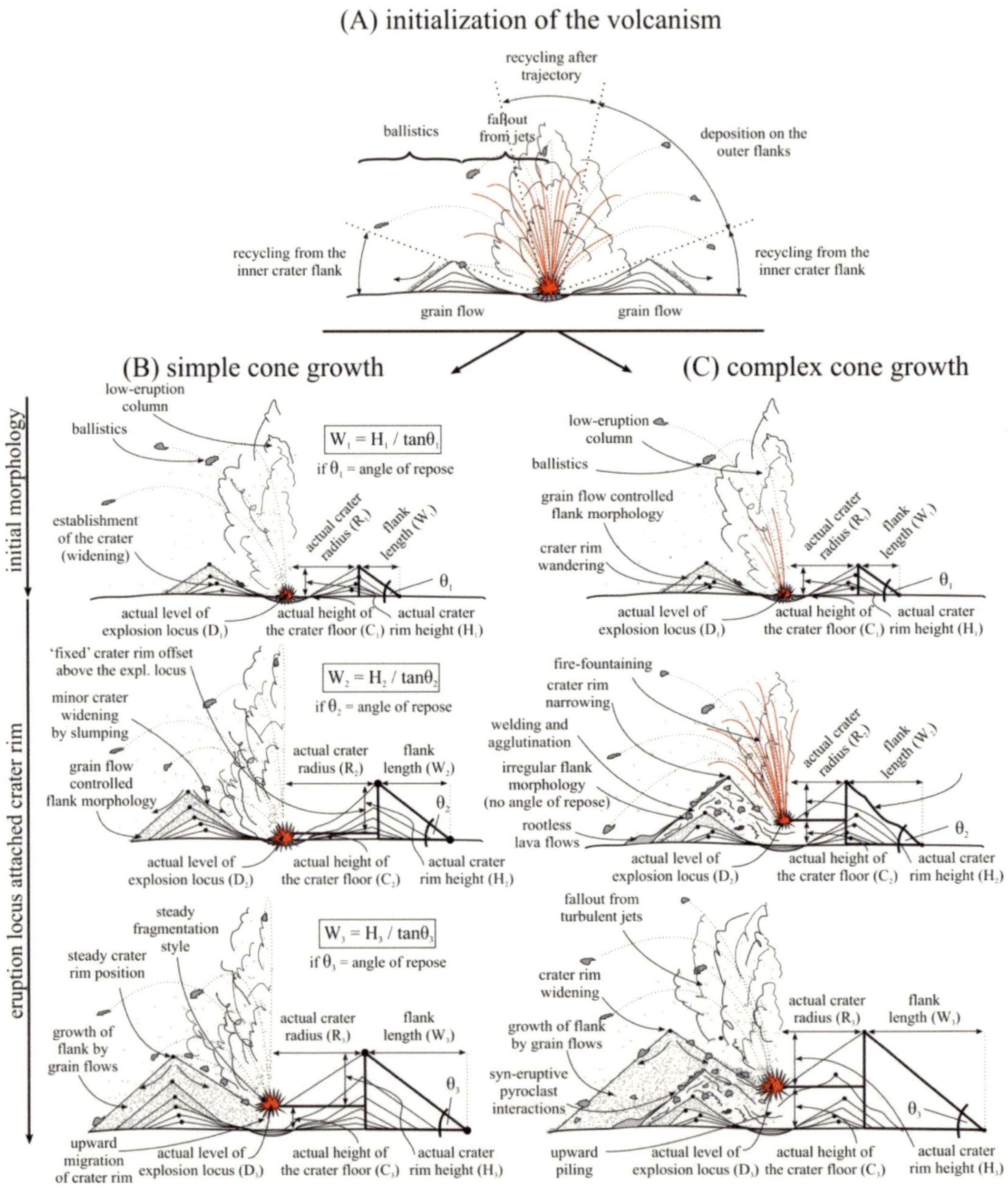

Figure 9. Growth of cone-type volcanoes (e.g. scoria cones) through ballistics' emplacement and jet fallout models. After the initialization of the volcanism (A), there are two different types of cone growth: simple and complex. The simple cone growth model (B) supposes a steady fragmentation mechanism and associated eruptive style and sedimentary processes, thus the angle of repose is near constant, $\theta_1=\theta_2=\theta_3$ over the eruption history. The variation of the relative height of the crater rim (H) above the location of the explosion locus and the radius of the crater (R) is 'fixed' or varies in a narrow range ($H_1 \leq H_2 = H_3$ and $R_1 \leq R_2 = R_3$), after the initial rapid growth of rim height and crater width. The simple cone growth model implies that constructional and flank morphologies are the results of a major pyroclast-transport mechanism (e.g. grain flows in the case of a scoria cone or welding, formation of rootless lava flows in the

case of spatter cones). On the other hand, the complex cone growth model (C) involves gradual or abrupt changes in eruption style, triggering multiple modes of pyroclast transport, and possible changes in the relative height and diameters of the crater rim ($H_1 \neq H_2 \neq H_3$). The granulometric diversity allows for post-emplacement pyroclast interactions which permits or blocks the free-avalanching of the particle on the outer flanks. Therefore, the angle of repose is constant and may not always be reached ($\theta_1 \neq \theta_2 \neq \theta_3$), especially when clast accumulation rate and temperature is higher, causing welded and agglutinated spatter horizons.

$$W_i = H_i / \tan \theta_i \tag{6}$$

When the eruption style is constant during the eruption and produces pyroclasts of the same characteristics, the pyroclasts behave as granular media (i.e. they are controlled by the angle of repose); the aspect ratio between the height and flank width should be in a narrow range until the 'tandem' relationship is established between the crater rim and explosion loci. Once the eruption is in progress, the explosion locus stay either at the same depth (e.g. at the pre-eruptive ground level) or migrates upwards if enough material is piled up within the crater resulting in a relative up-migration of the crater floor over time (Figure 9B). Consequently, this upward migration of the explosion locus should result in an elevation increase of the crater rim, if the eruption style remains the same. Once the crater rim rises, the majority of the pyroclasts avalanche downward from higher position, which creates a wider flank and increases the overall basal width of the edifice. The repetition of such crater rim growth and flank formation takes place until the magma supply is exhausted. Due to the dominance of loose and brittle scoria in the edifice, the H_{co}/W_{co} ratio is in a narrow range in accordance with Eq. 3. Although, there is a slight difference between H_{co}/W_{co} and H_i/W_i, because the former contains the crater. The crater is possibly the most sensitive volcanic feature of a cone that could be modified easily (e.g. shifting in eruption style in the course of the eruptions and/or vent migration). Finally, this growing process is in agreement with the earlier documented narrow ranges of the H_{co}/W_{co} ratio that are governed by the angle of repose of the ejecta [84, 301].

The complex cone growth model assumes the cone is the result of many distinctive eruption styles and changes between them, which trigger a complex cone growth mechanism from various eruptive and sedimentary processes (Figure 9C). Such changes usually relate to changes from one eruption style to another, which have consequences for the morphological evolution of the growing cone. The switching in efficiency of magma fragmentation can be triggered by the relative influence of externally and internally-governed processes or reorganization of internal or external controls without shifting from one to the other. An example for the first change is a gradual alteration in the abundance of ground water and consequent shift from phreatomagmatism to magmatic eruption styles. On the other hand, the reorganization of processes in either in the internally- or externally-driven eruption styles could be related to the changes in degree of vesiculation and efficiency of gas segregation in the conduit system. Each of these changes could modify the dominant eruption style that determines the grain size, pyroclast transport and in turn edifice growth processes. An internal gas-driven magma fragmentation leading to a Hawaiian eruption produces larger (up to a few meters) magma clots that are emplaced ballistically, while fines are deposited from turbulent jets and a low-eruption column, in agreement with the processes observed at

Kilauea Iki in Hawaii [e.g. 176]. In this eruption style, the dominance of coarse particles (e.g. lava clots up to 1–2 m in diameter) are common. These large lava clots cannot solidify during their ballistic transport, and therefore after landing they could deform plastically, weld and/or agglutinate together or with other smaller pyroclasts [171, 380], depending on the accumulation rate and the clast temperature [172]. As a result of the efficient welding processes, the landing is not usually followed by free-avalanching, unlike loose, sufficiently cooled, brittle particles from other eruption styles, e.g. normal Strombolian styles. At the other end of the spectrum, energetic eruptions styles, such as violent Strombolian or phreatomagmatic eruptions, tend to generate localized sedimentation from pyroclastic density currents such as base surges. Similar to spatter generation, pyroclasts from pyroclastic density currents do not conform to the angle of repose. In contrast, the pyroclasts from sensu stricto Strombolian-eruptions are sufficiently fragmented to suffer rapid cooling during transport either as ballistics or fallout from turbulent jets and eruption columns (Figure 9C). Thus, the particle with sufficient kinetic energy post-emplacement can fuel grain avalanches on the outer flanks of a growing pyroclastic pile. The active grain flows deposit pyroclasts on flanks in accordance with the syn-eruptive depositional surface properties and granulometric characteristics that provide the kinetic and static angle of repose of the granular pile. This is sustained until the ejected material is characterized by the same granulometric characteristics. With increasing degree of magma fragmentation, the dominant grain size of the tephra decreases. The increased efficiency of the magma fragmentation (e.g. violent Strombolian eruptions) commonly results in a higher eruption column, and therefore broader dispersion of tephra. The high variability and fluctuating eruption styles form a wide range of textural varieties of edifice such as Lathrop Wells in Southwest Nevada Volcanic Field, Nevada [60, 188], Pelagatos in Sierra Chichinautzin, Mexico [14, 378] or Los Morados, Payún Matru, Argentina [273]. Due to the variability in eruption styles, particles have different surfaces or granulometric characteristics. These differences induce some fine-scale post-emplacement pyroclast interactions. The effect of these syn-eruptive pyroclast interactions could prevent effective grain flow processes, 'reset' or delay (previous) sedimentary processes on the flanks of a growing conical volcano. Thus it could be a key control to determine the flank morphology of the resultant volcanic edifice. Examples for syn-eruptive granulometric differences can be found in the pyroclastic succession of the Holocene Rangitoto scoria cone in Auckland, New Zealand (Figure 10). During deposition of beds with contrasting dominant particle sizes, the angle of repose is not always a function of granulometric properties, but can be a result of the mode of pyroclast interactions with the syn-eruptive depositional surface. There is wide range of combinations of pyroclast interactions among ash, lapilli, block and spatter (Figure 10). Such transitions in eruption style and resulting pyroclast characteristics are important due to their blocking of freshly landed granular particles conforming (immediately) to the angle of repose expected if they were circular, dry and hard grains. For instance, the deposition of an ash horizon on a lapilli-dominated syn-eruptive surface must fill the inter-particle void, causing ash 'intrusions' into the lapilli media (Figure 10). These pyroclast interactions possibly slow the important cone growing mechanisms (e.g. grain flow efficiency) down, creating a sedimentary delay when the angle of repose is established on the syn-eruptive surface. In the complex cone growth model, the changes of eruptions style have an effect on the relative position of the crater rim shifting the position of the maximum sedimentation and the mode of pyroclast transport [182]. For example, switching from normal Strombolian to violent Strombolian style eruptions could increase the initial exit velocity from the usual 60–80 m/s [180, 183, 372, 386] to higher values, ≥150–200 m/s [111, 382, 387-388]. This shift in eruption style causes further decrease in the grain size of the ejecta [163, 367, 389]. This change could altogether cause migration of the location of the maximum sedimentation point further towards the crater rim if the tephra is transported by jet instead of pure ballistic trajectories [182]. Such changes in eruption styles introduce significant horizontal and vertical crater rim wandering during the eruption history of a monogenetic volcano (Figure 9C). This wandering modifies, in turn, the sedimentary environment leading the formation of cones with complex inner facies architectures, which may or may not be reflected by the morphology. Consequently, the major control in this cone growth model is on the eruption styles and their fluctuations over the eruption history.

In both cone growth models, lava flows can occur as a passive by-product of the explosive eruptions 'draining' the degassed magma away from the vent. Once the magma reaches the crater without major fragmentation it can either form lava lakes and/or (later) intrude into the flanks, increasing the stress [111, 298, 369]. When the magma pressure exceeds the strength of the crater walls, the crater wall may collapse or be rafted outwards [111, 369]. Sometimes, the magma flows out from foot of the cone, possibly fed from dykes [188, 273, 367, 369, 388, 390-391]. In this case the flank collapse is initiated by the inflation of the lava flow by discrete pulsation of magma injected beneath the cooler dyke margins [392-393] beneath a certain flank sector of a cone [111]. If the lava yield strength is reached and overtakes the pressure generated by the total weight of a certain flanks sector, flank collapse and subsequent rafting of remnants are common [111, 273, 390], leading to complex morphologies with breached craters and overall horse-shoe shape [322, 394]. The direction of crater breaching of scoria cones may not always be the consequence of effusion activity, but may coincide with regional/local principal stress orientations or fault directions [298, 315, 395-396]. If the magma supply is sufficient, the edifice that has been (partially) truncated by slope failure could be (partially) rebuilt or 'reheal' as documented from Los Morados scoria cone in Payun Matru, Argentina [273] or Red Mountain, San Francisco Volcanic Field, Arizona [397]. The direction of lava outflow from a cone commonly overlaps with the overall direction of background syn-eruptive surface inclination. This overall terrain tilt could cause differences in the tension in the downhill flank sector [82, 298, 391, 398]. Any kinds of changes during basaltic monogenetic eruptions could cause sudden decompression of the conduit system, leading to a change in the eruption style [273, 399]. Consequently, these destructive processes are more likely to occur during complex edifice growth, instead of simple cone growth. These changes could account for the fine-scale morphometric variability and architectural diversity observed in granular pile experiments and field observations [68, 182, 298, 314, 398]. Evidence of fine-scale morphometric variability due to lava outflow and crater breaching is observed through systematic morphometric analysis on young (≤4 ka) scoria cones in Tenerife [398]. Two types of morphometric variability were found: intra-cone and inter-cone variability. Intra-cone variability was characterised among individual flank facets. The slope angle variability was calculated to be as high as 12° between flanks sectors along (±45°) and perpendicular (±45°) to the main axis of the tilt direction of the syn-eruptive terrain [398]. This is about a third of the entire range of the natural spectrum of angles of repose of loose, granular material, i.e. scoria-dominated flanks [84, 182]. Inter-cone variability was detected on cones of the same age. According to Wood [71], these fresh cones should be in a narrow morphometric range (e.g. slope angle of 30.8±3.9°) based on fresh and pristine scoria cones analysed from the San Francisco Volcanic Field, Arizona [71, 324]. In contrast to this expected high value, the average slope angles of the studied cones from Tenerife turned out to vary from 22° to 30°, which has significant impact on many traditionally used interpretations of morphometric data including morphometric-based dating [e.g. 338], the morphometric signature concepts and erosion rate calculations. Both inter- and intra-cone variability were interpreted as a sign of differences in syn-eruptive processes coupled between internal, such as changes in efficiency of fragmentation, magma flux, effusive activity and associated crater breaching [398], and external controlling factors, such as interaction between pre-existing topography and the eruption processes [398]. All of these diverse eruptive and sedimentary processes are somehow integrated into the fresh morphology that is the subject of morphometric parameterization. A few of these processes could be detected while others could not, using morphometric parameters at one or multiple scales. For instance, some of this morphometric variability is usually undetectable using topographic maps and manual geomorphic analysis. This narrow variability, possibly associated with syn-eruptive differences in cone growth rates and trends, is in some instances in the range of the accuracy of the morphometric parameterization technique. For example, a manual calculation of slope angle from a 1:50 000 topographic maps with contour intervals of 20 m is ±5° [e.g. 312]. The fine-scale morphometric variability cannot be assessed accurately with this high analytical error range. On a DEM (either contour-based or airborne-based) with high vertical and horizontal accuracies (e.g. Root Mean Square Error under a few meters), this small-scale variability can be detected [e.g. 398]. An important consequence of this variability is that the initial geometry of the cone-type volcanoes, such as scoria cones, is not in a narrow range as previously expected [e.g. 71]. In other words, the morphometric signature of cone-type volcanoes are wider than described before, limiting the possibility of morphometric comparisons of individual edifices (especially eroded edifices) due to the lack of control on their initial geometries. It seems on the basis of the presented eruptive diversity, comparative morphologic studies should be focused on comparing cones that have similar processes involved in their formation (i.e. E_{simple}, $E_{compound}$ or $E_{complex}$) and limited post-eruptive surface modifications (i.e. younger than a few ka in age). The morphology of a fresh (≤ a few ka) cone-type volcano is the result of primary eruptive processes; therefore, the morphometric parameters should be interpreted as the numerical integration of such eruptive diversity and mode of edifice growth. As stated by Wood [84], only fresh cones must be used for detecting causes and consequences of changes in morphology. When cone-type volcanoes from a larger age spectrum, e.g. up to a few Ma [e.g. 312, 337], are studied, the primary, volcanic morphometric signatures are modified by post-eruptive processes. Thus, they contain a mixture of syn- and post-eruptive morphometric signatures. Hence, interpretation of large morphometric datasets should be handled with care. Furthermore, it is also evident that not all geomorphic changes experienced by the edifice during the eruption history are preserved in the final volcano morphology. This could be due to, for instance, rehealing of the edifice after a collapse event, or changing eruption style, reflecting the complex nature of cone growth. Future research should focus on finding the link between the eruptive processes and morphology, as well as finding out how syn-eruptive constructive and destructive processes can be discriminated from each other on an 'unmodified', fresh cone.

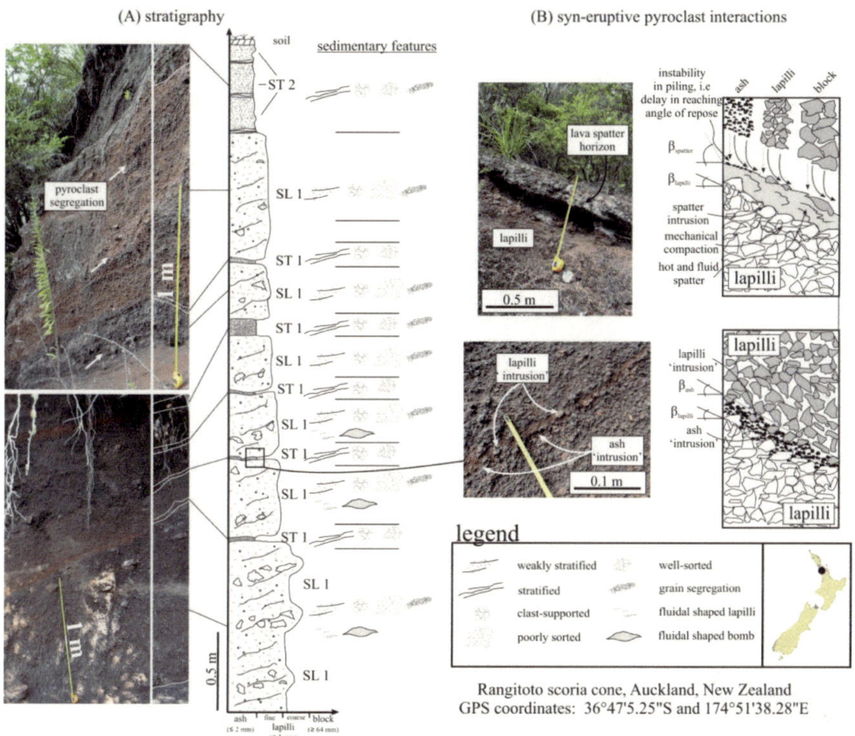

Figure 10. Stratigraphic log (A) of Rangitoto scoria cone, Auckland Volcanic Field, New Zealand showing the typical alteration of scoriaceous lapilli (SL1) and scoriaceous ash (ST1-2). Some examples are for syn-eruptive pyroclast interactions during growth of cone-type volcanoes such as scoria cones (B).

5. Degradation of monogenetic volcanoes

Once the eruption ceases, a bare volcanic surface is created with all of the primary morphologic attributes that have been determined by the temporal and spatial organization of the internally- and externally-controlled eruptive and sedimentary processes during the eruption history (Figure 6). The fresh surface is usually 'unstabilized' and highly permeable due to the unconsolidated pyroclasts, but there is often some degree of welding/agglutination, the presence of compacted ash, or lava flow cover. The degradation processes of a volcanic landform have a significant effect on the alteration of primary, volcanic geomorphic attributes (e.g. lowering of H_{co}/W_{co} or slope angle values). The significant transition from primary (pristine) volcanic to erosion landforms fundamentally starts when, for example, soil is formed, vegetation succession is developed or the surface is dissected over the primary eruptive products. However, modification of the pristine, eruption-controlled morphology could happen by non-erosion processes. For example, rapid, post-eruptive subsidence of the crater of a phreatomagmatic volcano due to diagenetic compaction, or lithification of the underlying diatreme infill during and immediately after the eruptions [400-401]. This usually leads to deepening of the crater or thickening of the sediments accumulated within the crater. Some compaction and post-eruptive surface fracturing, due to the gradual cooling down of the conduit and fissure system, is also expected at cone-type volcanoes, such as after the formation of Laghetto scoria cones at Mt. Etna, Italy [184] or Pu'u 'O'o spatter/scoria cone in Hawaii [173]. These processes could cause some geomorphic modification that may affect the morphometric parameters. On the other hand, the long-term surface modification of a monogenetic volcanic landform is related to degradation and aggradation processes over the erosion history. The structure of the degradation processes that operate on volcanic surfaces can be classified into two groups based on their frequency of occurrence and efficiency:

1. long-term (ka to Ma), slow mass movements, called 'normal degradation', as well as
2. short-term (hours to days), rapid mass movements, called 'event degradation'.

In the following section a few common degradation and aggradation processes are discussed briefly.

5.1. Long-term, normal degradation of monogenetic volcanoes

Normal degradation is a long-term (ka to Ma) mass wasting process that occurs by a combination of various sediment transport mechanisms and erosion processes such as rill and gully erosion, raindrop splash erosion, abrasion or deflation. Normal degradation requires initiation of the erosion agent that is usually the 'product' of the actual balance between many internal and external degradation controls at various levels, such as the climate or inner architecture (Figure 11). The external environment (e.g. annual precipitation, temperature or dominant wind direction etc.) is recognized as a major control on degradation [68, 71, 324-325, 337-338], influencing the chemical weathering and rates of CO_2 consumption [e.g. 402]. However, there is no single control on chemical weathering rates; the actual weathering rates are often a function of many controlling conditions [e.g. 403]. Important controls on weathering rate could be the climatic settings (e.g. surface runoff, moisture availability, temperature, atmospheric CO_2 level, rates of evaporation [e.g. 404]), the tectonics (e.g. the post-orogenic increase in chemical weathering that decreases the atmospheric CO_2 concentration [e.g. 405]), the geomorphology (e.g. age of the surface, surface drainage system, rates of sediment transport, relief, soil cover, sediment composition [e.g. 406]) or the biology (e.g. microorganism, plant cover, animal activity [e.g. 407]).

A combination of the abovementioned controls and processes on chemical weathering interacts in many ways depending on the internal composition and characteristics of the volcanics exposed to the environment (Figure 11). The importance of internal controls on degradation seems to be neglected by earlier studies on monogenetic volcanic edifices [e.g.

71] in contrast with recent studies [e.g. 68]. The facies architecture and granulometric characteristics of a volcanic surface govern how the edifice reacts to the environmental impacts, e.g. the flanks drain the rainwater 'overground' leading to the formation of rills and gullies or allow infiltration [71, 408-412]. The pyroclast-scale properties are determined by fluctuation of eruption styles during the eruption history, leading to accumulation of pyroclasts with contrasting geochemical, textural and granulometric characteristics. This pyroclast diversity will be responsible for the various rates of chemical weathering. Additionally, this diversity has an effect on the mode and efficiency of sediment transport during the course of degradation. For instance, the 'stability' (or amount of loose particles on the flanks) causes slight differences in rates, styles and susceptibility for erosion. The stability could increase with the formation of mature/immature soil, thick accumulation of weathering products, denudation of a lava-spatter horizon and/or heavy vegetation cover, which altogether help to stabilize the landscape. These changes on the mineral- to pyroclast-scales lead to transitions from 'unstabilized' to 'stabilized' stages. The duration of the transition depends on many factors (e.g. Figures 12A and B), such as the initial surface morphology, granulometric characteristics, volcanic environment and climatic settings [408-410, 413-417]. The transition could be as short as a couple of years if the volcanic surface is characterized by the dominance of fines, e.g. ejecta ring around a tuff ring, and typically exposed to a humid, tropical climate [408]. In arid climates the lag time between soil formation is significantly longer (if it takes place at all), up to 0.1–0.2 Ma [188, 413]. There are extreme environments where the soil/vegetation cover can barely be developed due to the high rates of volcanic degassing and acid rain, e.g. the intra-caldera environment in Ambrym, Vanuatu [e.g. 408] or cold polar regions, e.g. Deception Island, Antarctica [e.g. 418, 419]. Changes in surface stability could be governed by gradual denudation of inner, texturally compacted (e.g. welded or agglutinated spatter horizons or zones) pyroclastic units. This leads to rock selective erosion and higher preservation potential of an edifice in the long-term [e.g. 68]. Consequently, the degradation processes cannot be separated from the architecture of the degrading volcanic edifice, and therefore the erosion history is strongly attached to the eruption history. In this respect, the erosion history and rates seem to be governed (at least on one hand) by the time-lagged denudation of pyroclastic beds with varying susceptibility to erosion. In other words, the rate and style of degradation are theoretically the 'inverse' of the eruption history if the external controls are steady over the erosion history.

The actual balance between the internal and external controls determines the dominant rates and mode of sediment transport mechanism at a given point on the flanks of a monogenetic volcano (Figure 11). The mode and style of erosion of monogenetic volcanic landforms can be subdivided into 'overground' and 'underground' erosion. The long-term, overground degradation of volcanic surfaces can be accounted for through water-gravity (including rainfall, sea or freshwater, underground water or ice or various lateral movements of sediment/soil cover due to gravitation and water), and wind erosion agents. The sediment transport fluxes of such erosion agents and their time-scales are highly variable. The most significant, normal degradation processes on bare unstable (volcanic) surfaces are possibly triggered by the presence of water. Rainfall erosion causes small-scale movement of particles up to a couple of cm in diameter or chunks of soil when the rain drops impact on the

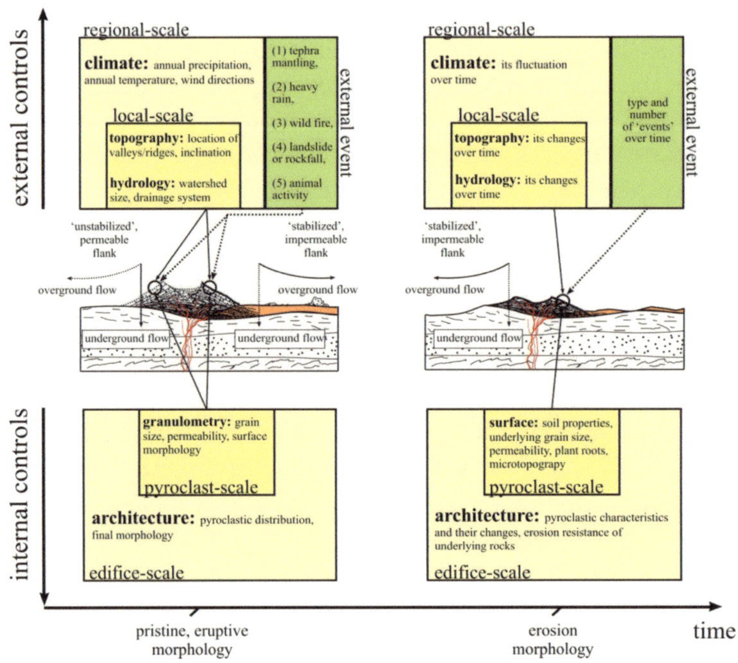

Figure 11. Conceptualized model for the configuration of internal and external degradation controls on determining erosion agents and sediment transport processes at a given point on 'unstabilized' and 'stabilized' volcanic surfaces over the erosion history. Degradation of a monogenetic volcano takes place by both long-term (ka to Ma; dark and light yellow boxes) and short-term, event-degradation processes (hours to years; green boxes) with different rates over the erosion history. The effects of such degradation processes take place both 'overground' and 'underground'.

surface [e.g. 420]. Besides the impact-triggered (or raindrop splash) sediment detachments, there are associated processes such as minor raindrop-induced surface wash [420-421]. On volcanic surfaces, the raindrop introduced and/or wind-driven rain splash erosion and related downward sediment transport on steep flanks (20–25°) is about 15.8 cm/yr for pyroclasts with diameter of 1.7 cm [323]. The regional-scale lowering of the landscape (over a 100 m length with 10° of slope angle) associated with rain-splash could be in the range of 0.1–1 mm/yr [e.g. 422]. This equivalent to a long-term erosion rates of 0.1 t/km/yr, calculating with 2.5 mm median rain drop diameter on a flank with 10° of slope angle [422]. If rainfall intensity exceeds the soil's infiltration capacity at any time during a rainfall event, overground flow, such as unchannelized sheet flow can be generated [e.g. 423]. The erosion capacity of sheet flow is higher than the rain-splash, but it is significantly lower than mass-wasting once the rill and gully network is developed (e.g. Figures 12E and F). Drainage system development on the flanks of monogenetic volcanoes is found to correlate with the age of the volcanic edifice [70, 321, 324, 424]. Although, the required time for their formation could be as short as a couple of months or years and it could develop on the gentle flanks (e.g. a tuff ring with slope angle of 5–10° in Figures 12E and F) which anomalously short period of time could introduce error in the relative morphology-based dating [425]. The fluvial erosion could remove sediment in a range of 10–100 000 t/km^2/yr [e.g. 426]. These overground surface processes, such as rill and gullies, and various soil/sediment creep [70-71, 323, 408, 425] have been accounted as a major mass wasting mechanism over the degradation history of a monogenetic volcanoes. The effect of these overground degradation processes were modelled mostly on scoria cones [324-325, 327, 337].

All varieties of soil and sediment creep and solifluction processes, such as soil and frost creep and gelifluction, are usually slow processes [e.g. 427], in comparison with surface runoff. Thus, these processes modify the volcano flanks' morphology constantly and over a longer period of time. The rates of erosion vary depending on the topography (e.g. slope gradient), sediment/soil properties (e.g. proportion of fines, moist content) and predominant climate (e.g. amount and type of precipitation, annual temperature), but rarely exceed the downhill movement rates of 1 m/yr and the volumetric velocity of between 1×10^{-10}–1×10^{-8} $km^3/km/yr$ [e.g. 427]. The sediment transport rates introduced by solifluction are many orders of magnitude smaller than the erosion loss of a volcano by fluvial processes. In cold semi-arid and arid regions, the ice plays the major role on the sediment transport [e.g. 428]. Consequently, the surface modification and movements are related to diurnal and annual frost-activity, such as ground freezing and thawing cycles. On scoria cones at the periglacial Marion Island, South Indian Ocean, dominant sediment transportation processes on scoria cone flanks are the needle-ice-induced frost creep related to the diurnal and possibly annual frost cycles [428]. The frost creep rates are 53.2 cm/yr for ash (≥70% of grains ≤2 mm), 16.1 cm/yr for lapilli (≥30% of grains between 2–60 mm), and 2.6 cm/yr for bomb/blocks (≥70% of grains ≤60 mm), based on measurements on painted rocks [428]. The rates are primarily controlled by the predominant grain-size of the sediment, the slope angle of the underlying terrain and altitude [428]. Based on the transportation rates, the processes are the same or an order of magnitude faster than rain-splash-induced pyroclast transport in a semi-arid environment, such as San Francisco volcanic field in Arizona [323].

Probably the most effective overground degradation process on a freshly created volcanic surface is the wave-cut erosion. In rocky, coastal regions, the wave cut notch moving back and sideward removes mass by hydraulic action and abrasion [e.g. 429]. Wave-cut erosion mostly affects tuff cones that are located on or offshore (Figures 12C, D, E and F). In this environment abrasion is a common syn-eruptive [430] or post-eruptive erosion process, e.g. Surtsey [431-432] or the early formation Jeju Island, Korea [433]. For instance, post-eruptive coastal modification through abrasion generated an area-loss of about 0.2 km^2 between 1975 and 1980 on Surtsey island in Iceland [431]. The rate of volume-loss of non-volcanic coastal regions is in the range of 10 000 to 100 000 t/km/yr [e.g. 434]. This enhanced rate of mass-removal is in agreement with advanced states of erosion on a recently formed tuff cone within Lake Vui in the caldera of Ambae volcano, Vanuatu, which formed in 2005 [263]. The initial surface has been intensively modified by wave-cut erosion and slumping from the crater walls, leading to an enormous enlargement of the crater and crater breaching over 10 months (Figures 12C and D).

The effect of wind deflation is often limited, especially in humid climates. However, there are examples in volcanic environments where, in spite of the high annual precipitation, the sediment transport rates are still significantly high due to wind action, e.g. in some parts of south Iceland, around 600 t/km/yr [435]. Expressing the long-term effect and rate of sediment transport by wind is complicated due the high variability of wind intensities (i.e. storm events versus normal background intensities) and directions [e.g. 436]. Sediment transport fluxes can vary in a wide range as a function of wind energy and surface characteristics (e.g. sediment availability, vegetation cover or water saturation). Long-term sediment transport by wind in volcanic areas (e.g. Iceland) is in the range of 100 to 1000 t/km/yr [435-436]. However, this sediment transport rate is in relation, but it is not equivalent to the erosion rates. Furthermore, efficient wind transportation as bed load by creep and saltation, and suspended load, is limited to particles generally ≤8 mm in diameter [e.g. 436]. This granulometric limit is crucial for the long-term erosion of volcanic landforms that built up from coarser pyroclasts, such as coarse lapilli-dominated scoria cones. Significant increase is observed during storm events, when these sediment transport rates could reach as high as a couple of percentage of the annual fluxes within an hour [435]. Thus, the long-term approximation of the sediment transport rates could be interpreted as cumulative values of normal, background and increased, storm/related erosion rates [e.g. 435]. In addition, there are a few examples such as Surtsey, where the wind deflation is considerable. In Surtsey, the strong wind is responsible for the polishing of palagonitized tephra surfaces and transporting and redistributing unconsolidated tephra on the freshly created island [431]. Direct observation of short-term volumetric change of a young scoria cones (e.g. Laghetto or Monte Barbagallo, ca. 2700–2800 m asl) is through deflation by wind in the summit region of Mt. Etna, Sicily [326]. Surface modification is inferred to occur on the windward side of the Monte Barbagallo cones [326]. The wind likely induces some minor pyroclast disequilibrium on the flanks that may lead to minor rock fall events or initiate grain flows [326]. In real semi-arid areas, wind-erosion is an important transport agent and surface modificator over unstabilized volcanic surfaces such as the Carapacho tuff ring in the Llancanelo Volcanic Field, Mendoza, Argentina (Figure 12A). The layer-by-layer stripping of the volcanic edifice is completely visible on the windward side of the erosion remnant facing the Andes. On the other hand, the wind-blown sediments can accumulate over time leading to sometimes expressible aggradation on volcanic surfaces (Figure 13A). Accumulation of aeolian addition could significantly contribute to the soil formation by gaining excess material, e.g. quartz or mica [415, 437]. Due to the generally high roughness of pyroclast- or lava rock-dominated surfaces (e.g. highly vesicular scoria or a'a lava flow), the wind slows down, leading to sedimentation and later accumulation of wind-transported particles [415-416, 437-438]. The wind-induced aggradation helps to reduce the transition time between an 'unstabilized' to a 'stabilized' surface by developing desert pavement in semi-arid/arid desert environments [415-416].

The previously mentioned, generally long-term overground degradation processes often account for most of the volumetric loss and surface modification of monogenetic volcanoes [e.g. 71]. In the case of the underground degradation, the surface water leaves the system through the groundwater if the actual soil infiltration capacity exceeds the rainfall intensity [423]. This underground water can remove weathering products (e.g. leaching of cations from the regolith) as dissolved sediment fluxes [e.g. 439]. The rate of chemical weathering could be extremely high in humid [e.g. 439] and lower in moderate climates [e.g. 440], based on chemical and solute-derived weathering data from rivers draining mafic to intermediate igneous rocks. The

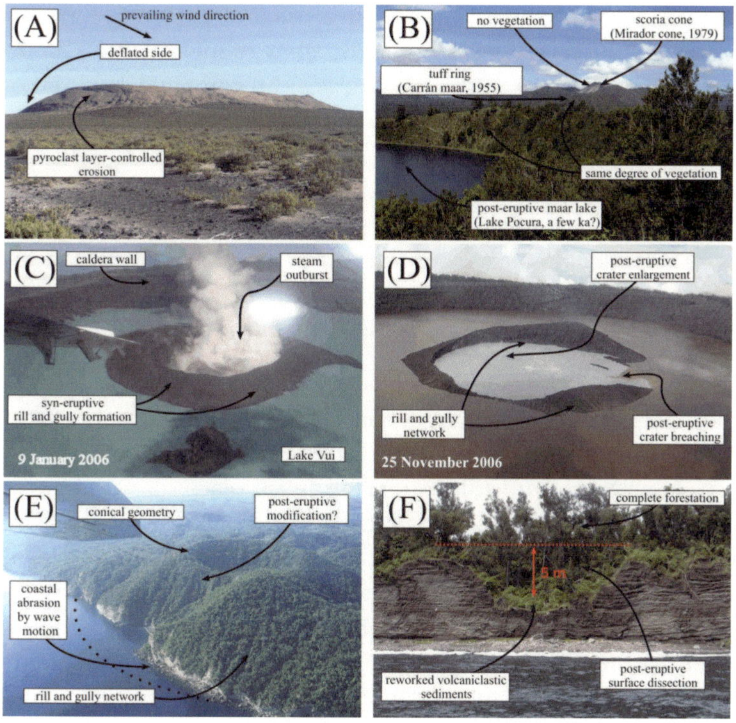

Figure 12. A) Contrasting rates of wind erosion on Pleistocene Carapacho tuff ring [61] under a semi-arid climate in the Llancanelo Volcanic Field, Mendoza, Argentina. It is interesting to note that the wind-erosion is strongly controlled by the resistance of individual beds to wind deflation. Lack of vegetation cover helps to maintain the long-term, slow erosion on the windward side of the volcanic edifice. (B) Contrasting style and rates of revegetation of volcanic surfaces on the flanks of maar/tuff ring and scoria cones. Lake Pocura (Ranco Province, Chile) is a few ka old maar crater and is characterised by the same degree of vegetation cover as the recently formed Carran maar (1955). The Mirador scoria cone (1979) lacks of vegetation cover. In comparison with Carapacho tuff ring (in A), and, the surface stabilization in semi-arid/arid climates, the time for vegetation to develop is much shorter, in a few decade time-scales. (C and D) Rapid syn- and post-eruptive erosion is observed on a freshly formed tuff cone in the caldera of Ambae, Vanuatu in 2006. (E) A tuff cone (≤2 ka old) located along the coastal region in Ambrym, Vanuatu. Note the geomorphic similarities of this tuff cone with scoria cones. (F) Cross-section through a post-eruptive, well-developed, gully exposed by intensive wave-cut erosion since 1913. The gully developed on the gentle-flanks of a tuff ring formed in the phreatomagmatic eruption in Ambrym, Vanuatu [425].

rates of underground chemical weathering could be in the range of 10 to 1000 t/km^2/year in a humid climate [e.g. 439], which is in the same order of magnitude as the erosion rate by surface runoff. This efficient removal of weathering products by infiltrating rainwater and subsequent underground flow has an important, previously unaccounted for effect on degradation of monogenetic volcanic landforms. It is also worth noting that volcanoes with conical geometry are suspected to have different weathering regimes in accordance with microclimatic and topographic conditions [e.g. 437]. Taking the Rangitoto scoria cone (Auckland, New Zealand) as an example, the area of the cone is about 0.41 km^2, while the bulk DEM-based volume is about 0.022 km^3 above 140 m asl [363, 441]. The edifice has a basal diameter of about 600 m and a crater diameter of 200 m. Considering the abovementioned ranges of subsurface weathering and erosion rates, the time scales of complete erosion of the scoria cone can be calculated. Assuming a 1200 kg/m^3 average density for moderately to highly vesiculated scoria deposits (i.e. 3.4 to 341.6 m^3/yr volume loss rates), the time scale of complete degradation would be between 6.4 Ma and 0.06 Ma, using the constant degradation rates mentioned above. Of course, the rates of chemical weathering tend to slow down when the soil coverage becomes thicker and the weatherable parental rocks are reduced [e.g. 442]. Apart from this, the underground erosion of volcanic edifices by infiltrating surface water (e.g. initial stages of scoria cone degradation) and groundwater flow could be very important and effective long-term degradation process that should have some influence on the morphology of monogenetic volcanoes, especially for volcanic areas with strong chemical weathering rates (e.g. humid, tropical areas with high annual temperature).

In summary, the degradation of a monogenetic volcano is many orders of magnitude longer ($\geq$100 ka to $\leq$ 50 000 ka) than their formation ($\leq$0.01 ka). For example, the degradation of a small-volume ($\leq$0.1 km^3) volcanic edifice usually takes place in a couple or 10s of Ma for welded and/or spatter-dominated edifices, such as in the Bakony-Balaton Highland Volcanic Field in Hungary [68] or in Sośnica hill volcano is Lower Silesia, Poland [335]. Phreatomagmatic volcanoes, especially those with diatremes, could degrade over a longer period of time due to their significant vertical extent, e.g. the Oligocene Kleinsaubernitz maar–diatreme volcano in Eastern Saxony, Germany [401]. During such a long degradation time, the rates and style of post-eruptive surface modification of monogenetic volcanic landforms are generally vulnerable to changes in the configuration and balance between internal and external degradation controls (Figure 11). These could be triggered internally, e.g. denudation of a spatter-dominated or a fine ash horizon (e.g. Figure 13B), or externally, such as long-term climate change or climate oscillation [39], initializing a gradual shift in the dominant mode of sediment transport. Each of these gradual changes (e.g. soil formation, granulometric and climatic changes etc.) causes a partial or complete reorganization of the controls on degradation. This adjustment of erosion settings could result in a change in erosion agent that may or may not increase or decrease sediment yield on the flanks of a volcano. All of these changes over the long erosion history open systemically new potential 'pathways' for erosion, leading to diverse erosion scenarios. The long-term degradation seems to be an iterative process, repeating a constant erosion agent adjustment that is triggered by many gradual changes over the erosion history of a volcano. In many previous erosion studies, the edifices are usually treated as individuals sharing the same internal (i.e. configuration of pyroclastic successions) and initial geometry [e.g. 71] and degrading in accordance with the climate of the volcanic field [324-325, 337]. Of course the climate is in general a important control on degradation, but the climatic forces are in continuous interactions with the volcanic surface, promoting the importance of architecture and granulometric characteristics of the exposed pyroclasts and lava rocks in the volcanic edifice. In extreme cases such as Pukeonake scoria cone (having typical monogenetic edifice dimensions of 150 m in height and 900 m in basal width) in the Tongariro Volcanic Complex in New Zealand, there is an unusual wide granulometric contrast within the pyroclastic succession (Figure 13B). Additionally, the trends and processes in degradation are in close relationship with the exposed pyroclast characteristics which determine the rates of chemical weathering, soil characteristics, surface permeability and, in turn, the mode of sediment transport on the flanks (e.g. Figures 13C and D).

5.2. Short-term, event degradation of monogenetic volcanoes

The event degradation processes take place in a short time frame (hours to days), but they could cause sudden disequilibrium in the degradation and sedimentary system. A monogenetic magmatic system tends to operate inhomogeneously both spatially, forming volcanic clusters, and temporally, forming volcanic cycles. Additionally, there are monogenetic volcanoes that can be found as parasitic or satellite vents on the flanks of larger, polygenetic volcanoes, such as Mt. Etna in Sicily [317] or Mauna Kea in Hawaii [314]. The spatial and temporal closeness of volcanic events, however, pose a generally overlooked problem related to the degradation of monogenetic volcanic landforms, such as tephra mantling or geomorphic truncation by eruptive processes of a surrounding volcano. Tephra mantling is considered to be an important process for the degradation of volcanic edifices as stated by Wood [71] and White [412]. The average distance between neighbouring volcanoes in intraplate settings, such as Auckland in New Zealand, is about 1340 m (i.e. 5.6 km^2), while on the flank of a polygenetic volcanic/volcanic island, such as Tenerife in Canary Islands, that average is about 970 m (i.e. 2.9 km^2). On the other hand, the typical area of a tephra blanket 1–2 cm thick ranges from 10 km^2 for Hawaiian eruptions [162] to 10^3 km^2 for violent Strombolian eruptions [163, 443-444] and for phreatomagmatic eruptions [358]. Consequently, the individual edifices commonly overlap each other's eruption footprint (i.e. area affected by the primary sedimentation from the eruptions), showing the importance of tephra mantling. Furthermore, there are monogenetic volcanic edifices that are developed on flanks of larger polygenetic volcanoes where the mantling by tephra could be more frequent and more significant than in intraplate volcanic fields (e.g. Mt. Roja in the southern edge of Tenerife in the Canary Islands, Figure 13E). A few cm thick tephra cover could cause complete or partial damage to the vegetation canopy [411, 445-447]. Mantling could reset all dominant surface processes, including sediment transport systems, erosion agents, vegetation cover or soil formation processes. This leads to similar reorganization of the degradation controls to those seen with the long-term gradual changes of external or internal factors during normal degradation, but in much shorter time-scales (hours to years). The sedimentary responses to mantling could occur instantly or with a slight delay. Increased erosion rates of older cones were documented instantly after the tephra mantling by fine/coarse ash from Paricutin, Michoacán-Guanajuato, Mexico between 1943 and 1952 [411]. The tephra that mantled the topography was fine ($Md\phi$ = 0.1–0.5 mm) and relatively impermeable, which led to the formation of new, extensive incisions by rill channels and significant deepening of older gullies by the increased sediment yield [411]. In contrast, the sedimentary response for the Tarawera eruption in New Zealand was delayed by the well-sorted, coarse and high permeability of the tephra accumulated over the landscape [448]. The mantling may have an effect on vegetation coverage (e.g. cover or burn the vegetation) and the erosional agent responsible for shaping the morphology of the volcano. The long-term effect of this may be the longer preservation of the landform, or increased dissection which temporarily enhances the overall rates of erosion. These changes will have an important influence on the majority of the morphometric parameters and their pattern of changes over the erosion history.

Surface modification of an already formed monogenetic volcanic edifice could also be triggered by the formation of another monogenetic vent close by [52, 412, 449]. The amalgamated or nested volcanic complexes that have some time delay between their formation are common in volcanic fields, for example Tihany in the Bakony-Balaton Highland, Hungary [41], Rockeskyllerkopf volcanic complex in the Eifel, Germany [450] or Songaksan in Jeju Island, Korea [9, 278]. The eruption of nearby volcano(es) seems to be a common process that may lead to the minor truncation of surfaces, bomb/block-dominated horizons and discordances in the stratigraphic log. This type of 'event' degradation by monogenetic eruption could modify the previously formed topography instantly.

On pyroclast surfaces with limited permeability (e.g. fine ash, lava spatter horizon or lava flows), the rain water tends to simply runoff depending on the actual infiltration rate and the rain fall intensity rate [409-410, 447, 451-452]. On this fine ash surface, the infiltration rates are an order of magnitude lower than on a loose, lapilli-covered flank. This is visible on the flank of La Fossa cone in Vulcano, Aeolian Islands (Figure 13C). The La Fossa cone is not a typical monogenetic volcano, but it has similar geometry and size to a typical monogenetic volcanic edifice. Erosion on the La Fossa cone is characterized by surface runoff on the upper steeper flanks (≥30°) built up by fine indurated ash (Mdϕ = 100 μm), while the erosion of the lower flanks (≥28° and Mdϕ = 1–2 mm) is usually due to debris flows forming levees and terminal lobes [410]. The strikingly different style of mass wasting mechanism is interpreted to be the result of the lack of vegetation cover and strong contrast in permeability and induration of the underlying pyroclastic deposits [410]. Erosion by debris flows forms deep and wide gullies even on a flank built up by permeable rocks, e.g. La Fossa in Vulcano [410] or Benbow tuff cone in Ambrym, Vanuatu [408]. The triggering mechanism for a volcaniclastic debris flow is limited to a period of intense, heavy rainfall [408, 410, 451, 453]. Thus it operates infrequently and it tends to typically redistribute a pocket of a few tens of m^3 of sediments [410].

Similarly to the volcaniclastic debris flows, landslides could also be part of the event degradation processes, especially on steep flanks (e.g. cone-type morphology). The susceptibility for landslides that remove large chunks from the original volcanic edifice, increases by diversity of the pyroclast in the succession. In other words, the layer-cake, usually bedded, inner architecture of either the ejecta ring around a phreatomagmatic volcano or a scoria cone, is extremely susceptible to landsliding triggered, for instance by heavy rain, earthquake, animal activity or surface instability of freshly deposited mantling tephra [e.g. 411].

Another event degradation process is the wild fire that is responsible for the temporal increase (by 100 000 times the 'background' sediment yield) in erosion rates and sediment yields on steep flanks [e.g. 456, 457]. The major effects on a surface by a wild-fire include accumulation of ash, partial or complete damage of vegetation, and organic matter in the soil, and modification of soil structure, if any, and its nutrient content [e.g. 456]. These changes of the surface properties lead to modifications of porosity, bulk density and infiltration rates of the surface, promoting overground flow which is able to carry the increased

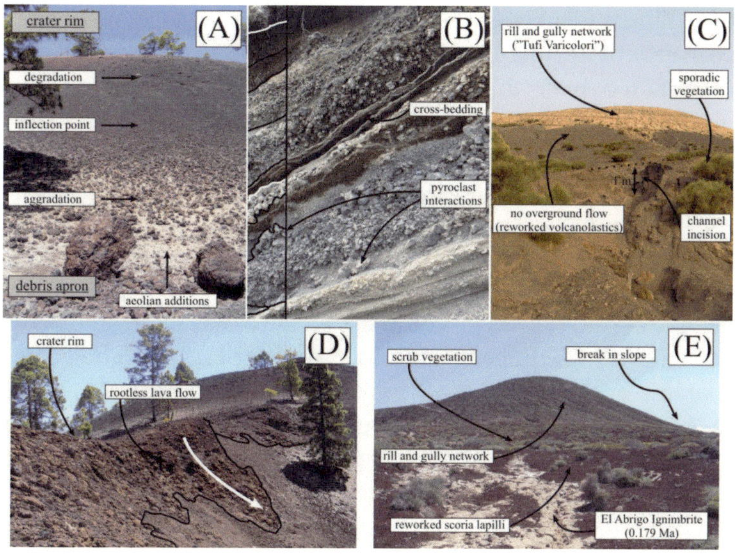

Figure 13. A) Degradation/aggradation on the outer flanks of Mt. Cascajo volcano (a few ka old?) in the NW rift zone, Tenerife, Canary Island. (B) Textural and granulometric inhomogeneity of the Pukeonake scoria cone, at the foot of the Tongariro Volcanic Complex in New Zealand. It is important to note that each of these units has different erosion resistance that may have an effect on the trends and patterns in the erosion history. (C) Contrasting granulometry and permeability caused difference in erosion surface modification on the flanks of La Fossa cone at Vulcano, Aeolian Islands. (D) Spatter accumulated on the crater rim feeds a small-volume rootless lava flow at the Mt. Cascajo volcano in Tenerife, Canary Island. The spatter is relatively impermeable in comparison with its environment (e.g. loose, scoria lapilli), thus will have an influence on the subsequent erosion patterns. (E) View of Mt. Roja in the southern edge of Tenerife, Canary Island. The Mt. Roja edifice was partly or completely covered by the El Ambrigo Ignimbrite (0.18 Ma) sourced from the former Las Cañadas edifice [454-455]. This mantling resulted in development of a rill and gully system on the flanks facing toward north.

sediment yield [e.g. 456]. The overground flow removes the fine sediment (e.g. volcanic ash and lapilli and non-volcanic ash) and the topsoil, causing enrichment of coarser sediment on the surface.

Animal activity is a commonly recognized erosion type due to its effect on the compaction of the uppermost soil [e.g. 458, 459] and/or linear dissection of the surface by trampling [e.g. 460, 461]. As a result of compaction, the rainwater cannot penetrate through the soil cover easily, leading to overground flow that increases the erosion rate and sediment yield [e.g. 458]. Wild animal tramping could be a source of rill and gully formation on flanks, mostly in semi-arid and arid environments [e.g. 460], particularly in, the crater-type volcanoes that commonly host post-eruptive lakes within the crater basins, such as Laguna Potrok Aike in Pali Aike Volcanic Field, Patagonia, Argentina [462] or Pula maar, Bakony-Balaton Highland, Western Hungary [195, 463]. These maar lakes create special habitats that could increase animal activity, creating more opportunity for animal activity-induced erosion.

The event degradation processes, such as heavy rain-induced debris flow, tephra mantling, landslide, post-wild fire runoff or animal activity, could individually trigger rapid geomorphic modifications that affect the long-term degradation rates of the volcanic edifices (Figure

11). The rates of sediment yield could be a thousand times larger in response to event degradation than in the case of normal degradation. These events are usually randomly or inhomogeneously distributed over the erosion history of a volcanic landform, making the quantification to the total erosion-loss complicated. Due to significant surface modification, their effect on the morphometric parameters could be large and difficult to quantify.

5.3. Post-eruptive erosion of monogenetic volcanoes by normal and event degradation processes

The geomorphic state of (monogenetic) volcanoes is commonly expressed by various morphometric parameters, including edifice height, slope angle or H_{co}/W_{co} ratios. The values of these morphometric parameters usually show a decreasing trend over the course of the erosion history [70-71, 290, 299, 317, 320, 464-465]. Consequently, the morphometric parameters show strong time-dependence. This systematic change in morphology was observed mostly on scoria cones in many classical volcanic fields, such as San Francisco volcanic field in Arizona [71] or Cima Volcanic Field in California [70]. This recognition led to the most obvious interpretation: the morphology is dependent on the degree of erosion, which is a function of time and the climate [e.g. 71]. Therefore, morphometric parameters may be used as a dating tool for volcanic edifices if the final geometry and internal architecture are similar among the volcanic edifices being compared [e.g. 71, 324]. These fundamental assumptions (regardless if stated or not) are valid for all comparative morphometric studies targeted volcanoes.

The concept outlined above is, however, sometimes oversimplified and the assumptions are not always fulfilled. The concern about the classical interpretation of morphometric parameters and their change over time is derived from various sources.

- The architecture of a single monogenetic edifice is commonly inhomogeneous in terms of internal facies characteristics. This architectural diversity usually results in diversity in erosion-resistance and susceptibility to chemical weathering of the pyroclastic rocks exposed to the external environment (e.g. Figures 10 and 13B or [e.g. 68]. Due to the continuous denudation of internal beds, the internal architectural irregularities could cause different rates of weathering and erosion leading to hardly predictable trends in degradation.

- The previous concept of monogenetic volcanism implied that the morphology of a volcanic landform is linked only to the specific eruption styles (i.e. Strombolian-type scoria cone). However, this is an oversimplification and belies the complex pattern in edifice growth (e.g. Figures 6–9).

- The final, pristine edifice morphology is mostly controlled by syn-eruptive processes (e.g. explosion energy, substrate stability, mass wasting and mode of pyroclast transport, e.g. Figures 7–9). Any change in either the internal or external controls during the course of an eruption could modify the final morphology of the edifice partially or dramatically. This is in agreement with the measured high variability of slope angle [e.g. 398] or aspect ratios [e.g. 314] on relatively fresh edifices. This supposedly eruptive process-related morphometric variability is observed on both the intra-edifice scale (e.g. between various parts of an edifice) and inter-edifice scale [398].

- There is a large difference in the rates and mode of chemical weathering and sediment transport operating on different types of pyroclastic deposits or lava rock surfaces (e.g. Figures 12 and 13). For instance, there is a contrast between mass wasting rates by under- and overground flow processes, e.g. spatter or a higher degree of welding/agglutination could cause asymmetric patterns in the permeability and, therefore, the subsequent initialization of the erosion on a freshly created surface.
- One theoretical concern about morphometric parameters, such as edifice height, aspect ratio or slope angle, is that they are intra-edifice, 'static' descriptors. Thus, they only express the current geomorphic state of the volcano. In contrast, they are often used to reveal and describe 'dynamic' processes, such as erosion patterns over time. It is obvious that trends in erosion processes cannot be seen based on these intra-edifice, 'static' parameters unless they are compared with other edifice parameters, or measure direct geomorphic modification by erosion processes over short periods of time (e.g. Figures 12C and D or [e.g. 464]). The first seems an 'extrapolation' due to the assumption of that all volcanic edifices in the comparison have similar eruptive- and erosion histories. The second is more suitable, because it is the direct observation of erosion loss and surface modification [e.g. 326].
- Comparative morphometric studies often lack or have limited age constraints (e.g. a few % of the total population of the studied edifices are dated) on the morphology, or inversely in special cases, the dating is the purpose of the comparison. There are just a few studies with complete age constraint [e.g. 68, 70, 326].
- The long-term surface modification is often believed as a result of the climate forces and climate-induced erosion processes. Wood [71] stated the importance of tephra mantling as a possible source of acceleration of erosion rates, but such event-degradation (e.g. tephra mantling, edifice truncation by eruption nearby, landsliding, wild fire, animal activity etc.), are usually neglected. They occur infrequently, but they could cause rapid and significant modification that may influence the patterns of future degradation.

Due to the concerns and arguments listed above, the morphometric parameters and their classical interpretations should be revised. Referring back to the complexity of construction of monogenetic volcanoes (Figure 6) and their primary geomorphic development (Figures 7–9), it is obvious that on a fresh volcanic landform the geomorphic feature is determined by syn-eruptive processes, which in turn are governed by the internally and externally-driven processes during the eruption history. Once the eruption ceases, the 'input' configuration of a monogenetic volcano in terms of architecture, pyroclast granulometric characteristics, geometry and geomorphology is given. The erosion agents at the start of the erosion history are determined by the interactions between the internal (e.g. pyroclastic rocks on the surface) and external processes (e.g. climate; Figure 11). The results of such series of interactions between these properties lead to surface and subsurface weathering, soil formation and development of vegetation succession over time. Each of these developments on the flanks of a monogenetic volcano has a feedback to the original controls modifying the actual balance towards one side. This leads to disequilibrium in the system and subsequent adjusting mechanism. These processes are called normal degradation, operating at a longer-time scale (ka to Ma). However, the degradation mechanism sometimes does not function as 'normal'. During the erosion history of a monogenetic volcano, there are some environmental effects called 'events' such as tephra mantling or heavy-rain-induced grain flows. These 'events' are documented to cause orders of magnitude larger surface modification and possibly initialize new rates and trends in the dominant sediment-transport system and increase the sediment yield [e.g. 411-412, 447-448, 457]. Consequently, the erosion history of a monogenetic volcano comprises both normal ('background') and event degradation processes (Figure 11). The cumulative result of many interactions, reorganization of erosion agents, and effects of event degradation processes over the erosion history, are integrated into the geomorphic state at the time of examination.

The degradation of the volcanic edifice leads to aggradation at the foot of the edifice and the development of a debris apron (Figure 13A). Based on the behaviour and changes of intensity of the abovementioned major sediment transport processes, it is evident that the individual contribution of such erosion processes is not constant over time. It is more likely that they are enhanced or eased by each other at certain stages of the degradation. The gradual changes in style, rate and mode of sediment transport on the flank of a monogenetic volcanic edifice are likely triggered by the shifting of dominant external (e.g. climate change) and internal environments (e.g. variability of erosion resistant layer within the edifice as observed in Figure 14A). Consequently, the degradation of the monogenetic edifice as a whole cannot be linear (or maybe just certain parts of the erosion history) and must erode faster at the beginning and slower at the end of the degradation [71, 324] in accordance with the wide range of rates and time-scales of sediment transport processes. Consequently, a single geomorphic agent cannot account for a volcano's degradation. Instead, it seems to be the result from the overall contribution of all processes with complex temporal distribution. Without event degradation processes, given the fact that the erosion history lasts at least over a time-scale of a couple of ka for a typical monogenetic volcanoes, this increases the likelihood of some changes in the external environment that could modify the degradation trend.

These surface modifications and degradation processes should be in correlation with the values of morphometric parameters, but their interpretation is possibly not a straightforward process. The pristine unmodified geomorphic stage of a monogenetic volcano is predominantly controlled by the processes that occurred in the eruption history. Once the degradation proceeds (e.g. erosion surface modification, soil formation or development of vegetation cover), these primary geomorphic attributes are gradually replaced by excess 'signatures' of the various post-eruptive processes. This will result in 'noise' in the original syn-eruptive state of morphometric parameters extracted from the topographic attributes. The soil cover on the surface creates a buffer zone between the pyroclastic deposits and the environment. In this buffer zone, most of the weathering and erosion processes take place (e.g. overground flow). During the degradation the actual erosion surface, regardless of whether it is 'unstabilized' or 'stabilized', could contain pyroclasts with contrasting granulometric and textural characteristics (e.g. Figure 13B). For instance, the rates of weathering, weathering product transport and soil formation could be different at the base of a volcanic cone than at the crater rim, due to the differences in flank morphology, aspect or microclimate. These differences are demonstrated for various sectors of a cone-type volcano by the variation in microclimatic setting, e.g. insolation, freeze-thaw cycles or snow cover [437]. If there are a couple of meter difference in sediment accumulation/loss, chemical weathering and soil formations, it could cause a variation of a few degrees in the slope angle values. In extreme cases, these differences could cause misinterpretation of the morphometric parameters, thus these should be taken into account or stated as an assumption of the interpretation. The increase of post-eruptive 'noise' of the morphometric parameters will possibly increase over the erosion history, and possibly the largest in the late stage degradation of the edifice (e.g. Ma after its formation). In the case of older scoria cones, the architectural control could increase, as the well-compacted and welded units are exposed, leading to rock-selective erosion styles and longer preservation potential for a volcanic landform. This is found to be important to the good preservation of the Pliocene (2.5–3.8 Ma) scoria cones such as Agár-tető or Bondoró at the Bakony-Balaton Highland in Hungary [68]. For instance, these scoria cones are old but they resemble considerably younger cone morphologies, due to their higher morphometric values (e.g. height about 40–80 m, slope angle of 10–15°). These parameters could be similar to the degradation signatures of a much younger cone, e.g. Early Pleistocene cones (slope angle of 13±3.8°) from Springerville volcanic field, Arizona [324].

Many lines of evidence suggest that the neglected internal architecture, initial variability in geomorphic state or effect of 'event' degradation processes play an important role on edifice degradation rates and trends. Once the degradation histories for various edifices are characterized by

1. different 'input' morphometric conditions (e.g. Figure 14B), and
2. large variability of rates and trends in mass wasting processes in accordance with the susceptibility of chemical weathering of the underlying volcanic rocks and the total capacity of sediment transport, it is possible that the same geomorphic state can be reached not only by ageing of the edifice, but via a combination of other processes.

This further implies that the monogenetic volcanic edifice has a unique eruptive (e.g. Figure 6) and erosion history (e.g. Figures 12–14). As a result of the eruptive diversity, the erosion history is not independent from the eruption history (i.e. the complexity of the monogenetic edifices). In this interpretation, there is a chance to have edifices showing the same 'geomorphic state' (in terms of the basic geometric parameters) reached through different 'degradation paths'. An example for this could be the case of the two scoria cones in Figures 14C and D. In Figure 14C, the geometry of the edifice is strongly attached to the erosion-resistance and to the position of the spatter-dominated collar along the crater rim. In this eruptive history and subsequent erosion, the slope angle can be increased due to the undermining of the flanks. Consequently, the morphology of the cone is becoming 'younger' over time, that is the slope angle or H_{co}/W_{co} ratio will increase rather than decrease. On the other hand, a classical-looking cone (Figure 14D) that has a homogenous inner architecture, experiences different rates and degrees of erosion over different time scales. Therefore, both cones degrade through different patterns and rates. To confidently say that the decreasing trend in morphometric parameters is associated with age, it is important to reconstruct the likely environment where the edifice degradation has taken place, including the number of 'event' and major changes in the degradation controls. This includes understanding the combination and diversity of facies architecture [68, 468], the stratigraphic position of the edifice within the stratigraphic record of the volcanic field [412, 469], the approximate likelihood of aggradation by, for example, tephra mantling [411], and spatial and temporal combination and fluctuation of 'normal' and 'event' degradation processes over the erosion history.

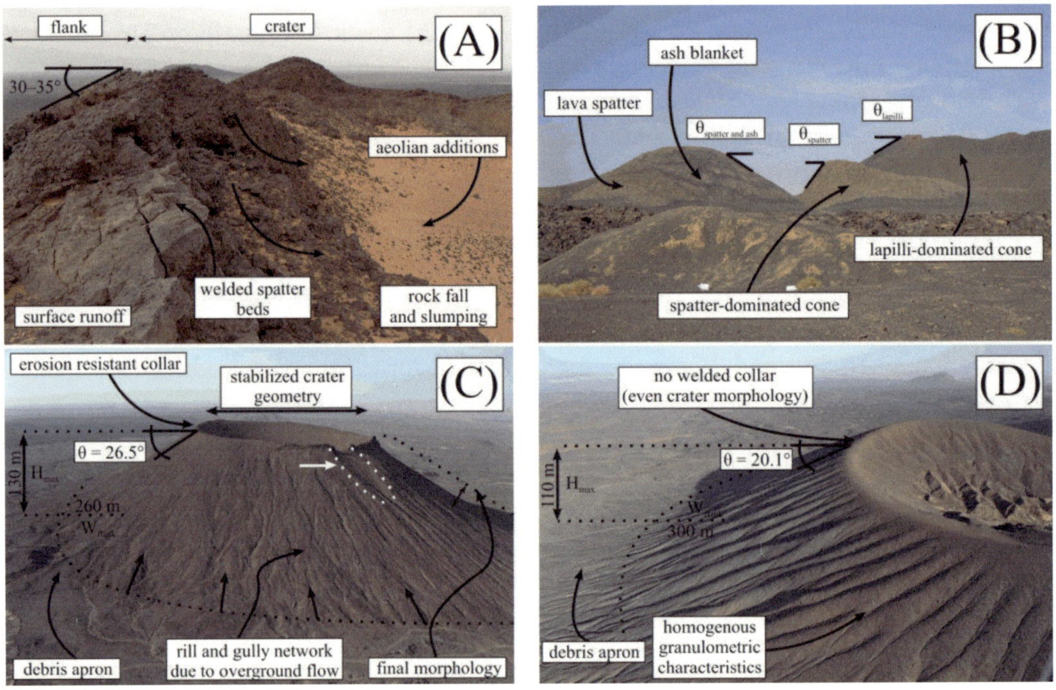

Figure 14. A) Difference in mode of erosion (rock fall or surface runoff) due to spatter accumulation on the crater rim of a 1–3 Ma old scoria cone in the Al Haruj Volcanic Field in Libya [288, 466]. (B) Variability in slope angle on the flanks of spatter-dominated and lapilli-dominated cones (1256 AD) in the last eruptions at the Harrat Al-Madinah Volcanic Field, Saudi Arabia [467]. Due to the young ages, these differences could be the results of differences in syn-eruptive processes (e.g. fragmentation mechanism, degree of welding and granulometric properties). These different 'input' geomorphic states alone can also lead to the large variability of degradation paths of monogenetic volcanic landforms. (C and D) Architecturally-controlled erosion pattern on Pleistocene scoria cones in the Harrat Al-Madinah Volcanic Field, Saudi Arabia. The ages are between 1.2 and 0.9 Ma for the cone in Figure C, and only a couple of ka for the cone in Figure D [467]. The geomorphic contrast between the edifices is striking in the slope angles, θ, calculated as $\theta = \arctan(H_{max}/W_{flank})$ from basic morphometric data. The erosion resistant collar on the crater rim changes the erosion patterns by keeping the crater rim at the same level over even Ma. This results in the 'undermining' of the flanks (small black arrows at the foot of the cones represented by a dashed line in Figure D) leading to a gradual increase of the slope angles in contrast to all previously proposed erosion models for cone-type monogenetic volcanoes. The white arrow near the rim (Figure C) indicates the significant surface modification by event degradation (e.g. mass wasting of the erosion-resistant spatter collar). It is speculative, but the consequence of this irreversible and possibly 'random' event may have initialized the formation of a deeper gully (white dashed lines) leading to crater breaching over a longer time-scale.

6. Conclusions: towards understanding the complexity of monogenetic volcanoes

A typical monogenetic volcanic event begins at the magma source region, usually in the mantle, and ends when the volcanics have been fully removed by, for example, erosion processes. Within this conceptualized life cycle of a monogenetic volcano, there is an active stage (e.g. propagation of the magma towards the surface feeding a monogenetic eruption; Figure 15) and a passive stage (e.g. post-eruptive degradation until the feeder system is exposed; Figure 15). The active stage of evolution is dependent on many interactions between internally- or externally-driven factors. The magma (left hand side on Figure 15) intrudes into shallow parts of the crust that can be fragmented in accordance with the actual balance between the magmatic and external conditions at the time of the fragmentations. This could result in 6 varieties of volcanic eruptions if the composition is dominantly basaltic, which is responsible for the construction of a monogenetic volcanic edifice with a simple eruption history (E_{simple}, that is 6^1 combinations of eruption styles). Once there is some disequilibrium in the system during the course of the eruption that will result in changing eruption styles adjusting the balance in the system, forming compound eruption histories ($E_{compound}$ that is 6^2 combinations of eruption styles). Each number of shifts in dominant eruption style opens a new phase of edifice growth and therefore increases the complexity of the eruption history towards $E_{complex}$ (that is 6^3 or more combinations of eruption styles). There could be even thousands of theoretical combinations of eruption styles if the volcano is built up by more than 4 phases with different eruption styles, until the magma supply is completely exhausted or new vent is established by migration of the magma focus. With increasing complexity of the eruption history, the complexity of the facies architecture of the volcanic edifice increases. Conceptually, these eruption histories can be numerically described by matrices, based on spatial and temporal characteristics of eruption styles (e.g. Figure 6). The coding of eruption styles could be

1. Hawaiian,
2. Strombolian,
3. violent Strombolian,
4. phreatomagmatic,
5. Surtseyan and
6. effusive activity

, if the erupting melt is characterised by basaltic to basaltic andesitic in composition. This systems can be modified by adding further eruption styles such as sub-Plinian. The syn-eruptive geomorphology of a volcano is, however, not only the result of the eruption style and associated pyroclast transport mechanism, but there are stages of destructive processes, such as flank collapse during scoria cone growth (e.g. Figure 3) or wall rock mass wasting during excavation of a maar crater (e.g. Figure 4). These common syn-eruptive processes (constructive and destructive phases during the eruption history) have an important role on the resulting morphology, but they are not always visible/detectable in the morphology of the edifice.

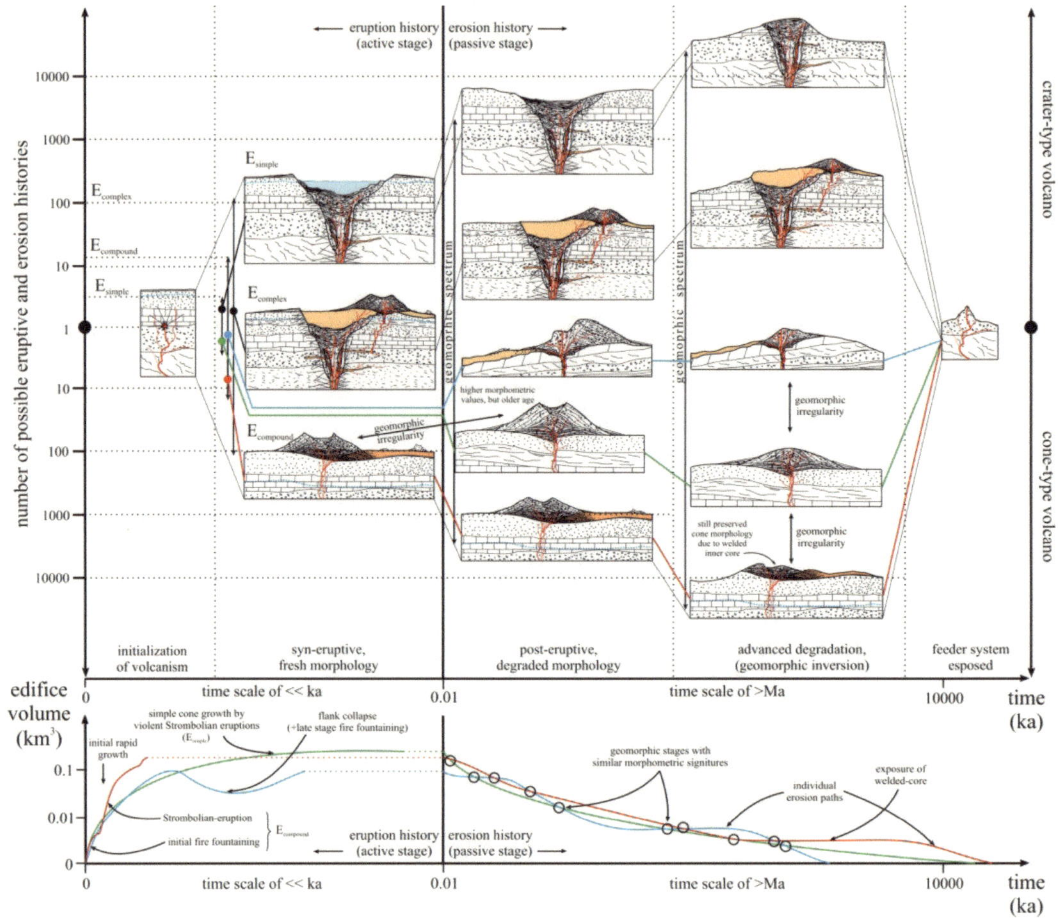

Figure 15. Conceptualized model to understand causes and consequences in the life cycle of crater- and cone-type monogenetic volcanoes. See the text for a detailed explanation. Note that the colour coding in the bottom graph corresponds the conceptualized degradation paths for the three cone-type volcanoes in the top graph. The double-headed arrows (top graph) and black circles (bottom graph) show the time slices when geomorphic states of the edifices are similar, expressed by the morphometric parameters.

On the other hand, after the eruption ceases a passive stage of surface modification takes place (right hand side on Figure 15). In the passive stage, the erosion history is also governed significantly by a series of interactions between the exposed pyroclastic deposits and lava rocks (and their textural and granulometric characteristics determining permeability) and external influences such as climate, location or hydrology of the area. These interactions determine the long-term (ka to Ma) degradation processes and rates. However, it is important to note that the erosion history is often a function of 'normal' and 'event' degradation. The effect of event degradation is expected to be larger, in some cases, than the cumulative surface modification by normal degradation processes. The relationship between event and normal degradation should be a subject for future studies. Due to the large number of combinations of eruption styles that can generate edifices with different pyroclastic successions and different initial geometries (at least a broader range than previously thought), volcanoes can have very different susceptibilities for erosion. This implies that degradation trends and patterns of monogenetic volcanoes should be individual volcano-specific (right hand side on Figure 15). In addition, the combination of erosion path of individual monogenetic volcanoes is an order of magnitude larger than during the eruption history due to the larger number of controlling factors (6 eruption styles versus varieties of 'normal' and 'event' mass wasting processes) and the longer time-scale of degradation ($\ll$ka versus >Ma). This has an important practical conclusion: there are certain stages during the degradation when some morphometric irregularity occurs if two or more volcanic edifices are compared. The morphometric irregularity refers to the state when two volcanoes appear similar through morphometric parameters such as H_{co}/W_{co} ratio or slope angle, but they have different absolute ages (black double-headed arrows of the top graph and black circles on the bottom graph in Figure 15). An important practical application of the volcano-specific degradation is that the correlation between the morphology of the edifice is not always a function of the time elapsed since formation of the volcanic edifice. As a consequence of the diverse active and passive evolution of a volcanic edifice, age grouping based on geomorphic parameters, such as H_{co}/W_{co} ratio or average or maximum cone slope angle, should be avoided. In terms of interpretation of the morphometric data, the post-eruptive surface modification causes unfortunate 'noise' in the primary morphometric signatures, which can be only reduced by using edifices with absolute age constraints. Due to the long-lived evolution of monogenetic volcanic fields (Ma-scale), there are usually volcanoes that are freshly formed, sometimes close to volcanoes with no primary morphological features at the time of the examination. The large contrasting and dynamic geological environment of such monogenetic volcanoes makes the interpretation of available topographic information more complicated than previously thought. Future studies should target this particular issue and define the meaning of morphology of these monogenetic volcanic edifices at many scales.

Acknowledgements

Constructive discussion about various aspects of monogenetic volcanism and remote sensing with Shane Cronin, Jon Procter, Jan Lindsay, Javier Agustín-Flores, Marco Brenna, Natalia Pardo, Nicola Le Corvec, Mark Bebbington and Mike Tuohy is acknowledged. Remarks and suggestion by Bob Stewart, Adrian Pittari, Marco Brenna, Kate Arentsen and Javier Agustín-Flores and an anonymous reviewer are gratefully appreciated. This study was supported by the NZ MSI-IIOF Grant "Facing the challenge of Auckland's Volcanism" (MAUX0808). The authors would like to thank the Auckland City Council for the LiDAR dataset and the Department of Conservation (Te Papa Atawhai) for the logistical help during field trips.

Author details

Gábor Kereszturi[1*] and Károly Németh[1,2]

*Address all correspondence to: kereszturi_g @yahoo.com

1 Volcanic Risk Solutions, Massey University, Palmerston North, New Zealand

2 King Abdulaziz University, Jeddah, Kingdom of Saudi Arabia

References

[1] Canon-Tapia, E., & Walker, G. P. L. (2004). Global aspects of volcanism: the perspectives of "plate tectonics" and "volcanic systems. *Earth-Science Reviews*, 66, 163-82.

[2] Morgan, W. J. (1971). Convection Plumes in the Lower Mantle. *Nature*, 230, 42-3.

[3] Tarduno, J., Bunge-P, H., Sleep, N., & Hansen, U. (2009). The Bent Hawaiian-Emperor Hotspot Track: Inheriting the Mantle Wind. *Science*, 324, 50-3.

[4] De Paolo, D. J., & Manga, M. (2003). Deep Origin of Hotspots- The Mantle Plume Model. *Science*, 300, 920-1.

[5] Németh, K. (2010). Monogenetic volcanic fields: Origin, sedimentary record, and relationship with polygenetic volcanism. *The Geological Society of America Special Paper*, 470, 43-66.

[6] Conrad, CP, Bianco, T. A., Smith, E. I., & Wessel, P. (2011). Patterns of intraplate volcanism controlled by asthenospheric shear. *Nat Geosci*, 4, 317-21.

[7] Bebbington, M. S., & Cronin, S. J. (2011). Spatio-temporal hazard estimation in the Auckland Volcanic Field, New Zealand, with a new event-order model. *Bull Volcanol*, 73(1), 55-72.

[8] Keating, G. N., Valentine, G. A., Krier, D. J., & Perry, F. V. (2008). Shallow plumbing systems for small-volume basaltic volcanoes. *Bull Volcanol*, 70, 563-82.

[9] Brenna, M., Cronin, S. J., Németh, K., Smith, I. E. M., & Sohn, Y. K. (2011). The influence of magma plumbing complexity on monogenetic eruptions, Jeju Island, Korea. *Terra Nova*, 23, 70-5.

[10] Connor, C. B., Stamatakos, J. A., Ferrill, D. A., Hill, B. E., Ofoegbu, G., Conway, F. M., et al. (2000). Geologic factors controlling patterns of small-volume basaltic volcanism: Application to a volcanic hazards assessment at Yucca Mountain, Nevada. *J Geophys Res*, 105(1), 417-32.

[11] Smith, I. E. M., Blake, S., Wilson, C. J. N., & Houghton, B. F. (2008). Deep-seated fractionation during the rise of a small-volume basalt magma batch: Crater Hill, Auckland, New Zealand. *Contrib Mineral Petrol*, 155(4), 511-27.

[12] Gencalioglu-Kuscu, G. (2011). Geochemical characterization of a Quaternary monogenetic volcano in Erciyes Volcanic Complex: Cora Maar (Central Anatolian Volcanic Province, Turkey). *Int J Earth Sci*, 100(8), 1967-85.

[13] Browne, B., Bursik, M., Deming, J., Louros, M., Martos, A., & Stine, S. (2010). Eruption chronology and petrologic reconstruction of the ca. 8500 yr B.P. eruption of Red Cones, southern Inyo chain, California. *Geol Soc Am Bull*, 122(9-10), 1401-1422.

[14] Agustín-Flores, J., Siebe, C., & Guilbaud-N, M. (2011). Geology and geochemistry of Pelagatos, Cerro del Agua, and Dos Cerros monogenetic volcanoes in the Sierra Chichinautzin Volcanic Field, south of México City. *J Volcanol Geotherm Res*, 201(1-4), 143-162.

[15] Strong, M., & Wolff, J. (2003). Compositional variations within scoria cones. *Geology*, 31(2), 143-6.

[16] Németh, K., White, J. D. L., Reay, A., & Martin, U. (2003). Compositional variation during monogenetic volcano growth and its implication for magma supply to continantal volcanic fields. *J Geol Soc London*, 160, 523-30.

[17] Brenna, M., Cronin, S. J., Smith, I. E. M., Sohn, Y. K., & Németh, K. (2010). Mechanisms driving polymagmatic activity at a monogenetic volcano, Udo, Jeju Island, South Korea. *Contrib Mineral Petrol*, 160, 931-50.

[18] Cebriá, J. M., Martiny, BM, López -Ruiz, J., & Morán-Zenteno, D. J. (2011). The Parícutin calc-alkaline lavas: New geochemical and petrogenetic modelling constraints on the crustal assimilation process. *J Volcanol Geotherm Res*, 201(1-4), 113-125.

[19] Shaw, C. S. J. (2004). The temporal evolution of three magmatic systems in the West Eifel volcanic field, Germany. *J Volcanol Geotherm Res*, 131, 213-40.

[20] Hasenaka, T., Ban, M., & Granados, H. D. (1994). Contrasting volcanism in the Michoacán-Guanajuato Volcanic Field, central Mexico: Shield volcanoes vs. cinder cones. *Geofisica Internacional*, 33(1), 125-38.

[21] Connor, C. B., & Conway, F. M. (2000). Basaltic volcanic fields. *Sigurdsson H, Houghton BF, McNutt SR, Rymer H, Stix J, editors. Encyclopedia of Volcanoes. San Diego: Academic Press*, 331-343.

[22] Hoernle, K., White, J. D. L., van den Bogaard, P., Hauff, F., Coombs, DS, Werner, R., et al. (2006). Cenozoic intraplate volcanism on New Zealand: Upwelling induced by lithospheric removal. *Earth Planet Sci Lett*, 248, 350-67.

[23] Németh, K., Cronin, S. J., Haller, MJ, Brenna, M., & Csillag, G. (2010). Modern analogues for Miocene to Pleistocene alkali basaltic phreatomagmatic fields in the Pan-

nonian Basin: "soft-substrate" to "combined" aquifer controlled phreatomagmatism in intraplate volcanic fields. *Cent Eur J Geosci*, 2(3), 339-61.

[24] Condit, C. D., & Connor, C. B. (1996). Recurrence rates of volcanism in basaltic volcanic fields: An example from the Springerville volcanic field, Arizona. *Geol Soc Am Bull*, 108, 1225-41.

[25] Kshirsagar, P. V., Sheth, H. C., & Shaikh, B. (2011). Mafic alkalic magmatism in central Kachchh, India: a monogenetic volcanic field in the northwestern Deccan Traps. *Bull Volcanol*, 73(5), 595-612.

[26] Sheth, H. C., Mathew, G., Pande, K., Mallick, S., & Jena, B. (2004). Cones and craters on Mount Pavagadh, Deccan Traps: Rootless cones? *Earth and Planetary Sciences (Proceedings of the Indian Academy of Sciences)*, 113(4), 831-8.

[27] Márquez, A., Verma, S. P., Anguita, F., Oyarzun, R., & Brandle, J. L. (1999). Tectonics and volcanism of Sierra Chichinautzin: extension at the front of the Central Trans-Mexican Volcanic belt. *J Volcanol Geotherm Res*, 93(1-2), 125-150.

[28] Germa, A., Quidelleur, X., Gillot, P. Y., & Tchilinguirian, P. (2010). Volcanic evolution of the back-arc Pleistocene Payun Matru volcanic field (Argentina). *J S Am Earth Sci*, 29(3), 717-30.

[29] Baloga, S. M., Glaze, L. S., & Bruno, B. C. (2007). Nearest-neighbor analysis of small features on Mars: applications to tumuli and rootless cones. *J Geophys Res*, 112(E03002), 10.1029/2005JE002652.

[30] Bishop, M. A. (2007). Higher-order neighbor analysis of the Tartarus Colles cone groups, Mars: The application of geographical indices to the understanding of cone pattern evolution Icarus. 197(1), 73-83.

[31] Broz, P., & Hauber, E. (2012). A unique volcanic field in Tharsis, Mars: Pyroclastic cones as evidence for explosive eruptions. *Icarus*, 218, 88-99.

[32] Keszthelyi, L. P., Jaeger, W. L., Dundas, C. M., Martínez-Alonso, S., Mc Ewen, AS, & Milazzo, M. P. (2010). Hydrovolcanic features on Mars: Preliminary observations from the first Mars year of HiRISE imaging. *Icarus*, 205, 211-29.

[33] Wilson, L., & Head, J. W. (1994). Mars: Review and analysis of volcanic eruption theory and relationships to observed landforms. *Rev Geophys*, 32(3), 221-63.

[34] Sandri, L., Jolly, G., Lindsay, J., Howe, T., & Marzocchi, W. (2011). Combining long- and short-term probabilistic volcanic hazard assessment with cost-benefit analysis to support decision making in a volcanic crisis from the Auckland Volcanic Field, New Zealand. *Bull Volcanol*, 74(3), 705-23.

[35] Lindsay, J., Marzocchi, W., Jolly, G., Constantinescu, R., Selva, J., & Sandri, L. (2010). Towards real-time eruption forecasting in the Auckland Volcanic Field: application of BET_EF during the New Zealand National Disaster Exercise 'Ruaumoko' Bull Volcanol. 72(2), 185-204.

[36] Siebe, C., Rodriguez-Lara, V., Schaaf, P., & Abrams, M. (2004). Radiocarbon ages of Holocene Pelado, Guespalapa, and Chichinautzin scoria cones, south of Mexico City: implications for archaeology and future hazards. *Bull Volcanol*, 66, 203-25.

[37] Delgado, H., Molinero, R., Cervantes, P., Nieto-Obregón, J., Lozaro-Santa, Cruz. R., Macías-González, H. L., et al. (1998). Geology of Xitle volcano in southern Mexico City-a 2000- year-old monogenetic volcano in an urban area. *Rev Mex Cienc Geol*, 15(2), 115-31.

[38] Valentine, G. A., & Hirano, N. (2010). Mechanisms of low-flux intraplate volcanic fields-Basin and Range (North America) and northwest Pacific Ocean. *Geology*, 38(1), 55-8.

[39] Kereszturi, G., Németh, K., Csillag, G., Balogh, K., & Kovács, J. (2011). The role of external environmental factors in changing eruption styles of monogenetic volcanoes in a Mio/Pleistocene continental volcanic field in western Hungary. *J Volcanol Geotherm Res*, 201(1-4), 227-240.

[40] Brenna, M., Cronin, S. J., Smith, I. E. M., & Maas, R., Sohn, Y. K. (2012). How Small-volume Basaltic Magmatic Systems Develop: a Case Study from the Jeju Island Volcanic Field, Korea. *J Petrol*, 53(5), 985-1018.

[41] Németh, K., Martin, U., & Harangi, S. (2001). Miocene phreatomagmatic volcanism at Tihany (Pannonian Basin, Hungary). *J Volcanol Geotherm Res*, 111(1-4), 111-35.

[42] Valentine, G. A., & Perry, F. V. (2007). Tectonically controlled, time-predictable basaltic volcanism from a lithospheric mantle source (central Basin and Range Province, USA). *Earth Planet Sci Lett*, 261(1-2), 201-16.

[43] Geyer, A., & Martí, J. (2010). The distribution of basaltic volcanism on Tenerife, Canary Islands: Implications on the origin and dynamics of the rift systems. *Tectonophysics*, 483(3-4), 310-326.

[44] Németh, K. (2012). An Overview of the Monogenetic Volcanic Fields of the Western Pannonian Basin: Their Field Characteristics and Outlook for Future Research from a Global Perspective. *Stoppa F, editors. Updates in Volcanology- A Comprehensive Approach to Volcanological Problems. In-Tech*, 27-52, http://www.intechopen.com/books/updates-in-volcanology-a-comprehensive-approach-to-volcanological-problems/an-overview-of-the-monogenetic-volcanic-fields-of-the-western-pannonian-basin-their-field-characteri.

[45] Jankovics, M. É., Harangi, S., Kiss, B., & Ntaflos, T. (2012). Open-system evolution of the Füzes-tó alkaline basaltic magma, western Pannonian Basin: Constraints from mineral textures and compositions. *Lithos*, 140-141, 25-37.

[46] Bali, E., Zanetti, A., Szabó, C., Peate, D. W., & Waight, T. E. (2008). A micro-scale investigation of melt production and extraction in the upper mantle based on silicate melt pockets in ultramafic xenoliths from the Bakony-Balaton Highland Volcanic Field (Western Hungary). *Contrib Mineral Petrol*, 155, 165-79.

[47] Kovács, I., Falus, G., Stuart, G., Hidas, K., Szabó, C., Flower, M. F. J., et al. (2012). Seismic anisotropy and deformation patterns in upper mantle xenoliths from the central Carpathian-Pannonian region: Asthenospheric flow as a driving force for Cenozoic extension and extrusion? *Tectonophysics*, 514-517, 168-79.

[48] Yang, K., Hidas, K., Falus, G., Szabó, C., Nam, B., Kovács, I., et al. (2010). Relation between mantle shear zone deformation and metasomatism in spinel peridotite xenoliths of Jeju Island (South Korea): Evidence from olivine CPO and trace elements. *J Geodyn*, 50(5), 424-40.

[49] Hidas, K., Falus, G., Szabó, C., Szabó, P. J., Kovács, I., & Földes, T. (2007). Geodynamic implications of flattened tabular equigranular textured peridotites from the Bakony-Balaton Highland Volcanic Field (Western Hungary). *J Geodyn*, 43, 484-503.

[50] Rubin, A. M. (1995). Propagation of Magma-Filled Cracks. *Annual Review Of Earth And Planetary Sciences*, 23, 287-336.

[51] Németh, K., & Cronin, S. J. (2011). Drivers of explosivity and elevated hazard in basaltic fissure eruptions: The 1913 eruption of Ambrym Volcano, Vanuatu (SW-Pacific). *J Volcanol Geotherm Res*, 201(1-4), 194-209.

[52] Kereszturi, G., & Németh, K. (2011). Shallow-seated controls on the evolution of the Upper Pliocene Kopasz-hegy nested monogenetic volcanic chain in the Western Pannonian Basin, Hungary. *Geol Carpath*, 62(6), 535-46.

[53] Gaffney, E. S., Damjanac, B., & Valentine, G. A. (2007). Localization of volcanic activity: 2. Effects of pre-existing structure. *Earth Planet Sci Lett*, 263, 323-38.

[54] Valentine, G. A., & Krogh, K. E. C. (2006). Emplacement of shallow dikes and sills beneath a small basaltic volcanic center- The role of pre-existing structure (Paiute Ridge, southern Nevada, USA). *Earth Planet Sci Lett*, 246(3-4), 217-30.

[55] Kiyosugi, K., Connor, C. B., Wetmore, P. H., Ferwerda, B. P., Germa, A. M., Connor, L. J., et al. (2012). Relationship between dike and volcanic conduit distribution in a highly eroded monogenetic volcanic field: San Rafael, Utah, USA. *Geology*, 40(8), 695-698.

[56] Lefebvre, N. S., White, J. D. L., & Kjarsgaard, B. A. (2012). Spatter-dike reveals subterranean magma diversions: Consequences for small multivent basaltic eruptions. *Geology*, 40(5), 423-6.

[57] Hintz, A. R., & Valentine, G. A. (2012). Complex plumbing of monogenetic scoria cones: New insights from the Lunar Crater Volcanic Field (Nevada, USA). *J Volcanol Geotherm Res*, 239-240(0), 19-32.

[58] Sulpizio, R., De Rosa, R., & Donato, P. (2008). The influence of variable topography on the depositional behaviour of pyroclastic density currents: The examples of the Upper Pollara eruption (Salina Island, southern Italy). *J Volcanol Geotherm Res*, 175(3), 367-85.

[59] Kereszturi, G., Csillag, G., Németh, K., Sebe, K., Balogh, K., & Jáger, V. (2010). Volcanic architecture, eruption mechanism and landform evolution of a Pliocene intracontinental basaltic polycyclic monogenetic volcano from the Bakony-Balaton Highland Volcanic Field, Hungary. *Cent Eur J Geosci*, 2(3), 362-84.

[60] Valentine, G. A., Krier, D., Perry, F. V., & Heiken, G. (2005). Scoria cone construction mechanisms, Lathrop Wells volcano, southern Nevada, USA. *Geology*, 33(8), 629-32.

[61] Risso, C., Németh, K., Combina, A. M., Nullo, F., & Drosina, M. (2008). The role of phreatomagmatism in a Plio-Pleisotcene high-density scoria cone field: Llancanelo Volcanic Field, Argentina. *J Volcanol Geotherm Res*, 168, 61-86.

[62] Magill, C., & Blong, R. (2005). Volcanic risk ranking for Auckland, New Zealand. II: Hazard consequences and risk calculation. *Bull Volcanol*, 67(4), 340-9.

[63] Connor, L. J., Connor, C. B., Meliksetian, K., & Savov, I. (2012). Probabilistic approach to modeling lava flow inundation: a lava flow hazard assessment for a nuclear facility in Armenia. *Journal of Applied Volcanology*, 1, 3, http://www.appliedvolc.com/1//3.

[64] Harris, A. J. L., Favalli, M., Wright, R., & Garbeil, H. (2011). Hazard assessment at Mount Etna using a hybrid lava flow inundation model and satellite-based land classification. *Nat Hazards*, 58(3), 1001-27.

[65] Németh, K., & Cronin, S. J. (2009). Phreatomagmatic volcanic hazards where rift-systems meet the sea, a study from Ambae Island, Vanuatu. *J Volcanol Geotherm Res*, 180(2-4), 246-258.

[66] Cronin, S. J., Bebbington, M., & Lai, C. D. (2001). A probabilistic assessment of eruption recurrence on Taveuni volcano, Fiji. *Bull Volcanol*, 63, 274-89.

[67] Siebe, C., Arana-Salinasa, L., & Abrams, M. (2005). Geology and radiocarbon ages of Tláloc, Tlacotenco, Cuauhtzin, Hijo del Cuauhtzin, Teuhtli, and Ocusacayo monogenetic volcanoes in the central part of the Sierra Chichinautzin, México. *J Volcanol Geotherm Res*, 141, 225-43.

[68] Kereszturi, G., & Németh, K. (2012). Structural and morphometric irregularities of eroded Pliocene scoria cones at the Bakony-Balaton Highland Volcanic Field, Hungary. *Geomorphology*, 136(1), 45-58.

[69] Rodriguez-Gonzalez, A., Fernandez-Turiel, J. L., Perez-Torrado, F. J., Aulinas, M., Carracedo, J. C., Gimeno, D., et al. (2011). GIS methods applied to the degradation of monogenetic volcanic fields: A case study of the Holocene volcanism of Gran Canaria (Canary Islands, Spain). *Geomorphology*, 134(3-4), 249-259.

[70] Dohrenwend, J. C., Wells, S. G., & Turrin, B. D. (1986). Degradation of Quaternary cinder cones in the Cima volcanic field, Mojave Desert, California. *Geol Soc Am Bull*, 97(4), 421-7.

[71] Wood, C. A. (1980). Morphometric analysis of cinder cone degradation. *J Volcanol Geotherm Res*, 8(2-4), 137-60.

[72] Németh, K. (2004). Calculation of long-term erosion in Central Otago, New Zealand, based on erosional remnants of maar/tuff rings. *Z Geomorphol*, 47(1), 29-49.

[73] Büchner, J., & Tietz, O. (2012). Reconstruction of the Landeskrone Scoria Cone in the Lusatian Volcanic Field, Eastern Germany- Long-term degradation of volcanic edifices and implications for landscape evolution. *Geomorphology*, 151-152, 175-187.

[74] Németh, K., & Martin, U. (1999). Late Miocene paleo-geomorphology of the Bakony-Balaton Highland Volcanic Field (Hungary) using physical volcanology data. *Zeitschrift fr Geomorphologie*, 43(4), 417-38.

[75] Lister, J. R., & Kerr, R. C. (1991). Fluid-Mechanical Models of Crack Propagation and Their Application to Magma Transport in Dykes. *J Geophys Res*, 96(B6), 10049-77.

[76] O'Neill, C., & Spiegelman, M. (2010). Formulations for Simulating the Multiscale Physics of Magma Ascent. *Dosseto A, Turner SP, van Orman JA, editors. Timescales of Magmatic Processes: From Core to Atmosphere. Chichester, UK. : John Wiley & Sons, Ltd*, 87-101.

[77] Gardine, M., West, M., & Cox, T. (2011). Dike emplacement near Parícutin volcano, Mexico in 2006. *Bulletin of Volcanology*, 73(2), 123-32.

[78] de la Cruz-Reyna, S., & Yokoyama, I. (2011). A geophysical characterization of monogenetic volcanism. *Geofisica Internacional*, 50(4), 465-84.

[79] Németh, K. (2010). Volcanic glass textures, shape characteristics and compositions of phreatomagmatic rock units from the Western Hungarian monogenetic volcanic fields and their implications for magma fragmentation. *Cent Eur J Geosci*, 2(3), 399-419.

[80] Houghton, B. F., Wilson, C. J. N., Rosenberg, M. D., Smith, I. E. M., & Parker, R. J. (1996). Mixed deposits of complex magmatic and phreatomagmatic volcanism: an example from Crater Hill, Auckland, New Zealand. *Bull Volcanol*, 58, 59-66.

[81] Lorenz, V. (2007). Syn- and posteruptive hazards of maar-diatreme volcanoes. *J Volcanol Geotherm Res*, 159, 285-312.

[82] Gutmann, J. T. (2002). Strombolian and effusive activity as precursors to phreatomagmatism: eruptive sequence at maars of the Pinacate volcanic fields, Sonora, Mexico. *J Volcanol Geotherm Res*, 113(1-2), 345-56.

[83] Foshag, W. F., & Gonzalez, J. R. (1956). Birth and development of Paricutin volcano, Mexico Geological Survey Bulletin. 965-D, 355-487.

[84] Wood, C. A. (1980). Morphometric evolution of cinder cones. *J Volcanol Geotherm Res*, 7(3-4), 387-413.

[85] Davey, F. J. (2010). Crustal seismic reflection measurements across the northern extension of the Taupo Volcanic Zone, North Island, New Zealand. *J Volcanol Geotherm Res*, 190(1-2), 75-81.

[86] Wilson, C. J. N., Houghton, B. F., Mc Williams, M. O., Lanphere, M. A., Weaver, S. D., & Briggs, R. M. (1995). Volcanic and structural evolution of Taupo Volcanic Zone, New Zealand: a review. *J Volcanol Geotherm Res*, 68(1-3), 1-28.

[87] Houghton, B. F., Wilson, C. J. N., Mc Williams, M. O., Lanphere, MA, Weaver, S. D., Briggs, R. M., et al. (1995). Chronology and dynamics of a large silicic magmatic system: Central Taupo Volcanic Zone, New Zealand. *Geology*, 23, 13-6.

[88] Wallace, L. M., Beavan, J., Mc Caffrey, R., & Darby, D. (2004). Subduction zone coupling and tectonic block rotations in the North Island, New Zealand. *J Geophys Res*, 109(B12), B12406. 10.1029/2004jb003241.

[89] Seghedi, I., & Downes, H. (2011). Geochemistry and tectonic development of Cenozoic magmatism in the Carpathian-Pannonian region. *Gondwana Res*, 20, 655-72.

[90] Lexa, J., Seghedi, I., Németh, K., Szakács, A., Konečný, V., Pécskay, Z., et al. (2010). Neogene-Quaternary Volcanic forms in the Carpathian-Pannonian Region: a review. *Cent Eur J Geosci*, 2(3), 207-70.

[91] Pécskay, Z., Lexa, J., Szakács, A., Seghedi, I., Balogh, K., Konecny, V., et al. (2006). Geochronlogy of Neogene magmatism in the Carpathian arc and intra-Carpathian area. *Geol Carpath*, 57(6), 511-30.

[92] Harangi, S., & Lenkey, L. (2007). Genesis of the Neogene to Quaternary volcanism in the Carpathian-Pannonian region: Role of subduction, extension, and mantle plume. *Beccaluva L, Bianchini G, Wilson M, editors. Cenozoic Volcanism in the Mediterranean Area. London: Geological Society of America*, 67-92.

[93] Szabó, C., Falus, G., Zajacz, Z., Kovács, I., & Bali, E. (2004). Composition and evolution of lithosphere beneath the Carpathian-Pannonian region: a review. *Tectonophysics*, 393(1-4), 119-37.

[94] Müller, R. D., Sdrolias, M., Gaina, C., & Roest, W. R. (2008). Age, spreading rates and spreading symmetry of the world's ocean crust. *Geochem Geophys Geosyst*, 9(Q04006).

[95] Katz, R. F., Spiegelman, M., & Holtzman, B. (2006). The dynamics of melt and shear localization in partially molten aggregates. *Nature*, 442, 676-9.

[96] Courtillot, V., Davaille, A., Besse, J., & Stock, J. (2006). Three distinct types of hotspots in the Earth's mantle. *Earth Planet Sci Lett*, 205, 295-308.

[97] King, S. D., & Anderson, D. L. (1998). Edge-driven convection. *Earth Planet Sci Lett*, 160, 289-96.

[98] Ballmer, MD, Ito, G., van Hunen, J., & Tackley, P. J. (2011). Spatial and temporal variability in Hawaiian hotspot volcanism induced by small-scale convection. *Nat Geosci*, 4, 457-60.

[99] Ballmer, M. D., van Hunen, J., Ito, G., Tackley, P. J., & Bianco, T. A. (2007). Non-hot-spot volcano chains originating from small-scale sublithospheric convection. *Geophysical Rerearch Letters*, 43(L23310), 10.1029/2007GL031636.

[100] King, S. D., & Ritsema, J. (2000). African Hot Spot Volcanism: Small-Scale Convection in the Upper Mantle Beneath Cratons. *Science*, 290, 1137-40.

[101] Foulger, G. R., & Natland, J. H. (2003). Is "Hotspot" Volcanism a Consequence of Plate Tectonics? *Science*, 300, 921-2.

[102] King, D. S., Zimmerman, M. E., & Kohlstedt, D. L. (2010). Stress-driven Melt Segregation in Partially Molten Olivine-rich Rocks Deformed in Torsion. *J Petrol*, 51(1-2), 21-42.

[103] Holtzman, B. K., Groebner, N. J., Zimmerman, M. E., Ginsberg, S. B., & Kohlstedt, D. L. (2003). Stress-driven melt segregation in partially molten rocks. *Geochem Geophys Geosyst*, 4(5), 8607.

[104] Mc Kenzie, D. (1984). The Generation and Compaction of Partially Molten Rock. *J Petrol*, 25(3), 713-65.

[105] Laporte, D., & Watson, E. B. (1995). Experimental and theoretical constraints on melt distribution in crustal sources: the effect of crystalline anisotropy on melt interconnectivity. *Chem Geol*, 124(3-4), 161-184.

[106] Kelemen, P. B., Hirth, G., Shimizu, N., Spiegelman, M., & Dick, H. J. (1997). A review of melt migration processes in the adiabatically upwelling mantle beneath oceanic spreading ridges. *Philosophical Transactions of the Royal Society A: Mathematical Physical & Engineering Sciences*, 355(1723), 283-318.

[107] Sleep, N. H. (1988). Tapping of Melt by Veins and Dikes. *J Geophys Res*, 93(B9), 10255-72.

[108] Valentine, G. A., & Perry, F. V. (2006). Decreasing magmatic footprints of individual volcanoes in a waning basaltic field. *Geophys Res Lett*, 33(L14305), 10.1029/2006GL026743.

[109] Takada, A. (1994). The influence of regional stress and magmatic input on styles of monogenetic and polygenetic volcanism. *J Geophys Res*, 99(B7), 10.1029/94JB00494.

[110] Ito, G., & Martel, S. J. (2002). Focusing of magma in the upper mantle through dike interaction. *J Geophys Res*, 107(B10), 10.1029/2001JB000251.

[111] Valentine, G. A., & Gregg, T. K. P. (2008). Continental basaltic volcanoes- Processes and problems. *J Volcanol Geotherm Res*, 177(4), 857-73.

[112] Dahm, T. (2000). On the shape and velocity of fluid-filled fractures in the Earth. *Geophys J Int*, 142(1), 181-92.

[113] Delaney, P. T., Pollard, D. D., Ziony, J. I., & Mc Kee, E. H. (1986). Field Relations Between Dikes and Joints: Emplacement Processes and Paleostress Analysis. *J Geophys Res*, 91(B5), 4920-38.

[114] Mazzarini, F., & D'Orazio, M. (2003). Spatial distribution of cones and satellite-detected lineaments in the Pali Aike Volcanic Field (southernmost Patagonia): insights into the tectonic settings of a Neogene rift system. *J Volcanol Geotherm Res*, 125, 291-305.

[115] Lara, L. E., Lavenu, A., Cembrano, J., & Rodríguez, C. (2006). Structural controls of volcanism in transversal chains: Resheared faults and neotectonics in the Cordón Caulle-Puyehue area (40.5°S), Southern Andes. *J Vol canol Geotherm Res*, 158(1-2), 70.

[116] Walker, G. P. L. (2000). Basaltic volcanoes and volcanic systems. *Sigurdsson H, Houghton BF, McNutt SR, Rymer H, Stix J, editors. Encyclopedia of Volcanoes. San Diego: Academic Press*, 283-290.

[117] Zernack, A. V., Procter, J. N., & Cronin, S. J. (2009). Sedimentary signatures of cyclic growth and destruction of stratovolcanoes: A case study from Mt. Taranaki, New Zealand. *Sediment Geol*, 220, 288-305.

[118] Carracedo, J. C., Rodríguez, Badiola. E., Guillou, H., Paterne, M., Scaillet, S., Pérez Torrado, F. J., et al. (2007). Eruptive and structural history of Teide Volcano and rift zones of Tenerife, Canary Islands. *Geol Soc Am Bull*, 119(9), 1027-51.

[119] Grosse, P., van Wyk de Vries, B., Petrinovic, I. A., Euillades, P. A., & Alvarado, G. E. (2009). Morphometry and evolution of arc volcanoes. *Geology*, 37(7), 651-4.

[120] Buck, R. W., Einarsson, P., & Brandsdóttir, B. (2006). Tectonic stress and magma chamber size as controls on dike propagation: Constraints from the 1975-1984 Krafla rifting episode. *J Geophys Res*, 111(B12404), 10.1029/2005JB003879.

[121] Marsh, B. D. (1989). Magma Chambers. *Annual Review of Earth and Planetary Sciences*, 17, 439-74.

[122] Burov, E., Jaupart, C., & Guillou-Frottier, L. (2003). Ascent and emplacement of buoyant magma bodies in brittle-ductile upper crust. *J Geophys Res*, 108(2177), 10.1029/2002JB001904.

[123] Hawkesworth, C. J., Blake, S., Evans, P., Hughes, R., Macdonald, R., Thomas, L. E., et al. (2000). Time scales of crystal fractionation in magma chambers- Integrating physical, isotopic and geochemical perspectives. *J Petrol*, 41(7), 991-1006.

[124] Alaniz-Alvarez, S. A., Nieto-Samaniego, A. F., & Ferrari, L. (1998). Effect of strain rate in the distribution of monogenetic and polygenetic volcanism in the Transmexican volcanic belt. *Geology*, 26(7), 591-4.

[125] Fedotov, S. (1981). Magma rates in feeding conduits of different volcanic centres. *J Volcanol Geotherm Res*, 9(4), 379-94.

[126] Suter, M., Contreras, J., Gómez-Tuena, A., Siebe, C., Quintero-Legorreta, O., García-Palomo, A., et al. (1999). Effect of strain rate in the distribution of monogenetic and polygenetic volcanism in the Transmexican volcanic belt: Comments and Reply. *Geology*, 27(6), 571-5.

[127] Mattsson, H., & Höskuldsson, Á. (2003). Geology of the Heimaey volcanic centre, south Iceland: early evolution of a central volcano in a propagating rift? *J Volcanol Geotherm Res*, 127(1-2), 55.

[128] Mc Gee, L. E., Beier, C., Smith, I. E. M., & Turner, S. P. (2011). Dynamics of melting beneath a small-scale basaltic system: a U-Th-Ra study from Rangitoto volcano, Auckland volcanic field, New Zealand. *Contrib Mineral Petrol*, 162(3), 547-63.

[129] Kiyosugi, K., Connor, C. B., Zhao, D., Connor, L. J., & Tanaka, K. (2010). Relationships between volcano distribution, crustal structure, and P-wave tomography: an example from the Abu Monogenetic Volcano Group, SW Japan. *Bull Volcanol*, 72(3), 331-40.

[130] Toprak, V. (1998). Vent distribution and its relation to regional tectonics, Cappadocian Volcanics, Turkey. *J Volcanol Geotherm Res*, 85(1-4), 55.

[131] Bacon, C. R. (1982). Time-predictable bimodal volcanism in the Coso Range, California. *Geology*, 10(2), 65-9.

[132] Frey, H. M., Lange, R. A., Hall, C. M., & Delgado-Granados, H. (2004). Magma eruption rates constrained by 40Ar/39Ar chronology and GIS for the Ceboruco-San Pedro volcanic field, western Mexico. *Geol Soc Am Bull*, 116(3-4), 259.

[133] Nakamura, K. (1977). Volcanoes as possible indicators of tectonic stress orientation-principal and proposal. *J Volcanol Geotherm Res*, 2(1), 1-16.

[134] Kuntz, M. A., Champion, D. E., Spiker, E. C., & Lefebvre, R. H. (1986). Contrasting magma types and steady-state, volume-predictable, basaltic volcanism along the Great Rift, Idaho. *Geol Soc Am Bull*, 97(5), 579-94.

[135] King-Y, C. (1989). Volume predictability of historical eruptions at Kilauea and Mauna Loa volcanoes. *J Volcanol Geotherm Res*, 38, 281-5.

[136] Parsons, T., & Thompson, G. A. (1991). The role of magma overpressure in suppressing earthquakes and topography: worldwide examples. *Science*, 253(5026), 1399-402.

[137] Ebinger, C. J., Keir, D., Ayele, A., Calais, E., Wright, T. J., Belachew, M., et al. (2008). Capturing magma intrusion and faulting processes during continental rupture: seismicity of the Dabbahu (Afar) rift. *Geophys J Int*, 174(3), 1138-52.

[138] Gudmundsson, A., & Loetveit, I. F. (2005). Dyke emplacement in a layered and faulted rift zone. *J Volcanol Geotherm Res*, 144(1-4), 311.

[139] Bebbington, M. (2008). Incorporating the eruptive history in a stochastic model for volcanic eruptions. *J Volcanol Geotherm Res*, 175, 325-33.

[140] Aranda-Gómez, J. J., Luhr, J. F., Housh, T. B., Connor, C. B., Becker, T., & Henry, C. D. (2003). Synextensional Pliocene-Pleistocene eruptive activity in the Camargo volcanic field, Chihuahua, México. *Geol Soc Am Bull*, 115(3), 298-313.

[141] Guilbaud-N, M., Siebe, C., Layer, P., Salinas, S., Castro-Govea, R., Garduño-Monroy, V. H., et al. (2011). Geology, geochronology, and tectonic setting of the Jorullo Volcano region, Michoacán, México. *J Volcanol Geotherm Res*, 201(1-4), 97-112.

[142] Wijbrans, J., Németh, K., Martin, U., & Balogh, K. (2007). 40Ar/39Ar geochronology of Neogene phreatomagmatic volcanism in the western Pannonian Basin, Hungary. *J Volcanol Geotherm Res*, 164(4), 193-204.

[143] Connor, C. B. (1990). Cinder Cone Clustering in the TransMexican Volcanic Belt: Implications for Structural and Petrologic Models. *J Geophys Res*, 95(B12), 19, 395-19,405.

[144] Hasenaka, T., & Carmichael, I. S. E. (1987). The Cinder Cones of Michoacán-Guanajuato, Central Mexico: Petrology and Chemistry. *J Petrol*, 28(2), 241-69.

[145] Huang, Y., Hawkesworth, C., van Calsteren, P., Smith, I., & Black, P. (1997). Melt generation models for the Auckland volcanic field, New Zealand: constraints from U-Th isotopes. *Earth Planet Sci Lett*, 149.

[146] Putirka, K. D., Kuntz, MA, Unruh, D. M., & Vaid, N. (2009). Magma Evolution and Ascent at the Craters of the Moon and Neighboring Volcanic Fields, Southern Idaho, USA: Implications for the Evolution of Polygenetic and Monogenetic Volcanic Fields. *J Petrol*, 50(9), 1639-65.

[147] Austin-Erickson, A., Ort, M. H., & Carrasco-Núñez, G. (2011). Rhyolitic phreatomagmatism explored: Tepexitl tuff ring (Eastern Mexican volcanic belt). *J Volcanol Geotherm Res*, 201(1-4), 325-41.

[148] Brooker, M. R., Houghton, B. F., Wilson, C. J. N., & Gamble, J. A. (1993). Pyroclastic phases of a rhyolitic dome-building eruption: Puketarata tuff ring, Taupo Volcanic Zone, New Zealand. *Bull Volcanol*, 55(6), 395-406.

[149] Walker, G. P. L. (1973). Explosive volcanic eruptions- a new classification scheme. *Geol Rundsch*, 62(2), 431-46.

[150] Parfitt, E. A. (2004). A discussion of the mechanisms of explosive basaltic eruptions. *J Volcanol Geotherm Res*, 134(1-2), 77-107.

[151] Wohletz, K. H. (1986). Explosive magma-water interactions: Thermodynamics, explosion mechanism and field studies. *Bull Volcanol*, 48(5), 245-64.

[152] Lorenz, V. (1985). Maars and diatremes of phreatomagmatic origin: a review. *S Afr J Geol*, 88(2), 459-70.

[153] Vergniolle, S., Brandeis, G., & Mareschal-C, J. (1996). Strombolian explosions: Eruption dynamics determined from acoustic measurements. *J Geophys Res*, 101, 20449-66.

[154] Parfitt, E. A., & Wilson, L. (1995). Explosive volcanic eruptions- IX. The transition between Hawaiian-style lava fountaining and Strombolian explosive activity. *Geophys J Int*, 121(1), 226-32.

[155] Blackburn, E. A., & Sparks, R. S. J. (1976). Mechanism and dynamics of Strombolian activity. *J Geol Soc*, 132, 429-40.

[156] Pioli, L., Azzopardi, B. J., & Cashman, K. V. (2009). Controls on the explosivity of scoria cone eruptions: Magma segregation at conduit junctions. *J Volcanol Geotherm Res*, 186(3-4), 407-415.

[157] Johnson, E., Wallace, P., Chashman, K., Granados, H. D., & Kent, A. (2008). Magmatic volatile contents and degassing-induced crystallization at Volcán Jorullo, Mexico: Implications for melt evolution and the plumbing systems of monogenetic volcanoes. *Earth Planet Sci Lett*, 269, 478-87.

[158] Walker, G. P. L., & Croasdale, R. (1972). Characteristics of some basaltic pyroclastics. *Bull Volcanol*, 35(2), 303-17.

[159] Houghton, B. F., & Gonnermann, H. M. (2008). Basaltic explosive volcanism: Constraints from deposits and models. *Chem Erde*, 68, 117-40.

[160] Wilson, L., & Head, J. W. (1981). Ascent and eruption of basaltic magma on the Earth and Moon. *J Geophys Res*, 86, 2971-3001.

[161] Sparks, R. S. J. (1978). The dynamics of bubble formation and growth in magmas: A review and analysis. *J Volcanol Geotherm Res*, 3(1-2), 1-37.

[162] Parfitt, E. A. (1998). A study of clast size distribution, ash deposition and fragmentation in a Hawaiian-style volcanic eruption. *J Volcanol Geotherm Res*, 84, 197-208.

[163] Pioli, L., Erlund, E., Johnson, E., Cashman, N. K., Wallace, P., Rosi, M., et al. (2008). Explosive dynamics of violent Strombolian eruptions: The eruption of Parícutin Volcano 1943-1952 (Mexico). *Earth Planet Sci Lett*, 271(1-4), 359-68.

[164] Coltelli, M., Del Carlo, P., & Vezzoli, L. (1998). Discovery of a Plinian basaltic eruption of Roman age at Etna volcano, Italy. *Geology*, 26(12), 1095-8.

[165] Houghton, B. F., Wilson, C. J. N., Del Carlo, P., Coltelli, M., Sable, J. E., & Carey, R. (2004). The influence of conduit processes on changes in style of basaltic Plinian eruptions: Tarawera 1886 and Etna 122 BC. *J Volcanol Geotherm Res*, 137, 1-14.

[166] Vergniolle, S., & Mangan, M. T. (2000). Hawaiian and Strombolian eruptions. *Sigurdsson H, Houghton B, McNutt SR, Rymer H, Stix J, editors. Encyclopedia of Volcanoes-San Diego: Academic Press*, 447-461.

[167] Stovall, W. K., Houghton, B. F., Gonnermann, H., Fagents, S. A., & Swanson, D. A. (2011). Eruption dynamics of Hawaiian-style fountains: the case study of episode 1 of the Kilauea Iki 1959 eruption. *Bull Volcanol*, 73(5), 511-29.

[168] Polacci, M., Corsaro, R. A., & Andronico, D. (2006). Coupled textural and compositional characterization of basaltic scoria: Insights into the transition from Strombolian to fire fountain activity at Mount Etna Italy. *Geology*, 34(3), 201-4.

[169] Duffield, W. A., Christiansen, R. L., Koyanagi, R., & Peterson, D. W. (1982). Storage, migration, and eruption of magma at Kilauea volcano, Hawaii, 1971-1972. *J Volcanol Geotherm Res*, 13, 273-307.

[170] Head, J. W., & Wilson, L. (1987). Lava fountain heights at Pu'u'O'o, Kilauea, Hawaii: Indicators of amount and variations of exsolved magma volatiles. *J Geophys Res*, 92(B13), 13715.

[171] Sumner, J., Blake, S., Matela, R., & Wolff, J. (2005). Spatter. *J Volcanol Geotherm Res*, 142, 49-65.

[172] Head, J. W., & Wilson, L. (1989). Basaltic pyroclastic eruptions: Influence of gas-release patterns and volume fluxes on fountain structure, and the formation of cinder cones, spatter cones, rootless flows, lava ponds and lava flows. *J Volcanol Geotherm Res*, 37(3-4), 261-271.

[173] Heliker, C., Kauahikaua, J., Sherrod, D. R., Lisowski, M., & Cervelli, P. F. (2003). The Rise and Fall of Pu'u'Ö'ö Cone, 1983-2002. *US Geological Survey Professional Paper 1676*, 29-52.

[174] Wolff, J. A., & Sumner, J. M. (2000). Lava fountains and their products. *Sigurdsson H, Houghton BF, McNutt SR, Rymer H, Stix J, editors. Encyclopedia of Volcanoes. San Diego: Academic Press*, 321-329.

[175] Capaccioni, B., & Cuccoli, F. (2005). Spatter and welded air fall deposits generated by fire-fountaining eruptions : Cooling of pyroclasts during transport and deposition. *J Volcanol Geotherm Res*, 145(3-4), 263 -280.

[176] Parfitt, E. A., & Wilson, L. (1999). A Plinian treatment of fallout from Hawaiian lava fountains. *J Volcanol Geotherm Res*, 88(1-2), 67-75.

[177] Head, J. W., Bryan, W. B., Greeley, R., Guest, J. E., Schultz, P. H., Sparks, R. S. J., et al. (1981). Distribution and morphology of basalt deposits on planets. *Project BVS, editors. Basaltic Volcanism on the Terrestrial Planets. New York: Pergamon Press*, 701-800.

[178] Kauahikaua, J., Sherrod, D. R., Cashman, K. V., Heliker, C., Hon, K., Mattox, T. N., et al. (2003). Hawaiian Lava-Flow Dynamics During the Pu'u 'Ö'ö-Küpaianaha Eruption: A Tale of Two Decades. *US Geological Survey Professional Paper 1676*, 63-88.

[179] Wood, C. A. (1979). Monogenetic volcanoes in terrestrial planets. *Proceedings of the 10th Lunar and Planetary Science Conference*.

[180] Patrick, M. R., Harris, A. J. L., Ripepe, M., Dehn, J., Rothery, D. A., & Calvari, S. (2007). Strombolian explosive styles and source conditions: insights from thermal (FLIR) video. *Bull Volcanol*, 69, 769-84.

[181] Vespermann, D., & Schmincke-U, H. (2000). Scoria cones and tuff rings. *Sigurdsson H, Houghton BF, McNutt SR, Rymer H, Stix J, editors. Encyclopedia of VolcanoesSan Diego: Academic Press*, 683-694.

[182] Riedel, C., Ernst, G. G. J., & Riley, M. (2003). Controls on the growth and geometry of pyroclastic constructs. *J Volcanol Geotherm Res*, 127(1-2), 121-52.

[183] Harris, A. J. L., Ripepe, M., & Hughes, E. A. (2012). Detailed analysis of particle launch velocities, size distributions and gas densities during normal explosions at Stromboli. *J Volcanol Geotherm Res*, 132, 109-131.

[184] Calvari, S., & Pinkerton, H. (2004). Birth, growth and morphologic evolution of the'Laghetto' cinder cone during the 2001 Etna eruption. *J Volcanol Geotherm Res*, 132, 225-39.

[185] Mc Getchin, T. R., Settle, M., & Chouet, BA. (1974). Cinder cone growth modeled after Northeast Crater, Mount Etna, Sicily. *J Geophys Res*, 79, 3257-72.

[186] Houghton, B. F., & Wilson, C. J. N. (1989). A vesicularity index for pyroclastic deposits. *Bull Volcanol*, 51(6), 451-62.

[187] Mangan, M. T., & Cashman, K. V. (1996). The structure of basaltic scoria and reticulite and inferences for vesiculation, foam formation, and fragmentation in lava fountains. *J Volcanol Geotherm Res*, 73, 1-18.

[188] Valentine, G. A., Krier, D. J., Perry, F. V., & Heiken, G. (2007). Eruptive and geomorphic processes at the Lathrop Wells scoria cone volcano. *J Volcanol Geotherm Res*, 161(1-2), 57.

[189] Settle, M. (1979). The structure and emplacement of cinder cone fields. *Am J Sci*, 279(10), 1089-107.

[190] Mac, Donald. G. A. (1972). Volcanoes. *Englewood Cliffs: Prentice-Hall.*

[191] Erlund, E. J., Cashman, K. V., Wallace, P. J., Pioli, L., Rosi, M., Johnson, E., et al. (2010). Compositional evolution of magma from Parícutin Volcano, Mexico: The tephra record. *J Volcanol Geotherm Res*, 197(1-4), 167-87.

[192] Zimanowski, B., & Büttner, R. (2002). Dynamic mingling of magma and liquefied sediments. *J Volcanol Geotherm Res*, 114, 37-44.

[193] Zimanowski, B., Büttner, R., & Lorenz, V. (1997). Premixing of magma and water in MFCI experiments. *Bull Volcanol*, 58(6), 491-5.

[194] White, J. D. L. (1996). Impure coolants and interaction dynamics of phreatomagmatic eruptions. *J Volcanol Geotherm Res*, 74(3-4), 155-70.

[195] Németh, K., Goth, K., Martin, U., Csillag, G., & Suhr, P. (2008). Reconstructing paleoenvironment, eruption mechanism and paleomorphology of the Pliocene Pula maar, (Hungary). *J Volcanol Geotherm Res*, 177(2), 441-56.

[196] Kokelaar, P. (1986). Magma-water interactions in subaqueous and emergent basaltic volcanism. *Bull Volcanol*, 48, 275-89.

[197] Zimanowski, B., & Wohletz, K. H. (2000). Physics of phreatomagmatism I. *Terra Nostra*, 6, 515-23.

[198] Lorenz, V., & Kurszlaukis, S. (2007). Root zone processes in the phreatomagmatic pipe emplacement model and consequences for the evolution of maar-diatreme volcanoes. *J Volcanol Geotherm Res*, 159, 4-32.

[199] Lorenz, V. (1986). On the growth of maar and diatremes and its relevance to the formation of tuff rings. *Bull Volcanol*, 48(5), 265-74.

[200] White, J. D. L., & Ross-S, P. (2011). Maar-diatreme volcanoes: A review. *J Volcanol Geotherm Res*, 201(1-4), 1-29.

[201] Büttner, R., Dellino, P., & Zimanowski, B. (1999). Identifying modes of magma/water interaction from the surface features of ash particles. *Nature*, 401, 688-90.

[202] Sottili, G., Palladino, D., Gaeta, M., & Masotta, M. (2012). Origins and energetics of maar volcanoes: examples from the ultrapotassic Sabatini Volcanic District (Roman Province, Central Italy). *Bull Volcanol*, 74(1), 163-86.

[203] Mattox, T. N., & Mangan, M. T. (1997). Littoral hydrovolcanic explosions: a case study of lava-seawater interaction at Kilauea volcano. *J Volcanol Geotherm Res*, 75(1-2), 1-17.

[204] Jurado-Chichay, Z., Rowland, S. K., & Walker, G. P. L. (1996). The formation of circular littoral cone from tube-fed pahoehoe: Mauna Loa, Hawaii. *Bull Volcanol*, 57(7), 471-82.

[205] Thorarinsson, S. (1953). The crater groups in Iceland. *Bull Volcanol*, 14, 3-44.

[206] Hamilton, C. W., Thordarson, T., & Fagents, S. A. (2010). Explosive lava-water interactions I: architecture and emplacement chronology of volcanic rootless cone groups in the 1783-1784 Laki lava flow, Iceland. *Bull Volcanol*, 72, 449-67.

[207] Hamilton, C. W., Fagents, S. A., & Thordarson, T. (2010). Explosive lava-water interactions II: self-organization processes among volcanic rootless eruption sites in the 1783-1784 Laki lava flow, Iceland. *Bull Volcanol*, 72, 469-85.

[208] Fagents, S. A., & Thordarson, T. (2007). Rootless volcanic cones in Iceland and on Mars. *Chapman MG, editors. The geology of Mars: evidence from Earth-based analogs.New York, USA: Cambridge University Press*, 151-177.

[209] Lorenz, V. (2003). Maar-Diatreme Volcanoes, their Formation, and their Setting in Hard-rock or Soft-rock environments. *Geolines*, 15, 72-83.

[210] Mattsson, H. B., & Tripoli, B. A. (2011). Depositional characteristics and volcanic landforms in the Lake Natron-Engaruka monogenetic field, northern Tanzania. *J Volcanol Geotherm Res*, 203, 23-34.

[211] Stoppa, F., Rosatelli, G., Schiazza, M., & Tranquilli, A. (2012). Hydrovolcanic vs Magmatic Processes in Forming Maars and Associated Pyroclasts: The Calatrava-Spain-Case History. *Stoppa F, editors. Updates in Volcanology- A Comprehensive Approach to Volcanological Problems. InTech*, 3-26, http://www.intechopen.com/books/updates-in-volcanology-a-comprehensive-approach-to-volcanological-problems/hydrovolcanic-vs-magmatic-processes-in-forming-maars-and-associated-pyroclasts-the-calatrava-spain-c.

[212] Stoppa, F. (1996). The San Venanzo maar and tuff ring, Umbria, Italy: eruptive behaviour of a carbonatite-melilitite volcano. *Bull Volcanol*, 57(7), 563-77.

[213] Lorenz, V., Mc Birney, A. R., & Williams, H. (1970). An investigation of volcanic depressions. *Part III. Maars, tuff-rings, tuff-cones, and diatremes. Houston, Texas.*

[214] Blaikie, T. N., Ailleres, L., Cas, R. A. F., & Betts, P. G. (2012). Three-dimensional potential field modelling of a multi-vent maar-diatreme- The Lake Coragulac maar, Newer Volcanics Province, southeastern Australia. *J Volcanol Geotherm Res*, 235-236(0), 70-83.

[215] Beget, J. E., Hopkins, D. M., & Charron, S. D. (1996). Largest known maars (Espenberg Maars) on earth, Seward Peninsula, northwest Alaska. *Arctic*, 49(1), 62-9.

[216] Grunewald, U., Zimanowski, B., Büttner, R., Phillips, L. F., Heide, K., & Büchel, G. (2007). MFCI experiments on the influence of NaCl-saturated water on phreatomagmatic explosions. *J Volcanol Geotherm Res*, 159(1-3), 126.

[217] Sheridan, M. F., & Wohletz, K. H. (1981). Hydrovolcanic Explosions: The Systematics of Water-Pyroclast Equilibration. *Science*, 212, 1387-9.

[218] Wohletz, K. H., & Valentine, G. A. (1990). Computer simulations of explosive volcanic eruptions. *Ryan MP, editors. Magma transport and storage*, 114-134.

[219] Büttner, R., Dellino, P., La Volpe, L., Lorenz, V., & Zimanowski, B. (2002). Thermohydraulic explosions in phreatomagmatic eruptions as evidenced by the comparison between pyroclasts and products from Molten Fuel Coolant Interaction experiments. *J Geophys Res*, 107(B11), 10.1029/2001JB000511.

[220] Büttner, R., & Zimanowski, B. (1998). Physics of thermohydraulic explosions. *Physical Review E*, 57(5), 5726-9.

[221] Wohletz, K. H., & Mc Queen, R. G. (1984). Experimental Studies of Hydrovolcanic Explosions. *editors. Explosive volcanism: Inception, evolution and hazards. Washington D.C., USA: National Academy Press*, 158-169.

[222] Kokelaar, B. P. (1983). The mechanism of Surtseyan volcanism. *J Geol Soc*, 140, 939-44.

[223] Wohletz, K. H., & Sheridan, M. F. (1983). Hydrovolcanic explosions II. Evolution of basaltic tuff rings and tuff cones. *Am J Sci*, 283, 385-413.

[224] Gençalioğlu-Kuşcu, G., Atilla, C., Cas, R. A. F., & Ilkay, K. (2007). Base surge deposits, eruption history, and depositional processes of a wet phreatomagmatic volcano in Central Anatolia (Cora Maar). *J Volcanol Geotherm Res*, 159(1-3), 198-209.

[225] Funiciello, R., Giordano, G., & De Rita, D. (2003). The Albano maar lake (Colli Albani Volcano, Italy): recent volcanic activity and evidence of pre-Roman Age catastrophic lahar events. J. *Volcanol Geotherm Res*, 123, 43-61.

[226] Vazquez, J. A., & Ort, M. H. (2006). Facies variation of eruption units produced by the passage of single pyroclastic surge currents, Hopi Buttes volcanic field, USA. *J Volcanol Geotherm Res*, 154(3-4), 222-36.

[227] Sulpizio, R., & Dellino, P. (2008). Chapter 2 Sedimentology, Depositional Mechanisms and Pulsating Behaviour of Pyroclastic Density Currents. *Joachim G, Joan M, iacute, editors. Developments in Volcanology. Elsevier.*, 57-96.

[228] Carrasco-Núñez, G., Ort, M. H., & Romero, C. (2007). Evolution and hydrological conditions of a maar volcano (Atexcac crater, Eastern Mexico). *J Volcanol Geotherm Res*, 159, 179-97.

[229] Auer, A., Martin, U., & Németh, K. (2007). The Fekete-hegy (Balaton Highland Hungary) „soft-substrate" and „hard-substrate" maar volcanoes in an aligned volcanic complex- Implications for vent geometry, subsurface stratigraphy and the paleoenvironmental setting. *J Volcanol Geotherm Res*, 159(1-3), 225-45.

[230] Pardo, N., Macias, J. L., Giordano, G., Cianfarra, P., Avellán, D. R., & Bellatreccia, F. (2009). The ~1245 yr BP Asososca maar eruption: The youngest event along the Nejapa-Miraflores volcanic fault, Western Managua, Nicaragua. *J Volcanol Geotherm Res*, 184(3-4), 292-312.

[231] Raue, H. (2004). A new model for the fracture energy budget of phreatomagmatic explosions. *J Volcanol Geotherm Res*, 129(1-3), 99.

[232] White, J. D. L., & Schmincke-U, H. (1999). Phreatomagmatic eruptive and depositional processes during the '49 eruption on La Palma (Canary Islands). *J Volcanol Geotherm Res*, 94(1-4), 283-304.

[233] Martin, U., & Németh, K. (2005). Eruptive and depositional history of a Pliocene tuff ring hat developed in a fluvio-lacustrine basin: Kissomlyó volcano (western Hungary). *J Volcanol Geotherm Res*, 147, 342-56.

[234] Sohn, Y. K. (1996). Hydrovolcanic processes forming basaltic tuff rings and cones on Cheju Island, Korea. *Geol Soc Am Bull*, 108(10), 1199-211.

[235] Chough, S. K., & Sohn, Y. K. (1990). Depositional mechanics and sequences of base surges, Songaksan tuff ring, Cheju Island, Korea. *Sedimentology*, 37, 1115-35.

[236] Aranda-Gómez, J. J., & Luhr, J. F. (1996). Origin of the Joya Honda maar, San Luis Potosí, Mexico. *J Volcanol Geotherm Res*, 74, 1-18.

[237] Lorenz, V. (1974). Vesiculated tuffs and associated features. *Sedimentology,* 21(2), 273-91.

[238] Alvarado, G. E., Soto, G. J., Salani, F. M., Ruiz, P., & de Mendoza, L. H. (2011). The formation and evolution of Hule and Río Cuarto maars, Costa Rica. *J Volcanol Geotherm Res,* 201(1-4), 342-356.

[239] Ngwa, C. N., Suh, C. E., & Devey, C. W. (2010). Phreatomagmatic deposits and stratigraphic reconstruction at Debunscha Maar (Mt Cameroon volcano). *J Volcanol Geotherm Res,* 192(3-4), 201-11.

[240] Dellino, P., Isaia, R., La Volpe, L., & Orsi, G. (2004). Interaction between particles transported by fallout and surge in the deposits of the Agnano-Monte Spina eruption (Campi Flegrei, Southern Italy). *J Volcanol Geotherm Res,* 133(1-4), 193-210.

[241] Németh, K., & White, C. M. (2009). Intra-vent peperites related to the phreatomagmatic 71 Gulch Volcano, western Snake River Plain volcanic field, Idaho (USA). *J Volcanol Geotherm Res,* 183(1-2), 30-41.

[242] Hetényi, G., Taisne, B., Garel, F., Médard, É., Bosshard, S., & Mattsson, H. (2012). Scales of columnar jointing in igneous rocks: field measurements and controlling factors. *Bull Volcanol,* 74(2), 457-82.

[243] Martí, J., Planagumà, L., Geyer, A., Canal, E., & Pedrazzi, D. (2011). Complex interaction between Strombolian and phreatomagmatic eruptions in the Quaternary monogenetic volcanism of the Catalan Volcanic Zone (NE of Spain). *J Volcanol Geotherm Res,* 201(1-4), 178-93.

[244] Aranda-Gómez, J. J., Luhr, J. F., & Pier, G. (1992). The La Brena- El Jagüey Maar Complex, Durango, México: I. Geological evolution. *Bull Volcanol,* 54(5), 393-404.

[245] Lorenz, V. (2003). Syn- and post-eruptive processes of maar-diatreme volcanoes and their relevance to the accumulation of port-eruptive maar creater sediments. *Földtani Kutatás,* 11(1-2), 13-22.

[246] Molloy, C., Shane, P., & Augustinus, P. (2009). Eruption recurrence rates in a basaltic volcanic field based on tephra layers in maar sediments: Implications for hazards in the Auckland volcanic field. *Geol Soc Am Bull,* 121(11-12), 1666-77.

[247] Pirrung, M., Fischer, C., Büchel, G., Gaupp, R., Lutz, H., & Neuffer-O, F. (2003). Lithofacies succession of maar crater deposits in the Eifel area (Germany). *Terra Nova,* 15(2), 125-32.

[248] Németh, K. (2001). Deltaic density currents and turbidity deposits related to maar crater rims and their importance for paleogeographic reconstructionof the Bakony-Balaton Highland Volcanic Field, Hungary. *Kneller B, McCaffrey B, Peakall J, Druitt T, editors. Sediment transport and deposition by particulate gravity currents. Oxford: Blackwell Sciences,* 261-277.

[249] Pastre-F, J., Gauthier, A., Nomade, S., Orth, P., Andrieu, A., Goupille, F., et al. (2007). The Alleret maar (Massif Central, France): A new lacustrine sequence of the early Middle Pleistocene in western Europe. *CR Geosci*, 339, 987-97.

[250] Zolitschka, B. (1993). Palaeoecological implications from the sedimentary record of a subtropical maar lake (Eocene Eckfelder Maar; Germany). *Negendank JW, Zolitschka B, editors. Paleolimnology of European Maar Lakes. Springer Berlin Heidelberg*, 477-484.

[251] Lorenz, V. (1974). Studies of the Surtsey tephra deposits. *Surtsey Research Progress Report VII*, 72-9.

[252] Kano, K. (1998). A shallow-marine alkali-basalt tuff cone in the Middle Miocene Jinzai Formation, Izumo, SW Japan. *J Volcanol Geotherm Res*, 87(1-4), 173.

[253] White, J. D. L. (2000). Subaqueous eruption-fed density currents and their deposits. *Precambrian Res*, 101(2-4), 87-109.

[254] White, J. D. L. (1996). Pre-emergent construction of a lacustrine basaltic volcano, Pahvant Butte, Utah (USA). *Bull Volcanol*, 58(4), 249-62.

[255] Cas, R. A. F., Landis, C. A., & Fordyce, R. E. (1989). A monogenetic, Surtla-type, Surtseyan volcano from the Eocene-Oligocene Waiareka-Deborah volcanics, Otago, New Zealand: A model. *Bulletin of Volcanology*, 51(4), 281-98.

[256] Martin, U., Breitkreuz, C., Egenhoff, S., Enos, P., & Jansa, L. (2004). Shallow-marine phreatomagmatic eruptions through a semi-solidified carbonate platform (ODP Leg 144, Site 878, Early Cretaceous, MIT Guyot, West Pacific). *Mar Geol*, 204(3-4), 251.

[257] Garvin, J. B., Williams, R. S., Frawley, J. J., & Krabill, W. B. (2000). Volumetric evolution of Surtsey, Iceland, from topographic maps and scanning airborne laser altimetry. *Surtsey Research*, 11, 127-34.

[258] Vaughan, R. G., & Webley, P. W. (2010). Satellite observations of a surtseyan eruption: Hunga Ha'apai, Tonga. *J Volcanol Geotherm Res*, 198(1-2), 177-86.

[259] Zanon, V., Pachecoa, J., & Pimentel, A. (2009). Growth and evolution of an emergent tuff cone: Considerations from structural geology, geomorphology and facies analysis of São Roque volcano, São Miguel (Azores). *J Volcanol Geotherm Res*, 180(2-4), 277-291.

[260] Kokelaar, B. P., & Durant, G. P. (1983). The submarine eruption and erosion of Surtla (Surtsey), Iceland. *J Volcanol Geotherm Res*, 19(3-4), 239-46.

[261] Thorarinsson, S. (1965). Surtsey eruption course of events and the developement of the new island. *Surtsey Research Progress Report I*, 51-5.

[262] Mattsson, H. B. (2010). Textural variation in juvenile pyroclasts from an emergent, Surtseyan-type, volcanic eruption: The Capelas tuff cone, São Miguel (Azores). *J Volcanol Geotherm Res*, 189(1-2), 81-91.

[263] Németh, K., Cronin, S. J., Charley, D., Harrison, M., & Garae, E. (2006). Exploding lakes in Vanuatu- "Surtseyan-style" eruption witnessed on Ambae Island. *Episodes*, 29(2), 87-92.

[264] Brand, B. D., & Clarke, A. B. (2009). The architecture, eruptive history, and evolution of the Table Rock Complex, Oregon: From a Surtseyan to an energetic maar eruption. *J Volcanol Geotherm Res*, 180(2-4), 203-24.

[265] Sohn, Y. K., & Chough, S. K. (1993). The Udo tuff cone, Cheju Island, South Korea: transformation of pyroclastic fall into debris fall and grain flow on a steep volcanic cone slope. *Sedimentology*, 40, 769-86.

[266] Mattsson, H. B., Höskuldsson, A., & Hand, S. (2005). Crustal xenoliths in the 6220 BP Saefell tuff-cone, south Iceland: Evidence for a deep, diatreme-forming, Surtseyan eruption. *J Volcanol Geotherm Res*, 145(3-4), 234-248.

[267] Suiting, I., & Schmincke-U, H. (2009). Internal vs. external forcing in shallow marine diatreme formation: A case study from the Iblean Mountains (SE-Sicily, Central Mediterranean). *J Volcanol Geotherm Res*, 186(3-4), 361-378.

[268] Martin, U., & Németh, K. (2004). Mio/Pliocene Phreatomagmatic Volcanism in the Western Pannonian Basin. *Budapest*.

[269] Cassidy, J., France, S. J., & Locke, CA. (2007). Gravity and magnetic investigation of maar volcanoes, Auckland volcanic field, New Zealand. *J Volcanol Geotherm Res*, 159(1-3), 153.

[270] Weinstein, Y. (2007). A transition from strombolian to phreatomagmatic activity induced by a lava flow damming water in a valley. *J Volcanol Geotherm Res*, 159, 267-84.

[271] Alvarado, G. E., Soto, G. J., Salani, F. M., Ruiz, P., & de Mendoza, L. H. (2011). The formation and evolution of Hule and Río Cuarto maars, Costa Rica. *Journal of Volcanology and Geothermal Research*, 201(1-4), 342-356.

[272] Luhr, J. F. (2001). Glass inclusions and melt volatile contents at Paricutin Volcano, Mexico. *Contrib Mineral Petrol*, 142(3), 261-83.

[273] Németh, K., Risso, C., Nullo, F., & Kereszturi, G. (2011). The role of collapsing and rafting of scoria cones on eruption style changes and final cone morphology: Los Morados scoria cone, Mendoza Argentina. *Cent Eur J Geosci*, 3(2), 102-18.

[274] Needham, A. J., Lindsay, J. M., Smith, I. E. M., Augustinus, P., & Shane, P. A. (2011). Sequential Eruption of Alkaline and Sub-Alkaline Magmas from a small Monogenetic volcano in The Auckland Volcanic Field, New Zealand. *J Volcanol Geotherm Res*, 201(1-4), 126-42.

[275] Cervantes, P., & Wallace, P. (2003). Magma degassing and basaltic eruption styles: a case study of 2000 year BP Xitle volcano in central Mexico. *J Volcanol Geotherm Res*, 120(3-4), 249-70.

[276] Johnson, E. R., Wallace, P. J., Cashman, K. V., & Delgado, Granados. H. (2010). Degassing of volatiles (H2O, CO2, S, Cl) during ascent, crystallization, and eruption of basaltic magmas. *J Volcanol Geotherm Res*, 197(1-4), 225-38.

[277] Houghton, B. F., & Schmincke-U, H. (1986). Mixed deposits of simultaneous strombolian and phreatomagmatic volcanism: Rothenberg volcano, East Eifel Volcanic Field. *J Volcanol Geotherm Res*, 30, 117-30.

[278] Sohn, Y. K., & Park, K. H. (2005). Composite tuff ring/cone complexes in Jeju Island, Korea: possible consequences of substrate collapse and vent migration. *J Volcanol Geotherm Res*, 141, 157-75.

[279] Ort, M. H., & Carrasco-Núñez, G. (2009). Lateral vent migration during phreatomagmatic and magmatic eruptions at Tecuitlapa Maar, east-central Mexico. *J Volcanol Geotherm Res*, 181(1-2), 67-77.

[280] Bishop, M. A. (2009). A generic classification for the morphological and spatial complexity of volcanic (and other) landforms. *Geomorphology*, 111(1-2), 104-9.

[281] Walker, G. (1971). Compound and simple lava flows and flood basalts. *Bull Volcanol*, 35(3), 579-90.

[282] Németh, K. (2004). The morphology and origin of wide craters at Al Haruj al Abyad, Libya: maars and phreatomagmatism in a large intracontinental flood lava field? *Z Geomorphol*, 48(4), 417-39.

[283] Solgevik, H., Mattsson, H., & Hennelin, O. (2007). Growth of an emergent tuff cone: Fragmentation and depositional processes recorded in the Capelas tuff cone, São Miguel, Azores. *J Volcanol Geotherm Res*, 159, 246-66.

[284] Gutmann, J. T. (1976). Geology of Crater Elegante, Sonora, Mexico. *Geol Soc Am Bull*, 87(12), 1718-29.

[285] Ross-S, P., & White, J. D. L. (2006). Debris jets in continental phreatomagmatic volcanoes: a field study of their subterranean deposits in the Coombs Hills vent complex, Antarctica. *J Volcanol Geotherm Res*, 149, 62-84.

[286] Sohn, Y. K., Cronin, S. J., Brenna, M., Smith, I. E. M., Németh, K., White, J. D. L., et al. (2012). Ilchulbong tuff cone, Jeju Island, Korea, revisited: A compound monogenetic volcano involving multiple magma pulses, shifting vents, and discrete eruptive phases. *Geol Soc Am Bull*, 10.1130/B30447.1.

[287] Shaw, C., & Woodland, A. (2012). The role of magma mixing in the petrogenesis of mafic alkaline lavas, Rockeskyllerkopf Volcanic Complex, West Eifel, Germany. *Bull Volcanol*, 74(2), 359-76.

[288] Martin, U., & Németh, K. (2006). How Strombolian is a "Strombolian" scoria cone? Some irregularities in scoria cone architecture from the Transmexican Volcanic Belt, near Volcán Ceboruco (Mexico), and Al Haruj (Libya). *J Volcanol Geotherm Res*, 155(1-2), 104-118.

[289] Karátson, D. (1996). Rates and factors of stratovolcano degradation in a continental climate: a complex morphometric analysis for nineteen Neogene/Quaternary crater remnants in the Carpathians. *J Volcanol Geotherm Res*, 73(1-2), 65.

[290] Inbar, M., Gilichinsky, M., Melekestsev, I., Melnikov, D., & Zaretskaya, N. (2011). Morphometric and morphological development of Holocene cinder cones: A field and remote sensing study in the Tolbachik volcanic field, Kamchatka. *J Volcanol Geotherm Res*, 201(1-4), 301-311.

[291] Rodriguez-Gonzalez, A., Fernandez-Turiel, J. L., Perez-Torrado, F. J., Gimeno, D., & Aulinas, M. (2009). Geomorphological reconstruction and morphometric modelling applied to past volcanism. *Int J Earth Sci*, 99(3), 645-60.

[292] Karátson, D., Telbisz, T., & Wörner, G. (2012). Erosion rates and erosion patterns of Neogene to Quaternary stratovolcanoes in the Western Cordillera of the Central Andes: An SRTM DEM based analysis. *Geomorphology*, 139-140, 122-135.

[293] Pike, R. J. (1978). Volcanoes on the inner planets: Some preliminary comparisons of gross topography. *Proceedings of the 10th Lunar and Planetary Science Conference*, 3, 3239-73.

[294] Karátson, D., Favalli, M., Tarquini, S., Fornaciai, A., & Wörner, G. (2010). The regular shape of stratovolcanoes: A DEM-based morphometrical approach. *J Volcanol Geotherm Res*, 193(3-4), 171-181.

[295] Grosse, P., van Wyk de Vries, B., Euillades, P. A., Kervyn, M., & Petrinovic, I. (2012). Systematic morphometric characterization of volcanic edifices using digital elevation models. *Geomorphology*, 136, 114-31.

[296] Procter, J. N., Cronin, S. J., Platz, T., Patra, A., Dalbey, K., Sheridan, M., et al. (2010). Mapping block-and-ash flow hazards based on Titan 2D simulations: a case study from Mt. Taranaki, NZ. *Nat Hazards*, 53, 483-501.

[297] Procter, J., Cronin, S. J., Fuller, I. C., Lube, G., & Manville, V. (2010). Quantifying the geomorphic impacts of a lake-breakout lahar, Mount Ruapehu, New Zealand. *Geology*, 38(1), 67-70.

[298] Tibaldi, A. (1995). Morphology of pyroclastic cones and tectonics. *J Geophys Res*, 100(B12), 24521-35.

[299] Inbar, M., & Risso, C. (2001). A morphological and morphometric analysis of a high density cinder cone volcanic field.- Payun Matru, south-central Andes, Argentina. *Z Geomorphol*, 45(3), 321-43.

[300] Colton, H. S. (1937). The basaltic cinder cones and lava flows of the San Francisco Mountain Volcanic Field. *Museum of Northern Arizona Bulletin*, 10, 1-58.

[301] Porter, S. C. (1972). Distribution, Morphology, and Size Frequency of Cinder Cones on Mauna Kea Volcano, Hawaii. *Geol Soc Am Bull*, 83(12), 3607-12.

[302] Moriya, I. (1986). Morphometry of Pyroclastic Cones in Japan. *The geographical reports of Kanazawa University*, 3, 58-76.

[303] Hasenaka, T., & Carmichael, I. S. E. (1985). A compilation of location, size, and geomophological parameters of volcanoes of the Michoacan-Guanajuato volcanic field, central Mexico. *Geofisica Internacional*, 24(4), 577-607.

[304] Hasenaka, T., & Carmichael, I. S. E. (1985). The cinder cones of Michoacán-Guanajuato central Mexico: their age, volume and distribution, and magma discharge rate. *J Volcanol Geotherm Res*, 25(1-2), 105-24.

[305] Bleacher, J. E., Glaze, L. S., Greeley, R., Hauber, E., Baloga, S. M., Sakimoto, S. E. H., et al. (2009). Spatial and alignment analyses for a field of small volcanic vents south of Pavonis Mons and implications for the Tharsis province, Mars. *J Volcanol Geotherm Res*, 185, 96-102.

[306] Mest, S. C., & Crown, D. A. (2005). Millochau crater, Mars: Infilling and erosion of an ancient highland impact crater. *Icarus*, 175, 335-59.

[307] Glaze, L. S., Anderson, S. W., Stofan, E. R., & Smrekar, S. E. (2005). Statistical distribution of tumuli on pahoehoe flow surfaces: analysis of examples in Hawaii and Iceland and potential applications to lava flows on Mars. *J Geophys Res*, 110(B08202), 10.1029/2004JB003564.

[308] Armstrong, J. C., & Leovy, C. B. (2005). Long term wind erosion on Mars. *Icarus*, 176, 57-74.

[309] Pike, R. J. (1976). Crater dimensions from Apollo data and supplemental sources. *The Moon*, 15, 463-77.

[310] Pike, R. J. (1977). Apparent depth/apparent diameter relation for lunar craters. *Proceedings of the 8th Lunar Science Conference*, 3, 3427-36.

[311] Ghent, R. R., Anderson, S. W., & Pithawala, T. M. (2012). The formation of small cones in Isidis Planitia, Mars through mobilization of pyroclastic surge deposits. *Icarus*, 217(1), 169-83.

[312] Bemis, K., Walker, J., Borgia, A., Turrin, B., Neri, M., & Swisher, III C. (2011). The growth and erosion of cinder cones in Guatemala and El Salvador: models and statistics. *J Volcanol Geotherm Res*, 201(1-4), 39-52.

[313] Sato, H., & Taniguchi, H. (1997). Relationship between crater size and ejecta volume of recent magmatic and phreato-magmatic eruptions: Implications for energy partitioning. *Geophys Res Lett*, 24(3), 205-8.

[314] Kervyn, M., Ernst, G. G. J., Carracedo-C, J., & Jacobs, P. (2012). Geomorphometric variability of "monogenetic" volcanic cones: Evidence from Mauna Kea, Lanzarote and experimental cones. *Geomorphology*, 136(1), 59-75.

[315] Corazzato, C., & Tibaldi, A. (2006). Fracture control on type, morphology and distribution of parasitic volcanic cones: an example from Mt. Etna, Italy. *J Volcanol Geotherm Res*, 158(1-2), 177-94.

[316] Tibaldi, A. (2003). Influence of cone morphology on dykes, Stromboli, Italy. *J Volcanol Geotherm Res*, 126(1-2), 79-95.

[317] Favalli, M., Karátson, D., Mazzarini, F., Pareschi, M. T., & Boschi, E. (2009). Morphometry of scoria cones located on a volcano flank: A case study from Mt. Etna (Italy), based on high-resolution LiDAR data. *J Volcanol Geotherm Res*, 186(3-4), 320-30.

[318] Conway, F. M., Connor, C. B., Hill, B. E., Condit, C. D., Mullaney, K., & Hall, C. M. (1998). Recurrence rates of basaltic volcanism in SP cluster, San Francisco volcanic field, Arizona. *Geology*, 26(7), 655-8.

[319] Rodríguez, S. R., Morales-Barrera, W., Layer, P., & González-Mercado, E. (2010). A quaternary monogenetic volcanic field in the Xalapa region, eastern Trans-Mexican volcanic belt: Geology, distribution and morphology of the volcanic vents. *J Volcanol Geotherm Res*, 197(1-4), 149-66.

[320] Aguirre-Díaz, G. J., Jaimes-Viera, M. C., & Nieto-Obregón, J. (2006). The Valle de Bravo Volcanic Field: Geology and geomorphometric parameters of a Quaternary monogenetic field at the front of the Mexican Volcanic Belt. *Geological Society of America Special Papers*, 402, 139-54.

[321] Dóniz, J., Romero, C., Carmona, J., & García, A. (2011). Erosion of cinder cones in Tenerife by gully formation, Canary Islands, Spain. *Physical Geography*, 32(2), 139-60.

[322] Doniz, J., Romero, C., Coello, E., Guillen, C., Sanchez, N., Garcia-Cacho, L., et al. (2008). Morphological and statistical characterisation of recent mafic volcanism on Tenerife (Canary Islands, Spain). *J Volcanol Geotherm Res*, 173(3-4), 185-95.

[323] Hooper, D. M. (1999). Cinder movement experiments on scoria cone slopes: Rates and direction of transport. *Landform Analysis*, 2, 5-18.

[324] Hooper, D. M., & Sheridan, M. F. (1998). Computer-simulation models of scoria cone degradation. *J Volcanol Geotherm Res*, 83(3-4), 241-67.

[325] Hooper, D. M. (1995). Computer-simulation models of scoria cone degradation in the Colima and Michoacán-Guanajuato volcanic fields, Mexico. *Geofisica Internacional*, 34(3), 321-40.

[326] Fornaciai, A., Behncke, B., Favalli, M., Neri, M., Tarquini, S., & Boschi, E. (2011). Detecting short-term evolution of Etnean scoria cones: a LIDAR-based approach. *Bull Volcanol*, 72(10), 1209-22.

[327] Pelletier, J. D., & Cline, M. L. (2007). Nonlinear slope-dependent sediment transport in cinder cone evolution. *Geology*, 35(12), 1067-70.

[328] Thouret-C, J. (1999). Volcanic geomorphology- an overview. *Earth-Science Reviews*, 47, 95-131.

[329] Karátson, D., Telbisz, T., & Singer, BS. (2010). Late-stage volcano geomorphic evolution of the Pleistocene San Francisco Mountain, Arizona (USA), based on high-resolution DEM analysis and 40Ar/39Ar chronology. *Bull Volcanol*, 72(7), 833-46.

[330] Székely, B., & Karátson, D. (2004). DEM-based morpometry as a tool for reconstructing primary volcanic landforms: examples from the Börzsöny Mountains, Hungary. *Geomorphology*, 63, 25-37.

[331] Lahitte, P., Samper, A., & Quidelleur, X. (2012). DEM-based reconstruction of southern Basse-Terre volcanoes (Guadeloupe archipelago, FWI): Contribution to the Lesser Antilles Arc construction rates and magma production. *Geomorphology*, 136(1), 148-64.

[332] Csatho, B., Schenk, T., Kyle, P., Wilson, T., & Krabill, W. B. (2008). Airborne laser swath mapping of the summit of Erebus volcano, Antarctica: Applications to geological mapping of a volcano. *J Volcanol Geotherm Res*, 177(3), 531-48.

[333] Parrot-F, J. (2007). Study of Volcanic Cinder Cone Evolution by Means of High Resolution DEMs. *MODSIM 2007 International Congress on Modelling and Simulation; 2007: Modelling and Simulation Society of Australia and New Zealand.*

[334] Parrot-F, J. (2007). Tri-dimensional parameterisation: an automated treatment to study the evolution of volcanic cones. *Géomorphologie: relief, processus, environnement*, 2007(3), 247-57.

[335] Awdankiewicz, M. (2005). Reconstructing an eroded scoria cone: the Miocene Sosnica Hill volcano (Lower Silesia, SW Poland). *Geological Quarterly*, 49, 439-48.

[336] Rapprich, V., Cajz, V., Kostak, M., Pécskay, Z., Ridkosil, T., Raska, P., et al. (2007). Reconstruction of eroded monogenetic Strombolian cones of Miocene age: A case study on character of volcanic activity of the Jicin Volcanic Field (NE Bohemia) and subsequent erosional rates estimation. *Journal of Geoscience*, 52(3-4), 169-80.

[337] Fornaciai, A., Behncke, B., Favalli, M., Neri, M., Tarquini, S., & Boschi, E. (2012). Morphometry of scoria cones, and their relation to geodynamic setting: A DEM-based analysis. *J Volcanol Geotherm Res*, 217-218, 56-72.

[338] Kereszturi, G., Geyer, A., Martí, J., Németh, K., & Dóniz-Páez, J. F. (in press) Evaluation of morphometry-based dating of monogenetic volcanoes – A case study from Bandas del Sur, Tenerife (Canary Islands). *Bull Volcanol.*

[339] Lanz, J. K., Wagner, R., Wolf, U., Kröchert, J., & Neukum, G. (2010). Rift zone volcanism and associated cinder cone field in Utopia Planitia, Mars. *J Geophys Res*, 115(E12), E12019.

[340] Pike, R. J. (1974). Craters on Earth, Moon, and Mars: Multivariate classification and mode of origin. *Earth Planet Sci Lett*, 22(3), 245-55.

[341] Frey, H., & Jarosewich, M. (1982). Subkilometer Martian Volcanoes: Properties and Possible Terrestrial Analogs. *J Geophys Res*, 87(B12), 9867-79.

[342] Avellán, D. R., Macías, J. L., Pardo, N., Scolamacchia, T., & Rodriguez, D. (2012). Stratigraphy, geomorphology, geochemistry and hazard implications of the Nejapa Volcanic Field, western Managua, Nicaragua. *J Volcanol Geotherm Res*, 213-214, 51-71.

[343] Taddeucci, J., Sottili, G., Palladino, D. M., Ventura, G., & Scarlato, P. (2009). A note on maar eruption energetics: current models and their application. *Bull Volcanol*, 72(1), 75-83.

[344] Goto, A., Taniguchi, H., Yoshida, M., Ohba, T., & Oshima, H. (2001). Effects of explosion energy and depth to the formation of blast wave and crater: Field Explosion Experiment for the understanding of volcanic explosion. *Geophys Res Lett*, 28(22), 4287-90.

[345] Martín-Serrano, A., Vegas, J., García-Cortés, A., Galán, L., Gallardo-Millán, J. L., Martín-Alfageme, S., et al. (2009). Morphotectonic setting of maar lakes in the Campo de Calatrava Volcanic Field (Central Spain, SW Europe). *Sediment Geol*, 222(1-2), 52-63.

[346] Valentine, G. A., Shufelt, N. L., & Hintz, A. R. L. (2011). Models of maar volcanoes, Lunar Crater (Nevada, USA). *Bull Volcanol*, 73(6), 753-65.

[347] Valentine, G. A. (2012). Shallow plumbing systems for small-volume basaltic volcanoes, 2: Evidence from crustal xenoliths at scoria cones and maars. *J Volcanol Geotherm Res*, 223-224, 47-63.

[348] Zimanowski, B. (1998). Phreatomagmatic explosions. *Freundt A, Rosi M, editors. From Magma to Tephra: Developments in Volcanology 4. Amsterdam: Elsevier*, 25-54.

[349] Leat, P. T., & Thompson, R. N. (1988). Miocene hydrovolcanism in NW Colorado, USA, fuelled by explosive mixing of basic magma and wet unconsolidated sediment. *Bull Volcanol*, 50(4), 229-43.

[350] Mc Clintock, M., & White, J. (2006). Large phreatomagmatic vent complex at Coombs Hills, Antarctica: Wet, explosive initiation of flood basalt volcanism in the Ferrar-Karoo LIP. *Bull Volcanol*, 68(3), 215-39.

[351] Hearn, B. C. (1968). Diatremes with Kimberlitic Affinities in North-Central Montana. *Science*, 159(3815), 622-5.

[352] Yokoo, A., Taniguchi, H., Goto, A., & Oshima, H. (2002). Energy and depth of Usu 2000 phreatic explosions. *Geophys Res Lett*, 29(24), 10.1029/2002GL015928.

[353] Nordyke, M. D. (1962). An Analysis of Cratering Data from Desert Alluvium. *J Geophys Res*, 67(5), 1965-74.

[354] Chabai, A. J. (1965). On Scaling Dimensions of Craters Produced by Buried Explosives. *J Geophys Res*, 70(20), 5075-98.

[355] Ross-S, P., Delpit, S., Haller, MJ, Németh, K., & Corbella, H. (2011). Influence of the substrate on maar-diatreme volcanoes- an example of a mixed setting from the Pali Aike volcanic field, Argentina. *J Volcanol Geotherm Res*, 201(1-4), 253-71.

[356] Jankowski, D. G., & Squyres, S. W. (1992). The Topography of Impact Craters in "Softened" Terrain. *Icarus*, 100, 26-39.

[357] Self, S., Kienle, J., & Hout-P, J. (1980). Ukinrek maar, Alaska, II Deposits and formation of the 1977 craters. *J Volcanol Geotherm Res*, 7, 39-65.

[358] Kienle, J., Kyle, P. R., Self, S., Motyka, R. J., & Lorenz, V. (1980). Ukinrek maar, Alaska, I April 1977 Eruption sequence, petrology and tectonic setting. *J Volcanol Geotherm Res*, 7, 11-37.

[359] Geshi, N., Németh, K., & Oikawa, T. (2011). Growth of phreatomagmatic explosion craters: A model inferred from Suoana crater in Miyakejima Volcano, Japan. *J Volcanol Geotherm Res*, 201(1-4), 30-8.

[360] Sparks, R. S. J., Baker, L., Brown, R. J., Field, M., Schumacher, J., Stripp, G., et al. (2006). Dynamical constraints on kimberlite volcanism. *J Volcanol Geotherm Res*, 155(1-2), 18-48.

[361] Walters, A. L., Phillips, J. C., Brown, R. J., Field, M., Gernon, T., Stripp, G., et al. (2006). The role of fluidisation in the formation of volcaniclastic kimberlite: Grain size observations and experimental investigation. *J Volcanol Geotherm Res*, 155(1-2), 119-37.

[362] Allen, S. R., & Smith, I. E. M. (1994). Eruption styles and volcanic hazard in the Auckland Volcanic Field, New Zealand. *Geoscience Reports of Shizuoka University*, 20, 5-14.

[363] Kereszturi, G., Németh, K., Cronin, J. S., Smith, I. E. M., Agustin-Flores, J., Lindsay, J., (in preparation). Eruptive magmatic volumes of the Quaternary Auckland Volcanic Field (New Zealand).

[364] Rout, D. J., Cassidy, J., Locke, CA, & Smith, I. E. M. (1993). Geophysical evidence for temporal and structural relationships within the monogenetic basalt volcanoes of the Auckland volcanic field, northern New Zealand. *J Volcanol Geotherm Res*, 57(1-2), 71-83.

[365] Murtagh, R. M., White, J. D. L., & Sohn, Y. K. (2011). Pyroclast textures of the Ilchulbong 'wet' tuff cone, Jeju Island, South Korea. *J Volcanol Geotherm Res*, 201(1-4), 385-96.

[366] Cole, P. D., Guest, J. E., Duncan, A. M., & Pacheco-M, J. (2001). Capelinhos 1957-1958, Faial, Azores: deposits formed by an emergent surtseyan eruption. *Bull Volcanol*, 63(2-3), 204-20.

[367] Luhr, J. F., & Simkin, T. (1993). Parícutin: the volcano born in a Mexican cornfield. *Geoscience Press*.

[368] Mannen, K., & Ito, T. (2007). Formation of scoria cone during explosive eruption at Izu-Oshima volcano, Japan. *Geophys Res Lett*, 34(L18302), 10.1029/2007GL030874.

[369] Gutmann, J. T. (1979). Structure and eruptive cycle of cinder cones in the Pinacate volcanic field and controls of strombolian activity. *J Geol*, 87, 448-54.

[370] Ripepe, M., & Harris, A. J. L. (2008). Dynamics of the 5 April 2003 explosive paroxysm observed at Stromboli by a near-vent thermal, seismic and infrasonic array. *Geophys Res Lett*, 35(7).

[371] Chouet, B., Hamisevicz, N., & Mc Getchin, T. R. (1974). Photoballistics of Volcanic Jet Activity at Stromboli, Italy. *J Geophys Res*, 79(32), 4961-76.

[372] Ripepe, M., Rossi, M., & Saccorotti, G. (1993). Image processing of explosive activity at Stromboli. *J Volcanol Geotherm Res*, 54(3-4), 335-51.

[373] Taddeucci, J., Scarlato, P., Capponi, A., Del Bello, E., Cimarelli, C., Palladino, D. M., et al. (2012). High-speed imaging of Strombolian explosions: The ejection velocity of pyroclasts. *Geophys Res Lett*, 39(2), L02301, 10.1029/2011GL050404.

[374] Rosi, M., Bertagnini, A., Harris, A. J. L., Pioli, L., Pistolesi, M., & Ripepe, M. (2006). A case history of paroxysmal explosion at Stromboli: Timing and dynamics of the April 5, 2003 event. *Earth Planet Sci Lett*, 243, 594-606.

[375] Sable, J., Houghton, B., Wilson, C., & Carey, R. (2006). Complex proximal sedimentation from Plinian plumes: the example of Tarawera 1886. *Bull Volcanol*, 69(1), 89-103.

[376] Kervyn, M., Ernst, G., Keller, J., Vaughan, R., Klaudius, J., Pradal, E., et al. (2010). Fundamental changes in the activity of the natrocarbonatite volcano Oldoinyo Lengai, Tanzania. *Bull Volcanol*, 72(8), 913-31.

[377] Patrick, M. R. (2007). Dynamics of Strombolian ash plumes from thermal video: Motion, morphology, and air entrainment. *J Geophys Res*, 112(B6), B06202-40, 10.1029/2006JB004387.

[378] Guilbaud-N, M., Siebe, C., & Agustín-Flores, J. (2009). Eruptive style of the young high-Mg basaltic-andesite Pelagatos scoria cone, southeast of México City. *Bull Volcanol*, 71(8), 859-880.

[379] Ernst, G. G. J., Sparks, R. S. J., Carey, S. N., & Bursik, M. I. (1996). Sedimentation from turbulent jets and plumes. *J Geophys Res*, 101(B3), 5575-89.

[380] Bertotto, G. W., Bjerg, E. A., & Cingolani, C. A. (2006). Hawaiian and Strombolian style monogenetic volcanism in the extra-Andean domain of central-west Argentina. *J Volcanol Geotherm Res*, 158(3-4), 430-44.

[381] Van Burkalow, A. (1945). Angle of repose and angle of sliding friction: an experimental study. *Geol Soc Am Bull*, 56(6), 669-707.

[382] Tokarev, P. (1978). Prediction and characteristics of the 1975 eruption of Tolbachik volcano, Kamchatka. *Bull Volcanol*, 41(3), 251-8.

[383] Fedotov, S., Enman, V., Nikitenko, Y., Maguskin, M., Levin, V., & Enman, S. (1980). Crustal deformations related to the formation of new tolbachik volcanoes in 1975-1976, Kamchatka. *Bull Volcanol*, 43(1), 35-45.

[384] Tokarev, P. I. (1983). Calculation of the magma discharge, growth in the height of the cone and dimensions of the feeder channel of Crater I in the Great Tolbachik Fissure

Eruption, July 1975. *Fedotov SA, Markhinin YK, editors. The Great Tolbachik Fissure Eruption: Geological and Geophysical Data, 1975-1976 Cambridge: Cambridge University Press*, 27-35.

[385] Fedotov, S., Chirkov, A., Gusev, N., Kovalev, G., & Slezin, Y. (1980). The large fissure eruption in the region of Plosky Tolbachik volcano in Kamchatka, 1975-1976. *Bull Volcanol*, 43(1), 47-60.

[386] Ripepe, M., Ciliberto, S., & Della Schiava, M. (2001). Time constraints for modeling source dynamics of volcanic explosions at Stromboli. *J Geophys Res*, 106(B5), 8713-27.

[387] Taddeucci, J., Pompilio, M., & Scarlato, P. (2004). Conduit processes during the July-August 2001 explosive activity of Mt. Etna (Italy): inferences from glass chemistry and crystal size distribution of ash particles. *J Volcanol Geotherm Res*, 137(1-3), 33-54.

[388] Self, S., Sparks, R. S. J., Booth, B., & Walker, G. P. L. (1974). The 1973 Heimaey Strombolian Scoria deposit, Iceland. *Geol Mag*, 111, 539-48.

[389] Doubik, P., & Hill, B. E. (1999). Magmatic and hydromagmatic conduit development during the 1975 Tolbachik Eruption, Kamchatka, with implications for hazards assessment at Yucca Mountain, NV. *J Volcanol Geotherm Res*, 91(1), 43-64.

[390] Holm, R. F. (1987). Significance of agglutinate mounds on lava flows associated with monogenetic cones: An example at Sunset Crater, northern Arizona. *Geol Soc Am Bull*, 99(3), 319-24.

[391] Valentine, G. A., Perry, F. V., Krier, D., Keating, G. N., Kelley, R. E., & Cogbil, A. H. (2006). Small-volume basaltic volcanoes: Eruptive products and processes, and post-eruptive geomorphic evolution in Crater Flat (Pleistocene), southern Nevada. *Geol Soc Am Bull*, 118(11-12), 1313-30.

[392] Self, S., Keszthelyi, L., & Thordarson, T. (1998). The importance of pahoehoe. *Annual Review of Earth and Planetary Sciences*, 26, 81-110.

[393] Calvari, S., & Pinkerton, H. (1999). Lava tube morphology on Etna and evidence for lava flow emplacement mechanisms. *J Volcanol Geotherm Res*, 90(3-4), 263-80.

[394] Inbar, M., Enriquez, A. R., & Graniel, J. H. G. (2001). Morphological changes and erosion processes following the 1982 eruption of El Chichón volcano, Chiapas, Mexico. *Géomorphologie: relief, processus, environnement*, 2001(3), 175-83.

[395] Lutz, T. M., & Gutmann, J. T. (1995). An improved method of determining alignments of point-like features and its implications for the Pinacate volcanic field, Mexico. *J Geophys Res*, 100(B9), 17659-70.

[396] Paulsen, T. S., & Wilson, T. J. (2010). New criteria for systematic mapping and reliability assessment of monogenetic volcanic vent alignments and elongate volcanic vents for crustal stress analyses. *Tectonophysics*, 482(1-4), 16-28.

[397] Riggs, N. R., & Duffield, W. A. (2008). Record of complex scoria cone eruptive activity at Red Mountain, Arizona, USA, and implications for monogenetic mafic volcanoes. *J Volcanol Geotherm Res*, 178, 763-76.

[398] Kereszturi, G., Jordan, G., Németh, K., & Dóniz-Páez, J. F. (2012). Syn-eruptive morphometric variability of monogenetic scoria cones. *Bull Volcanol*, 74(9), 2171-2185.

[399] Di Traglia, F., Cimarelli, C., de Rita, D., & Gimeno, Torrente. D. (2009). Changing eruptive styles in basaltic explosive volcanism: Examples from Croscat complex scoria cone, Garrotxa Volcanic Field (NE Iberain Peninsula). *J Volcanol Geotherm Res*, 180(2-4), 89-109.

[400] Suhr, P., Goth, K., Lorenz, V., & Suhr, S. (2006). Long lasting subsidence and deformation in and above maar-diatreme volcanoes- a never ending story. *Zeitschrift der Deutschen Gesellschaft für Geowissenschaften*, 157(3), 491-511.

[401] Lorenz, V., Suhr, P., & Goth, K. (2003). Maar-Diatrem-Vulkanismus- Ursachen und Folgen. Die Guttauer Vulkangruppe in Ostsachsen als Beispiel für die komplexen Zusammenhänge. *Zeitschrift für Geologische Wissenschaften*, 31(4-6), 267-312.

[402] Dessert, C., Dupré, B., Gaillardet, J., François, L. M., & Allègre, C. J. (2003). Basalt weathering laws and the impact of basalt weathering on the global carbon cycle. *Chem Geol*, 202(3-4), 257-73.

[403] West, A. J., Galy, A., & Bickle, M. (2005). Tectonic and climatic controls on silicate weathering. *Earth Planet Sci Lett*, 235(1-2), 211-28.

[404] Brady, P. V., & Carroll, S. A. (1994). Direct effects of CO2 and temperature on silicate weathering: Possible implications for climate control. *Geochim Cosmochim Acta*, 58(7), 1853-6.

[405] Raymo, M. E., & Ruddiman, W. F. (1992). Tectonic forcing of late Cenozoic climate. *Nature*, 359(6391), 117-22.

[406] Riebe, C. S., Kirchner, J. W., & Finkel, R. C. (2004). Erosional and climatic effects on long-term chemical weathering rates in granitic landscapes spanning diverse climate regimes. *Earth Planet Sci Lett*, 224(3-4), 547-62.

[407] Hinsinger, P., Fernandes, Barros. O. N., Benedetti, M. F., Noack, Y., & Callot, G. (2001). Plant-induced weathering of a basaltic rock: experimental evidence. *Geochim Cosmochim Acta*, 65(1), 137-52.

[408] Németh, K., Cronin, S. J., Stewart, R. B., & Charley, D. (2009). Intra- and extra-caldera volcaniclastic facies and geomorphic characteristics of a frequently active mafic island-arc volcano, Ambrym Island, Vanuatu. *Sediment Geol*, 220(3-4), 256-70.

[409] Major, J. J., & Yamakoshi, T. (2005). Decadal-scale change of infiltration characteristics of a tephra-mantled hillslope at Mount St Helens, Washington. *Hydrol Processes*, 19(18), 3621-30.

[410] Ferrucci, M., Pertusati, S., Sulpizio, R., Zanchetta, G., Pareschi, M. T., & Santacroce, R. (2005). Volcaniclastic debris flows at La Fossa Volcano (Vulcano Island, southern Italy): Insights for erosion behaviour of loose pyroclastic material on steep slopes. *J Volcanol Geotherm Res*, 145(3-4), 173-91.

[411] Segerstrom, K. (1950). Erosion Studies at Paricutin, State of Michoacán, Mexico. *Geological Survey Bulletin*, 965-A, 1-163.

[412] White, J. D. L. (1991). The depositional record of small, monogenetic volcanoes within terrestrial basins. *Fisher EV, Smith GA, editors. Sedimentation in Volcanic Settings*, 155-171.

[413] Valentine, G. A., & Harrington, C. D. (2006). Clast size controls and longevity of Pleistocene desert pavements at Lathrop Wells and Red Cone volcanoes, southern Nevada. *Geology*, 34(7), 533-6.

[414] Mc Fadden, L. D., Mc Donald, E. V., Wells, S. G., Anderson, K., Quade, J., & Forman, S. L. (1998). The vesicular layer and carbonate collars of desert soils and pavements: formation, age and relation to climate change. *Geomorphology*, 24(2-3), 101-45.

[415] Mc Fadden, L. D., Wells, S. G., & Jercinovich, M. J. (1987). Influences of eolian and pedogenic processes on the origin and evolution of desert pavements. *Geology*, 15(6), 504-8.

[416] Wells, S. G., Dohrenwend, J. C., Mc Fadden, L. D., Turrin, B. D., & Mahrer, K. D. (1985). Late Cenozoic landscape evolution on lava flow surfaces of the Cima volcanic field, Mojave Desert, California. *Geol Soc Am Bull*, 96(12), 1518-29.

[417] Kato, T., Kamijo, T., Hatta, T., Tamura, K., & Higashi, T. (2005). Initial Soil Formation Processes of Volcanogenous Regosols (Scoriacious) from Miyake-jima Island, Japan. *Soil Science & Plant Nutrition*, 51(2), 291-301.

[418] Melo, R., Vieira, G., Caselli, A., & Ramos, M. (2012). Susceptibility modelling of hummocky terrain distribution using the information value method (Deception Island, Antarctic Peninsula). *Geomorphology*, 155-156(0), 88-95.

[419] López-Martínez, J., Serrano, E., Schmid, T., Mink, S., & Linés, C. (2012). Periglacial processes and landforms in the South Shetland Islands (northern Antarctic Peninsula region). *Geomorphology*, 155-156(0), 62-79.

[420] Kinnell, P. I. A. (2005). Raindrop-impact-induced erosion processes and prediction: a review. *Hydrol Processes*, 19(14), 2815-44.

[421] Van Dijk, A. I. J. M., Bruijnzeel, L. A., & Wiegman, S. E. (2003). Measurements of rain splash on bench terraces in a humid tropical steepland environment. *Hydrol Processes*, 17(3), 513-35.

[422] Dunne, T., Malmon, D. V., & Mudd, S. M. (2010). A rain splash transport equation assimilating field and laboratory measurements. *J Geophys Res*, 115(F1), F01001.

[423] Horton, R. E. (1945). Erosional development of streams and their drainage basins; hydrophysical approach to quantitative morphology. *Geol Soc Am Bull*, 56(3), 275-370.

[424] Cronin, S. J., & Neall, V. E. (2001). Holocene volcanic geology, volcanic hazard and risk on Taveuni, Fiji New Zeal. *J Geol Geophys*, 44, 417-37.

[425] Németh, K., & Cronin, S. J. (2007). Syn- and post-eruptive erosion, gully formation, and morphological evolution of a tephra ring in tropical climate erupted in 1913 in West Ambrym, Vanuatu. *Geomorphology*, 86, 115-30.

[426] Poesen, J., Nachtergaele, J., Verstraeten, G., & Valentin, C. (2003). Gully erosion and environmental change: importance and research needs. *Catena*, 50(2-4), 91-133.

[427] Matsuoka, N. (2001). Solifluction rates, processes and landforms: a global review. *Earth-Science Reviews*, 55(1-2), 107-34.

[428] Holness, S. D. (2004). Sediment movement rates and processes on cinder cones in the maritime Subantarctic (Marion Island). *Earth Surf Processes Landforms*, 29(1), 91-103.

[429] Woodroffe, C. D. (2002). Coasts: Form, Process and Evolution. *Cambridge: Cambridge University Press*.

[430] Sohn, Y. K., & Yoon-H, S. (2010). Shallow-marine records of pyroclastic surges and fallouts over water in Jeju Island, Korea, and their stratigraphic implications. *Geology*, 38(8), 763-6.

[431] Calles, B., Lindé, K., & Norrman, J. O. (1982). The geomorphology of Surtsey island in 1980. *Surtsey Research Progress Report IX*, 117-32.

[432] Norrman, J. O., Calles, B., & Larsson, R. A. (1974). The geomorphology of Surtsey Island in 1972. *Surtsey Research Progress Report VII*, 61-71.

[433] Sohn, Y. K., Park, K. H., & Yoon-H, S. (2008). Primary versus secondary and subaerial versus submarine hydrovolcanic deposits in the subsurface of Jeju Island, Korea. *Sedimentology*, 55(4), 899-924.

[434] Rachold, V., Grigoriev, M. N., Are, F. E., Solomon, S., Reimnitz, E., Kassens, H., et al. (2000). Coastal erosion vs riverine sediment discharge in the Arctic Shelf seas. *Int J Earth Sci*, 89(3), 450-60.

[435] Arnalds, O., Gisladottir, F. O., & Orradottir, B. (2012). Determination of aeolian transport rates of volcanic soils in Iceland. *Geomorphology*, 167–168, 4–12.

[436] Thorarinsdottir, E. F., & Arnalds, O. (2012). Wind erosion of volcanic materials in the Hekla area, South Iceland. *Aeolian Research*, 4, 39-50.

[437] Rech, J. A., Reeves, R. W., & Hendricks, D. M. (2001). The influence of slope aspect on soil weathering processes in the Springerville volcanic field, Arizona. *Catena*, 43(1), 49-62.

[438] Dohrenwend, J. C., Mc Fadden, L. D., Turrin, B. D., & Wells, S. G. (1984). K-Ar dating of the Cima volcanic field, eastern Mojave Desert, California: Late Cenozoic volcanic history and landscape evolution. *Geology*, 12(3), 163-7.

[439] Rad, S. D., Allègre, C. J., & Louvat, P. (2007). Hidden erosion on volcanic islands. *Earth Planet Sci Lett*, 262(1-2), 109-24.

[440] Gaillardet, J., Millot, R., & Dupré, B. (2003). Chemical denudation rates of the western Canadian orogenic belt: the Stikine terrane. *Chem Geol*, 201(3-4), 257-79.

[441] Kereszturi, G., Procter, J., Cronin, J. S., Németh, K., Bebbington, M., & Lindsay, J. (2012). LiDAR-based quantification of lava flow hazard in the City of Auckland (New Zealand). *Remote Sens Environ*, 125, 198-213.

[442] Oliva, P., Viers, J., & Dupré, B. (2003). Chemical weathering in granitic environments. *Chem Geol*, 202(3-4), 225-56.

[443] Ort, M. H., Elson, M. D., Anderson, K. C., Duffield, W. A., & Samples, T. L. (2008). Variable effects of cinder-cone eruptions on prehistoric agrarian human populations in the American southwest. *J Volcanol Geotherm Res*, 176(3), 363-76.

[444] Rowland, S. K., Jurado-Chichay, Z., & Ernst, G. J. (2009). Pyroclastic deposits and lava flows from the 1759-1774 eruption of El Jorullo, México: Aspects of "violent strombolian" activity and comparison with Parícutin. *Hoskuldsson A, Thordarson T, Larsen G, Self S, Rowland S, editors. The Legacy of George PL Walker, Special Publications of IAVCEI 2. London, UK: Geological Society of London*, 105-128.

[445] Cronin, S. J., Hedley, M. J., Neall, V. E., & Smith, R. G. (1998). Agronomic impact of tephra fallout from the 1995 and 1996 Ruapehu Volcano eruptions, New Zealand. *Environ Geol*, 34(1), 21-30.

[446] Grishin, S., & del Moral, R. (1996). Dynamics of forests after catastrophic eruptions of Kamchatka's volcanoes. *Turner IM, Diong CH, Lim SSL, Ng PKL, editors. Biodiversity and Dynamics of Ecosystems. DIWPA Series*, 1, 133-146.

[447] Chinen, T., & Riviere, A. (1990). Post-eruption erosion processes and plant recovery in the summit atrio of Mt. Usu, Japan. *Catena*, 17(3), 305-14.

[448] White, J. D. L., Houghton, B. F., Hodgson, K. A., & Wilson, C. J. N. (1997). Delayed sedimentary response to the A.D. 1886 eruption of Tarawera, New Zealand. *Geology*, 25(5), 459-62.

[449] Kralj, P. (2011). Eruptive and sedimentary evolution of the Pliocene Grad Volcanic Field, North-east Slovenia. *J Volcanol Geotherm Res*, 201(1-4), 272-84.

[450] Shaw, S. J., Woodland, A. B., Hopp, J., & Trenholm, N. D. (2010). Structure and evolution of the Rockeskyllerkopf Volcanic Complex, West Eifel Volcanic Field, Germany. *Bull Volcanol*, 72, 971-90.

[451] Bullard, F. M. (1947). Studies on Paricutin volcano, Michoacan, Mexico. *Geol Soc Am Bull*, 58, 433-50.

[452] Chinen, T. (1986). Surface erosion associated with tephra deposition on Mt. Usu and other volcanoes. *Environmental Science, Hokkaido*, 9(1), 137-49.

[453] Valentine, G. A., Palladino, D. M., Agosta, E., Taddeucci, J., & Trigila, R. (1998). Volcaniclastic aggradation in a semiarid environment, northwestern Vulcano Island, Italy. *Geol Soc Am Bull*, 110(5), 630-43.

[454] Bryan, S. E., Marti, J., & Cas, R. A. F. (1998). Stratigraphy of the Bandas del Sur Formation; an extracaldera record of Quaternary phonolitic explosive eruptions from the Las Canadas edifice. *Tenerife (Canary Islands) Geol Mag*, 135(5), 605-36.

[455] Martí, J., Mitjavila, J., & Araña, V. (1994). Stratigraphy, structure and geochronology of the Las Cañadas caldera (Tenerife, Canary Island). *Geol Mag*, 131(6), 715-27.

[456] Shakesby, R. A. (2011). Post-wildfire soil erosion in the Mediterranean: Review and future research directions. *Earth-Science Reviews*, 105(3-4), 71-100.

[457] Inbar, M., Tamir, M., & Wittenberg, L. (1998). Runoff and erosion processes after a forest fire in Mount Carmel, a Mediterranean area. *Geomorphology*, 24(1), 17-33.

[458] Trimble, S. W., & Mendel, A. C. (1995). The cow as a geomorphic agent- A critical review. *Geomorphology*, 13(1-4), 233-53.

[459] Butler, D. R. (1995). Zoogeomorphology- Animals as Geomorphic Agents. *Cambridge, UK: Cambridge University Press.*

[460] Boelhouwers, J., & Scheepers, T. (2004). The role of antelope trampling on scarp erosion in a hyper-arid environment, Skeleton Coast, Namibia. *J Arid Environ*, 58(4), 545-57.

[461] Butler, D. R. (1993). The Impact of Mountain Goat Migration on Unconsolidated Slopes in Glacier National Park, Montana. *The Geographical Bulletin*, 35(2), 98-106.

[462] Zolitschka, B., Schäbitz, F., Lücke, A., Corbella, H., Ercolano, B., Fey, M., et al. (2006). Crater lakes of the Pali Aike Volcanic Field as key sites for paleoclimatic and paleoecological reconstructions in southern Patagonia, Argentina. *J S Am Earth Sci*, 21(3), 294-309.

[463] Hably, L., & Kvaček, Z. (1998). Pliocene mesophytic forests surrounding crater lakes in western Hungary. *Review of Palaeobotany and Palynology*, 101(1), 257-69.

[464] Inbar, M., Hubp, J. L., & Ruiz, L. V. (1994). The geomorphological evolution of the Paricutin cone and lava flows, Mexico, 1943-1990. *Geomorphology*, 9, 57-76.

[465] Sucipta, I. G. B. E., Takashima, I., & Muraoka, H. (2006). Morphometric age and petrological characteristic of volcanic rocks from the Bajawa cinder cone complex, Flores, Indonesia. *J Mineral Petrol Sci*, 101(2), 48-68.

[466] Németh, K., Suwesi, S. K., Peregi, Z., Gulácsi, Z., & Ujszászi, J. (2003). Plio/Pleistocene Flood Basalt Related Scoria and Spatter Cones, Rootless Lava Flows, and Pit Craters, Al Haruj Al Abiyad, Libya. *Geolines*, 15, 98-103.

[467] Moufti, M. R., Moghazi, A. M., & Ali, K. A. (2012). Geochemistry and Sr-Nd-Pb isotopic composition of the Harrat Al-Madinah Volcanic Field, Saudi Arabia. *Gondwana Res*, 21(2-3), 670-89.

[468] Cabrera, A. P., & Caffe, P. J. (2009). The Cerro Morado Andesites: Volcanic history and eruptive styles of a mafic volcanic field from northern Puna, Argentina. *J S Am Earth Sci*, 28(2), 113-31.

[469] Manville, V., Németh, K., & Kano, K. (2009). Source to sink: A review of three decades of progress in the understanding of volcaniclastic processes, deposits, and hazards. *Sediment Geol*, 220(3-4), 136-61.

Chapter 2

The Role of the Andesitic Volcanism in the Understanding of Late Mesozoic Tectonic Events of Bureya-Jziamysi Superterrain, Russian Far East

I.M. Derbeko

Additional information is available at the end of the chapter

http://dx.doi.org/10.5772/51908

1. Introduction

At the moment Bureya-Jziamysi superterrain is a very discussable geological object [1]. It is distinguished as a component of Amur plate or a part of a microcontinent [2] of the eastern part of Euroasia (Fig. 1a). Nowadays a kinematic model is obtained [3] that describe the dislocation of Euroasian and Amur plates as independent tectonic units (Fig. 1a). The GPS-calculations [4, 3] showed that the eastern border of Amur plate goes along of the one of the branches of Than-Lu fracture system (Fig. 1a). The branch is also an eastern border of Bureya-Jziamysi superterrain. The northern border is identified by its contact with Mongol-Okhotsk orogenic belt and correlates to the northern border of Amur plate [5]. On the west and south the superterrain is framed with Paleozoic and early Mesozoic orogenic belts: South Mongolian – Khingan, Solonkersky, Vundurmiao [5, 6]. South Mongol – Khingan orogenic belt separates it from Argun superterrain that is also a component of Amur microcontinent (Fig. 1b).

There are almost no controversies about the time of the connection of the researched area to the Argun superterrain in the literature. The authors [7, 2 et al.] agree that these tectonic events took place in the second half of Paleozoic. And than the newly formed Amur microcontinent, together with the north Chinese plate, moved to the north and accreted to Siberian platform at Early Cretaceous supported by the data of [8], at late Jurassic by the data of [9] or at the end of Paleozoic [10].

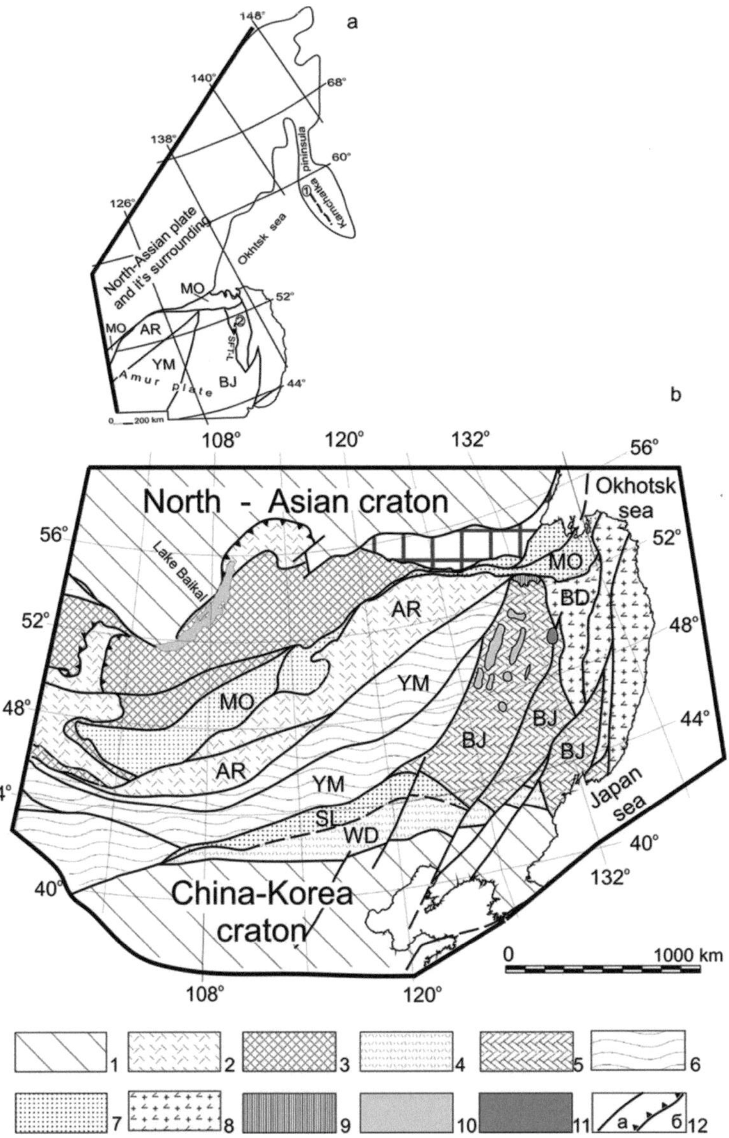

Figure 1. Schemes of the major tectonic structures dislocation. a). Regional scheme. The mountain ridges mentioned in the text are shown with the dotted line: 1 – Sredinnyi, 2 - Small Khingan. b). Bureya-Jziamysi superterrain and it's surrounding: Cratons (1): North Asian, Sino-Korean. Orogenic belts and the fragments of orogenic belts: Late Riphean (2), Late Cambrian – Early Ordovician (3), Silurian (4), Early Paleozoic (5), Late Paleozoic (6), Late Paleozoic – Early Mesozoic (7), Late Jurassic - Early Cretaceous (8). Volcanoes field complexes: Burunda (9), Pojarka (10), Stanolir (11). Tectonic contacts (12). a, b). Letter marks: YM – South Mongolian – Khingansky, SL Solonker, WD Vundurmiao; superterrains – BJ – Bureya-Jziamysi, A- Argunian, terrain – Badzhal terrain, SFT-L - the system faults Tan-Lu. The scheme is made by [5].

It is considered that the border between the Amur microcontinent and the Mongol-Okhotsk structure was amalgamated at the late Mesozoic period by volcano-plutonic complexes of early-late Cretaceous [6]. High precision geochronology and chemical composition of the complexes deny the late Mesozoic unity in the evolutional process of the superterrains that formed the Amur microcontinent. For the Argun superterrain and South Mongolian – Khingan orogenic belt the following stages of the volcanic activity are stated: 147 Ma – sub-alkaline rhyolitic intra-plate complex, 140-122 Ma – calc-alkaline volcano-plutonic complex of intermediate composition with subductional origin, 119-97 Ma – bimodal volcano-plutonic intra-plate complex [11]. Bimodal volcano-plutonic complexes accompany the closure of Mongol-Okhotsk basin in the frames of western link of Mongol-Okhotsk belt [12] and of the eastern link [13]. But the analogues of the rocks are absent in the zone of the connection of Mongol-Okhotsk belt and Bureya-Jziamysi superterrain.

2. Late Mesozoic volcanism of Bureya-Jziamysi superterrain

The volcanic complexes that are developed in the frames of Bureya-Jziamysi superterrain differ from the same formations that are developed in the frames of the Argun superterrain in the South Mongol-Khingan orogenic belt both by the time of the formation and by the material composition. Volcanites of Bureya-Jziamysi superterrain traditionally refer to the three different volcanogenic complexes: Low Zeya – central and western part of the investigated territory; Khingan-Okhots (Khingan-Olonoi zone) – east and south-east, Umlekan-Ogodzha (Ogodzha zone) – north. Volcanites of the Low Zeya volcanic zone are represented by Early Cretaceous rhyolites (137 Ma) and andesites of the Poyarka complex (117-105 Ma) [11, 14, 15]. Ogodzha zone is formed with the Burunda andesite complex (111 – 105 Ma). Its rocks are developed along the northern border of Bureya-Jziamysi superterrain. Khingan-Olonoi zone is represented by two Early Cretaceous complexes in the frames of the superterrain: the Stanolir andesites (111-105 Ma) and the rhyolite-alkaline dacite complexes (101.5 – 99 Ma) [16, 17, 18, 11]. The volcanites of acidic-alkaline composition correspond to typical intraplate formations by their petrochemical characteristics [11]. Thus, in the composition of each of the volcanic complexes of andesite formation is separated, such as: Poyarka, Burunda, and Stanolir andesites.

2.1. Poyarka andesite volcanic complex

Poyarka andesite volcanic complex [19, 11] was formed mostly on the tectonic stress-released zones, commonly referred as riftous. The beginning of their formation of these andesites coincides by the time period with the outpour of large volume rhyolites in the beginning of Early Cretaceous. The rocks of Poyarka complex are represented by small singular outcrops. They are mostly described on the drill logs uncovered by the deep boreholes. According to the open casts the main rock types of the volcanic complex are various andesites that form the covering and subvolcanic facies of volcanites that make more than 50% of the total volume of the volcanites of the complex. Volcanogenic-sedimentary rocks of the covering facies – eg. Poyarka suite – are divided into two parts by their chemical composition and by the floristic signatures indicating specific ages: 1) lower part and 2) upper part. The lower part has got a polyfacies composition but its genesis and sedimentary features both horizontally and vertically. Upwards along the open-cast proluvial deposits are changed into alluvial lake-swamp deposits. Non-volcanic sediment accumulation was parallel to the volcanic activity. As a result of this, the terrigenous formations are gradually replaced by volcanogenic rocks to the edge of the sedimentary basins indicating the proximity of the volcanic source. The base of Poyarka suite concordantly occurs on the covering volcanites of silicic composition. Where the coverings are absent, it lays on the Premesozoic foundation. The thickness of the volcanosedimentary succession is not more than 400 m.

The upper part of the suite consists primarily of primary volcanites. The volume of volcanites concordantly increases in the open-cast of the lower part of the suite. The volcanites are represented by the intermediate –to-basic lavas, pyroclastic rocks and clastogenic lavas. The top of the open-cast ends up with relatively thin (up to 20 m) alternating alleurolits and argillits with rare interlayers of sandstones, tuffaceous sandstones, tuffs, carbonaceous argillites with conglomerates and lens of the coals. Thickness of the volcanogenic component is laterally variable from 130 to 340 m.

The subvolcanic formations of the Poyarka complex are composed of andesites, basaltic andesites and diorite-porphyry bodies. They form laccolith, lopolith or sill bodies 20 km^2 or more. Petrochemical and geochemical compositions of the subvolcanic rocks correspond well to the composition of the covering part of Poyarka suite.

The biggest part of the Poyarka complex is mostly composed from andesites, rarely basalts, and rarely trachybasalt tu trachyandesites. These rocks are of black to dark gray, green-gray, sealing-wax color with a massive fluidal or almond-shaped texture; with a porphyric or serial-porphyric structure.

The sizes of the porphyric fragments are up to 4 mm, and total to an amount of about 5-60%. The structure of the main mass is – pilotaxite, hyalopilite, intersertal or cryptocrystalline. The spheroidal parting is characteristic for the basalts. The volcanites are divided into hornblende, hornblende-pyroxene, dipyroxene and biotite-hornblende. Olivine might be present in basalts and basaltic andesites. In all of the rocks types the main proportion of minerals are plagioclases with a composition for andesites - $An_{35\text{-}55}$, and for basalts or basaltic andesites - $An_{53\text{-}68}$. Dark colored minerals in these rocks are monoclinic and rombic pyroxenes or basaltic hornblende and rarely – biotite.

The main mass is formed by the lath-shaped plagioclase (up to 0.3 mm large), granules of pyroxenes, magnetits and volcanic glass, in different degree replaced by illite, chlorite and iron oxide. Accessory minerals are apatite, sphene, magnetite, ilmenite, and rarely zircon. Almonds are made by montmorillonite, chalcedony and calcite.

Tuffs of andesites and basaltic andesites are massive, stratified. The fragments make 20 – 80 % of the rock. Cement is almost fully replaced by the secondary minerals of chlorite, sericite, chalcedony, limonite, argillaceous minerals.

2.1.1. *Petrochemical and geochemical characteristics*

By the petrochemical data the volcanites of Poyarka complex relate to moderate silica, basic – intermediate silica rocks. They are low in alkali content that is in a range of 2.1-5.9 wt.% (Fig. 2a). Na_2O constantly prevails over K_2O (Fig. 2).

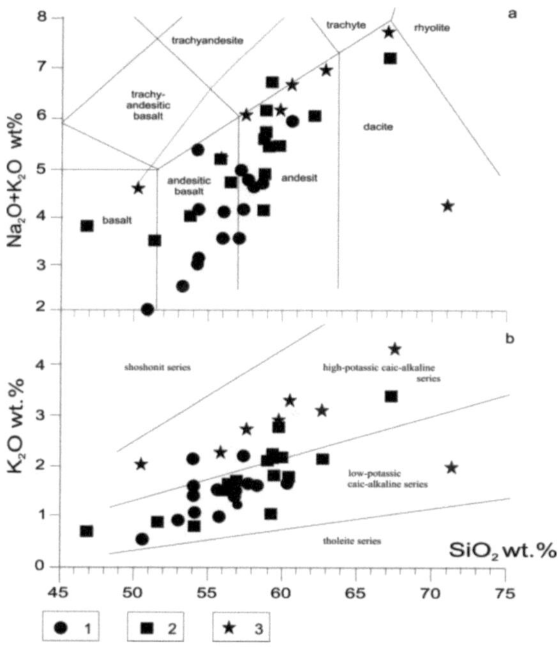

Figure 2. Classification diagrams for the rocks of volcano complexes of Bureya-Jziamysi supertarrain: a) (Na_2O+K_2O) - SiO_2; b) K_2O+SiO_2 [20]. The line of the separation of alkali and subalkali rocks by [21]. Complexes: Poyarka (1), Burunda (2); Stanolir (3).

The rocks belong to the low potassic, in rare cases – high potassic (K_2O = 0.9-1.6 wt.%) calc-alkali series (Fig 2b). The content of Na_2O is irregularly increasing with the growth of silica concentration. The basalts are alkali type. All the other types have potassium-alkali type (Na_2O/K_2O = 1.45 - 4.85). The MgO concentration changes from 9.37 wt.% (high-magnesial basalts, andesitic basalts, andesites) to 3.0 wt.% (moderate magnesial); all varieties are – moderate titanium (TiO_2 = 0.62-0.99 wt.%), ASI (with aluminum saturation index) = 0.9-1.4. By the content of MgO, CaO the volcanites of Poyarka complex are congruous to the volcanites of Burunda and Stanolir complexes and by their content of TiO_2 to the Burunda complex.

On the diagrams of the distribution of REE in the rocks of Poyarka complex (Fig. 3a) Eu anomaly absence or weak positive (Eu/Eu* = 0.89-1.05) and insignificantly prevalence of temporary over HREE (Gd/Lu)n = 2.5-4.5. On the diagram of the REE elements normalized to primitive mantle (Fig. 3 b) the Sr enrichment of the rocks (1029 ppm), Ba (443-642 ppm) is revealed by their impoverishment of Nb (>4-10 ppm), Ta (0.49 ppm), Rb (20.4-43.5 ppm), Th (1.70-4.97 ppm), Y (8-29 ppm), Ti (3100-3300 ppm).

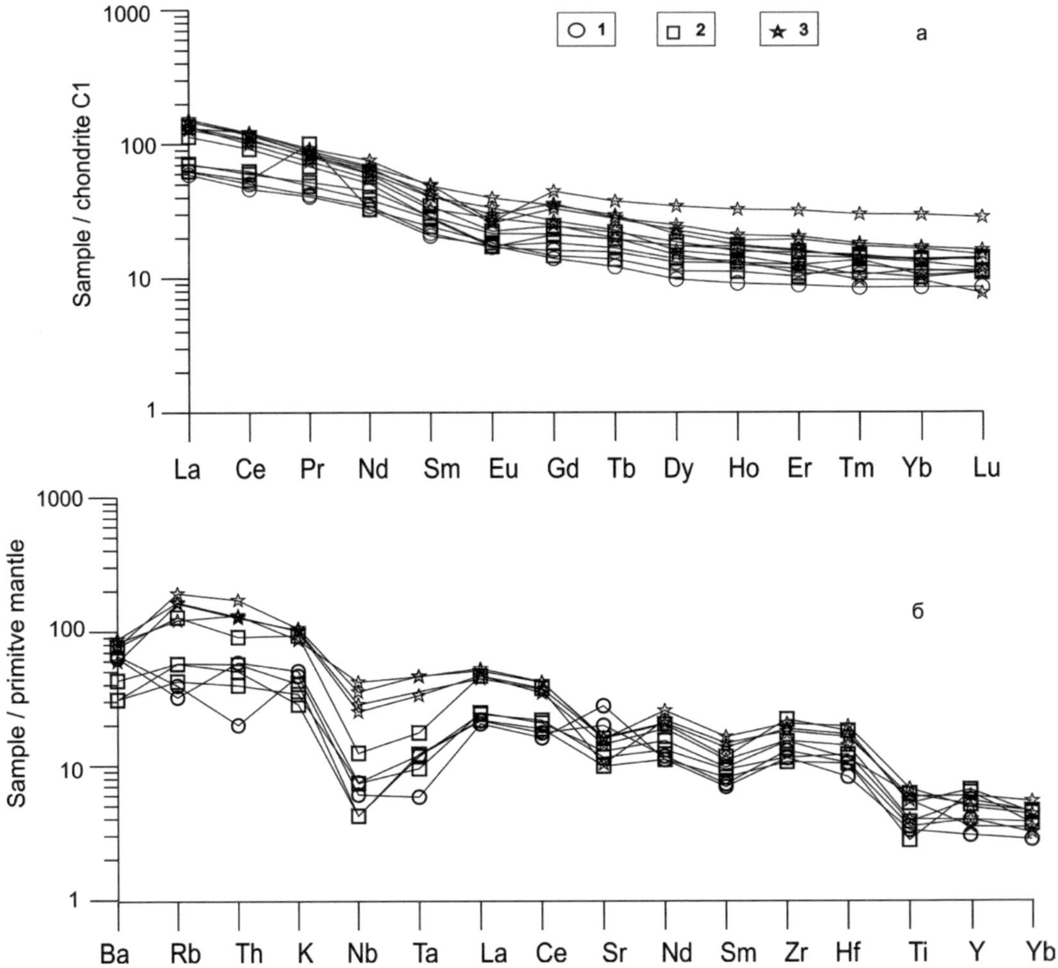

Figure 3. The concentration of the rare elements is normalized to chondrite composition (a) and to primitive mantle (b), in the formations of the volcanic complexes: 1 - Poyarka, 2 - Burunda, 3 - Stanolir. The composition of chondrite and primitive mantle are made by [41].

2.1.2. The age of the formation of the volcanic complex

For terrigenous formation of the Poyarka suite it is certain its Hauterivian-Barremian age based on the rich and complex fresh-water fauna and flora [22]. For the top part of the rock sequence it is characteristic an independent floristic complex which corresponds to an age of Aptian-Albian stage [19]. The similar age is given by palynology methods [19]. Thus, the age of the Poyarka suite is established as Hauterivian - Albian stage, and it displays of a volcanic activity, accordingly, occurred in an interval Aptian - Albian stage. The age is confirmed by radiometric geochronological datings as well (eg. $^{40}Ar/^{39}Ar$ a method) and yielded to an age of about 117 million years [15].

http://dx.doi.org/10.5772/51908

2.2. Burunda andesite volcanic complex

Burunda andesite volcanic complex is composed of tuffs and lavas mainly with intermediate composition and subordinate basic or more silicic volcanite types [23, 24, 13 et al.]. The rock types of the complex make variably broad lithological stripe from 3 to 30 km width on the border of the eastern flank of Mongol-Okhotsk orogenic belt and Bureya-Jziamysi superterraine (Fig. 4).

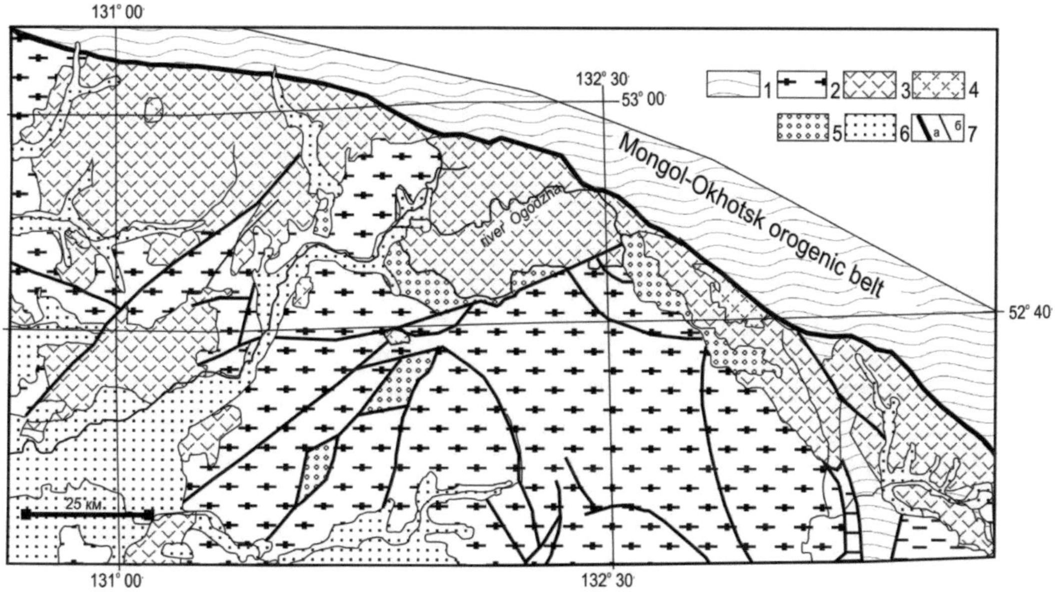

Figure 4. Scheme of dislocation of the rocks of Burunda complex. The rocks are: Paleozoic – Mesozoic of Mongol-Okhotsk orogenic belt (1), Paleozoic of Bureya-Jziamysi superterrain (2), integumentary volcanites of Burunda complex (3), and subvolcanic (4), terrigenic deposits of Ogodzha suite (5), friable deposits of quarter (6). Tectonic borders (7): a – a border between Mongol-Okhotsk orogenic belt and Bureya-Jziamysi superterrain, (b) other borders. Scheme is made by [14].

The complex is presented by covering facies, subvolcanic facies and vent facies which form a volcano-tectonic structure of about up to 40 km in diameter.

The open cast of the integumentary facies - *Burunda suite* – is represented by the lower undersuite that consists mostly of tufts in the base and in the top mostly of the lava rocks. The border between the suites goes symbolically by the beginning of the prevalence of lavas above tuffs in the rocks sequence. The estimated total thickness of Burunda is about 1050 m [22].

The volcanites inconsistently superpose Carboniferous to Early Cretaceous deposits of Ogodzha suite on the base of floristic evidences and have tectonic boundaries with the other undifferentiated Paleozoic rocks formations [23, 13, 24].

The lower part of the rock sequence is presented by tuffs and lava breccias of andesites and dacitic andesites, by tuff-terrigenous deposits with various dimensions of fragmental material, by argillites, by interbed and lenses of dacitic andesites, andesitic basalts and their lava breccias. Sometimes in the base there is a pack of tuffaceous conglomerate with the total thickness of more than 300 m. Tuffaceous conglomerates change into by tuffs of andesitic basalts - dacitic andesites. These tuffs have got various structures ranging from pelitic up to agglomerating, at prevalence psammitic varieties. The upper part of the rock sequence increases the lower part concordantly. It covers the lower part of the rock mass at less than 10 % of the area. Lavas are andesites and andesitic basalts. Dacitic andesites and dacites coexist in individual outcrops, and their underlying tuffs and lava breccias are rarely exposed.

Subvolcanic bodies of the Burunda complex have rather various morphology including stocks, laccoliths, lopoliths, sills with less than 2-3 km^2 surface area and dykes. These subvolcanic bodies are made of andesites, granodiorite-porphyries, diorite-porphyrites, and rarely dacites.

The main representatives of the complex are andesites hornblende - pyroxene, plagioclase-hornblende, bipyroxene and hornblende. These are massive dark grey or greenish rocks with porphyritic structure. Porphyritic minerals are formed by plagioclase ($An_{36\text{-}46}$), clino-and orthopyroxenes, and green hornblende. The main mass has got the hyalopilitic, microlitic, intersertal, and hyaline or pilotaxitic structure. Laths of plagioclase and fine grains of dark colored minerals, similar to phenocrystals are defined in the texture of the main mass of rocks. Accessory minerals are ilmenite, magnetite and apatite, and among secondary minerals sericite, chlorite, carbonate, epidote, zeolites and limonite prevail. Basalts contain olivine from 1 up to 15 %, an oligophyric structure and zoned plagioclase appear in them (An_{80} - a nucleus, $An_{36\text{-}46}$ - periphery). Sphene is added to accessory minerals, among secondary serpentine and iddingsite appear. Porphyritic texture in dacitic andesites are presented by zoned plagioclase $An_{30\text{-}65}$, clino-and (or) orthopyroxene, hornblende, biotite, quartz, olivine (singular minerals). The main mass consists of volcanic glass (up to 20 %) in which laths of plagioclase, grains of pyroxenes, hornblende, quartz, scales of biotite and accessory (ilmenite, magnetite, spinal) are defined. Secondary formations are similar to those in andesites, and on plagioclase additionally albite develops. Dacites are presented by light grey, greenish, lilac, massive or almond-shaped rocks with a fine or average porphyritic structure.

Porphyritic rocks contain plagioclase $An_{20\text{-}47}$, hornblende, biotite, quartz, muscovite, and in single cases clinopyroxene. The main mass has a microfelsitic, hyalopilitic or poikilitic structure and is combined with the quartz-feldspathic unit. Accessory minerals are presented by apatite, zircon, sphene and ore minerals. Comagmatic to integumentary volcanites and subvolcanic bodies differ by a greater degree of crystalline texture. The change of structure within the limits of one body is characteristic from thickly- to rarely-porphyric textures.

2.2.1. *Petrochemical and geochemical characteristics*

Rocks of the Burunda complex are characterized by wide fluctuations of the silica content, 47-66 wt.%, and they belong to moderate to low silica rock formations (Fig. 2a). Low-alkaline rocks are those of having Na_2O/K_2O = 1.1-3.5. Change of the Na_2O concentration with increase of SiO_2 fluctuates within the limits of 1.0 wt.%, and it maintains K_2O increases more than three times. The concentration of MgO in the rocks changes from 7.78 wt. % to 1.46 wt. %. The rocks are moderate and high titanium. According to the content of Al_2O_3, all varieties of the complex relate to the high aluminiferous rocks with ASI = 0.9-1.3, mainly low potassic calc -alkali series (at the content of SiO_2> 60 % - to high potassic calc -alkali series) (Fig. 2b).

For the REE distribution (Fig. 3a) the volcanic complex has the following characteristics: 1) poorly expressed Eu anomaly (Eu/Eu* = 0.74-0.85), 2) insignificant prevalence of the content of normalized LREE over the intermediate $(La/Sm)_n$ = 2.5-3.8, and 3) changeable prevalence of the intermediate elements over HREE $(Gd/Lu)_n$ = 1.0-5.0. Rocks are moderately enriched with Sr (230-910 ppm), Zr (121-301 ppm), Hf (178-212 ppm), Ti (2887-6190 ppm), Y (19-31 ppm), and are impoverished with Ta (0.39-0.72 ppm) and Nb (<5-13 ppm) (Fig. 3b).

2.2.2. *The age of the formation of the volcanic complex*

The age of the rocks of the Burunda volcanic complex was estimated to be as early as Cretaceous based on sporadic age data on fossil flora, spores and pollen from tuffaceous rocks from dispersed outcrops [19, 23]. Radiometric isotope dating ($^{40}Ar/^{39}Ar$) on samples of covering and subvolcanic facies of rocks of the volcanic complex resulted comparable ages with those inferred from paleontological data within the limits of technical errors. Magmatic lithoclasts from volcanogenic sediments and coherent magmatic bodies yielded an age of 108-105 Ma for the volcanic complex that represents the beginning of Albian in the Upper Cretaceous [25]. The Rb-Sr isochrone is revised for the subvolcanic dacites [23]. It defines the age of the rocks as 109.3±1.2 Ma. Age of 111 Ma was obtained by $^{40}Ar/^{39}Ar$ dating method for the andesites of Burunda suite recently [11].

2.3. Stanolir andesite volcanic complex

Stanolir andesite volcanic complex forms the fields of volcanites of north-eastern direction at the foot of Small Hingan range and it is spatially combined with younger (101-99 Ma) acidic (silicic) - alkaline volcanic formations. Therefore the rocks preserved in the surface is complex and unfortunately insignificant, as they are over covered by fields of younger volcanites making to understand the volcanic stratigraphy difficult (Fig. 5).

Stanolir volcanic complex is composed of rock formations of covering, subvolcanic and vent-filling clastogenic lavas and lava foot/top breccias) facies [26, 27, 17, 18, 11, et al.]. The basic rock formations in the structure of Stanolir volcanic complex belong to andesites, trachyandesites, seldom andesitic basalts, dacites and rhyolite dacites, as well as their subordinate pyroclastic rock types including various ignimbrites.

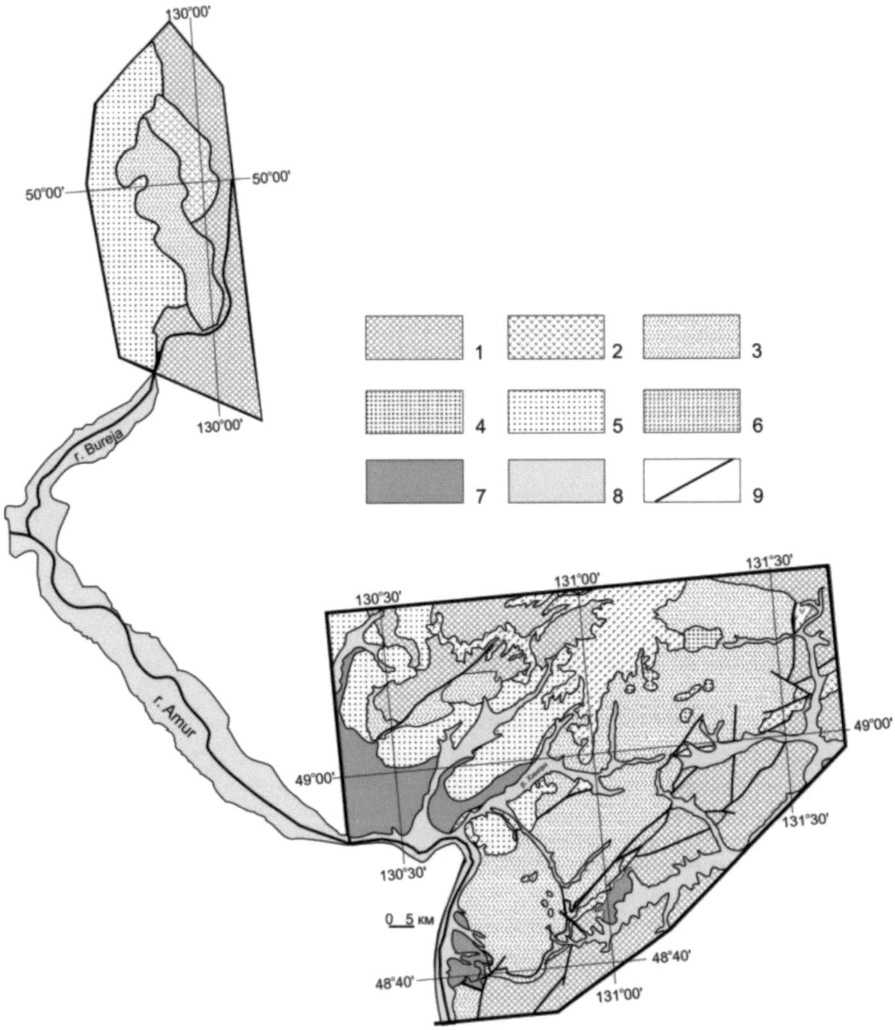

Figure 5. Scheme of the dislocation of the volcanites of Stanolir complex. Made on the base of [14] and by the authors data. The rocks are: Pre-Mesozoic magmatic and metamorphic formations (1); Early Cretaceous volcano-plutonic complexes: Stanolir complex – (2), acid-alkali composition (3), subvolcanic bodies of granitoids of the acid composition (4).Lower and upper Cretaceous sedimentary rocks (5), Cenozoic basalts (6), Lower Cenozoic sediments (7), modern sediments of the valleys of the river-bed (8), tectonic borders (9).

Covering facies - *Stanolir suite* - lies on Pre-Mesozoic crystalline basin rocks and Early Mesozoic granitoids. It composed of lavas and pyroclastic rocks of andesites, trachyandesites, andesitic basalts, trachybasalts, dacites, and also volcanogenetic and normal non-volcanic terrigenous rocks. Normal non-volcanic terrigenous rocks are located mainly in the base of the suite. Tuffs from aleurolite to coarse fragments are present in the rock sequences. Non-volcanic terrigenous rocks are relatively rare and small volume fraction of the entire volcanic complex. These terrigeneous rocks are dominantly arkose sandstones with less than 10 m in thickness commonly interbedded with coaly slates that contain up to 50 % charred vegetative detritus [26]. The general thickness of integumentary facies reaches 930 m, and it contains a cumulative lava flow units of an estimated thickness of about 150-460 m [22].

The basic representatives of the complex are andesites of plagioclase-pyroxene or plagioclase-pyroxene-amphibole. Plagioclase An_{45-48} forms grains up to 3 mm in the size. Secondary formations are widely developed. In andesitic basalts insets are presented by plagioclase An $_{46-53}$, monoclinic pyroxene - augite and olivine (up to 5 %). Olivine sometimes is completely replaced by iddingsite. Trachybasalts are characterized by greater crystallisation of the main mass. They are divided into pyroxene and olivine varieties. The porphyres of plagioclase in basalts correlate with plagioclases of An_{55-63}. In the main mass there are plagioclases with $An_{45-48.}$ Olivine is established both in porphyres and in the main mass.

2.3.1. Petrochemical and geochemical characteristics

Volcanites of Stanolir complex correspond with moderate silica concentration rock types interbedded with some, low SiO_2 varieties as well as some more acidic, silica-enriched rock varieties (Fig. 2) providing some petrochemical peculiarities to this volcanic complex. The rocks relate to the two groups by the content of the alkalias are characterized as the main-moderate of moderate alkalinity and moderate-acid of normal alkalinity (Fig. 2a) of potassic-natrium type (Na_2O/K_2O = 0.7-1.6). The sum of alkalis naturally increases from the basic varieties to the acidic rocks (Na_2O+K_2O = 4.88-7.37 wt. %), at almost constant content of Na_2O (3.05 - 3.73 wt %) and proportionally increasing K_2O (1.83-4.26 wt. %) toward the silicic rock types. All varieties of the rocks are representatives of calc-alkaline rocks (Fig. 2b) of high potassium content. The rocks are moderately magnesial, in occasional cases they are low magnesial by the content of TiO_2, but all the other varieties are high titanium formations except for moderately titanium trachybasalt, ASI = 1.04-1.31.

The rocks are characterized with moderate concentrations of Ba (430-696 ppm) and Rb (43-135 ppm) [16, 17, 18, 11]. The quantity of Rb increases from trachybasalt to dacite. The content of Sr has an opposite tendency of change (642 - 190 ppm). Moderate and moderately high concentrations are peculiar to the rocks which noticeably increase from the main rocks to moderate acid; eg. Zr (129 - 412 ppm), Hf (3 - 13 ppm), Nb (7 - 39 ppm).

REE (Fig. 3a) are characterized by inconstancy of display of negative Eu anomalies. For moderately alkaline main-moderate rocks exhibit an almost full absence of Eu-anomalies and an $(Eu/Eu^*)_n$ = 0.94-0.99 ratio is established. However Eu-anomalies are clearly shown in andesite-basalts, some andesites and dacites, where the amount of $(Eu/Eu^*)_n$ falls to a range of 0.56 - 0.70 (Fig. 3a). LREE slightly prevail above intermediate - $(La/Sm)_n$ = 2.6-4.0, at non-uniform prevalence intermediate above HREE - $(Gd/Lu)_n$ = 2.3-10.8.

The contents of Ba (430 - 700 ppm) and Rb (43 - 135 ppm) are moderate; and the contents of Zr (170 - 400 ppm), Hf (4 - 13 ppm), Nb (18 - 39 ppm), Ta (1.36 - 1.90 ppm) are moderately raising, with irregular growth of their concentration from the basic rocks to moderate acid. On the diagrams of normalization of the rocks composition to a primitive mantle (Fig. 3b) a clear Ta-Nb minimum is established, but with smaller amplitude, than on these diagrams for rocks of Poyarka and Burunda complexes, and poor expressed negative anomaly concerning Sr (190 - 642 ppm). The composition of the other elements matches with the elements of Poyarka and Burunda complexes almost completely.

2.3.2. The age of the formation of the volcanic complex

The values of the isotope plateau age, that were defined by the $^{40}Ar/^{39}Ar$ method for the matrix of andesites and dacite, yielded to a range of 109 – 105 Ma and when calculating by the isochrone line the values has changed slightly to an age of 104-111 Ma [16, 17, 18, 13].

Therefore, the interval of 105-111 Ma is the most suitable interval of the formation of the volcanic component of Stanolir complex. The radiometric ages correlate with the age estimates based on previous floristic data [28].

3. Evolution of the Late Mesozoic volcanism on the territory of Bureya-Jziamysi superterrain

The continental volcanism in the end of Late Mesozoic correlates to three age stages in the frames of the northern flank of Bureya-Jziamysi superterrain: 1) the beginning of Early Cretaceous (136 Ma), 2) Aptianian - Albian (117 – 105 Ma), 3) the end of Early Cretaceous – Albian (101 – 99 Ma). The spreading of the volcanic formations in the beginning of Early Cretaceous is timed to the contour of Amur-Zeya depression. The Amur-Zeya depression continues on south-western direction as Songliao depression on the territory of China. In the limits of Songliao depression the acid volcanites aged 136 ± 0.3 Ma [29] are stated. The belonging of the two volcanites to the intraplate formations is well confirmed by the petro-geochemical characteristics of the rocks of the volcanic complex [11].

Low potassic andesites of Poyarka volcanic complex are formed on the territory of the superterrain in the end of early Cretaceous (117 – 105 Ma). They are depleted by highly charged elements (Nb, Ta, Zr, Hf) and are enriched by Sr, Ba. Such geochemical characteristics are peculiar to the products of subduction-related volcanism, what is also confirmed by series of discrimination diagrams of major element oxides and minor element and element ratio values commonly used for geodynamical discriminations of magmatic suites (Fig. 6, 7, 8).

Judging by the presence of a spheroidal jointing of lavas and by the presence of the carbonaceous layers in the lower and upper part of the exposed covering rock facies, the outflow of lavas occurred under conditions of shallow coastal areas in a continental basin, which is in good concert with other researchers' interpretations [33].

The rocks are also compared with over subduction-related rock formations petrochemical characteristics (Fig. 6, 7). Correlation among incoherent elements the studied rocks are in close similarities with young island arc volcanites of Kamchatka (Fig. 8) that show strikingly similar values obtained from rocks especially from the Poyarka complex. The rocks of Burunda complex are the closest ones to the island arc formations lay on continental crust by the ratio La/Yb – Sc/Ni (Fig. 8).

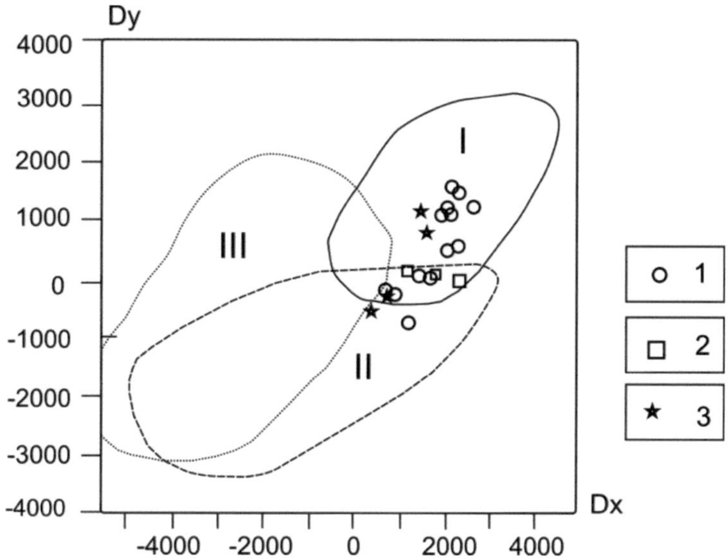

Figure 6. Discrimination diagrams for the definition of the tectonic situations. The rocks of Bureya-Jziamysi superterrain: Poyarka (1), Burunda (2), Stanolir (3). According to the data of: D_x/D_y [30] for the main rocks (D_x = (176.94* SiO_2) - (1217.77* TiO_2) + (154.51* Al_2O_3) - (63.1* FeOt) - (15.69* MgO) + (372.43* CaO) + (104.41* Na_2O) - (19.96* K_2O) - (873.69* P_2O_5) - 11721.488; D_y = (94.39* SiO_2) - (103.3* TiO_2) + (417.98* Al_2O_3) - (55.63* FeOt) + (57.61* MgO) + (118.42* CaO) + (502.02* Na_2O) + (6.37* K_2O) + (415.31* P_2O_5) - 13724.66). The fields of the basalts: I – island arcs, II – traps, III – continental rifts.

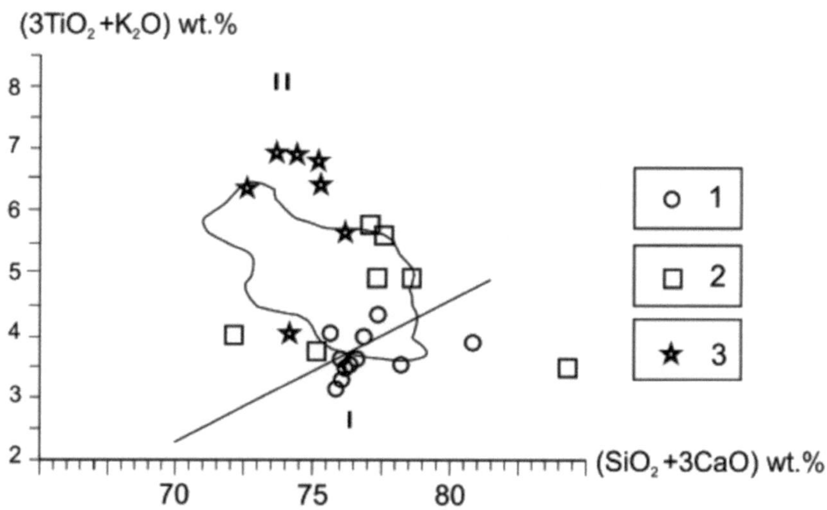

Figure 7. Discrimination petrochemical diagram for the rocks of the volcanic complexes based on the classification of [31].: Poyarka (1), Burunda (2), Stanolir (3). The situation of the volcanites of Okhotsk-Chukotica volcanogenic belt is marked with the contour on the diagram. Type of the associations: I – oceanic, II – continental-margin.

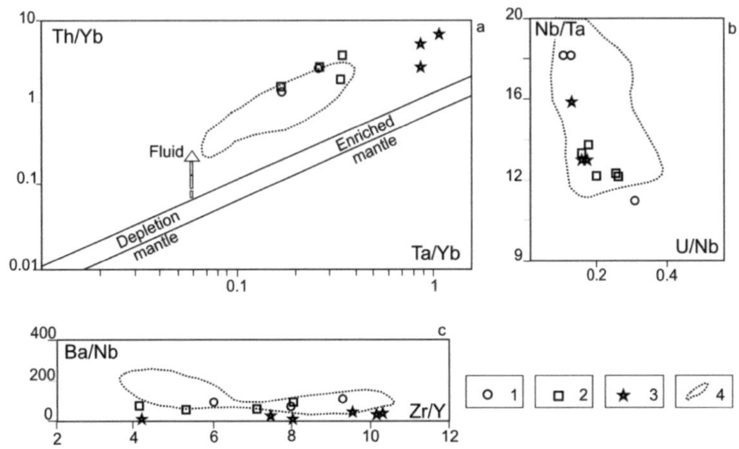

Figure 8. Diagram of the ratio of the incoherent element for the rocks of the volcano-plutonic complexes: Poyarka (1), Burunda (2), Stanolir (3), island arc type of Sredinniy of Kamchatka fault range island arc (4) by [32].

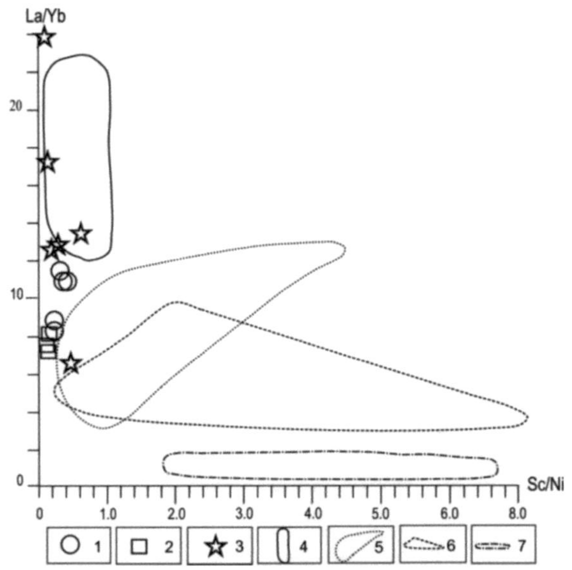

Figure 9. Ratio of the minor elements La/Yb – Sc/Ni in the volcanic complexes: 1 – Poyarka, 2 – Burunda, 3 – Stanolir. Fields of the rocks by the data of [34]: 4 – Andean active continental margin; 5 – island arcs laying on the continental crust, 6 – island arcs laying on the oceanic crust, 7 – low potassic oceanic basalts.

The values of the ratios of Burunda complex rocks are La/Yb<10; La/Ta = 30-102; Zr/Hf = 36.0-39.7 (almost constant). Such values are characteristic for the island arc rocks.

Along the eastern border of superterrain (by modern coordinates) during the period 108 – 105 Ma the andesites of Stanolir complex were formed. On the tectonic diagrams (Fig. 6, 7, 8, 9) they get into the fields of the subduction conditions of their formation. On the diagrams of the REE composition (Fig. 3b) they are characterized with higher content of Nb, Ta, Zr and lower content of Sr, with the conservation of the clear minimums of Ta and Nb, one of the typical values of subduction-related signatures [35, 36]. The proximity to boundary values of the ratios La/Ta = 18-23 [37] are characteristic for these rocks. By the correlations of such incoherent elements, as Nb/Ta - U/Nb (Fig. 8 b), it is inferred that the rocks are relate to the island arc formations. According to the correlation Th/Yb – Ta/Yb and Ba/Nb – Zr/Y (Fig. 8 a, c) the volcanites of the Stanolir complex are located on the continuation of the fields of the island arc formations.

Isotope-geochronological data for lavas and subvolcanic rock formations of the investigated volcanic complexes define the time of the formation of the magma component. But the beginning of their establishment has started prior the formation of the preserved coherent magmatic bodies as evidenced by the presence of basal thick volcanogenic terrestrial sedimentary successions part of the underlying sedimentary succession. It should be mentioned that the thickness of this component is almost the same for all complexes – 200 – 450 m. Tuffogenic – sedimentary successions of Poyarka volcanic complex was accumulated during Hauterivian – Barremian period which is more or les the same time frame for the Burunda volcanic complex which was inferred to have been accumulated during Barremian – Aptianian period [22].

It can be stated, that Poyarka volcanic complex started to form from the accumulation of the tuffogenic-sediment component at about 120 Ma. About 117 Ma large volume of lava outflow – part of the volcanic complex – took place. The analogical formations of Burunda complex started to form 111 Ma. Stanolir complex started to form about 108 Ma ago. The outflow of Poyarka volcanic complex were near-continues till Albian – 107 Ma. That is the time when lava outflow of Burunda and Stanolir volcanic complexes begins.. All the volcanites belong to calcareous-alkaline low and high potassic series. They are characterized with snbductional type origin based on the distribution pattern of the minor elements such as for instance the concentration of Nb and Ta is low while the concentration of Ba, Rb, K, Ti, Sr are relatively high

The diagram (Fig. 10) illustrates the formation of the initial melt for the three complexes occurred at the expense of the melt of peridotite.

By the correlation of Tb/Yb normalized to chondrite –C1 [39], that make less than 1.8 (except some of the trials of Stanolir complex) it can be stated that, spinel peridotites were the magma-forming substratum for the formation of the andesite of the volcanic complex. By all that the stage of the melt of the substratum of the spinel peridotite was decreasing from the volcanites of Poyarka complex to the volcanites of Stanolir complex (Fig. 32). The coefficient of REE = 2.5-4.3. (K_{REE} = 0.1La/Yb+Ho/Yb+(Dy+Ho)/(Yb+Lu) by [40], elements normalized to chondrite [41]). Such values confirm the presence of pyroxene in the melting substrata. By the ratio Ni/Co [42] the rocks of Poyarka (completely), Stanolir (mostly) and Burunda (subvolcanic) complexes are derivatives from the melts of the mantle. The derivatives of the melt of the lower crust formations are the lavas of Burunda complex and (rarely) Stanolir complex. Thus, the rocks of the three complexes can be partly examined as primary. It is confirmed by the absence (or a weak presence) of Eu anomalies, that is one of the criteria of the primary nature of magmas [43]. By the correlation of the incoherent elements: Nb/Ta - U/Nb (Fig. 7 b) the formations of the complexes are comparable with the rocks of the subduction type of the Sredinnii mountain ridge of Kamchatka. By the correlations Th/Yb-Ta/Yb (Fig. 7a) the volcanites of Stanolir complex are displaced to the side of the enriched mantle. The relations of the coherent elements (Ce/La, Zr/La, Nb/La, Th/La, Yb/La) are not only close to the constant values, but they also correlate with each other. This confirms the belonging of the rocks of the three complexes to a singular magmatic stage. The derivatives of the stage underwent the evolution because of the decay of the subduction processes in the frames of the researched region. Many authors connect the lowering of the concentration of Sr and growth of the concentration of Ce and Th with the "decay" of the subduction [44 - 49, 36]. It can be seen in the geochemical characteristics of the rocks of the researched complexes in the direction from the volcanites of Poyarka to the volcanites of Stanolir complex: Sr – from 1029 ppm to 153 ppm, Ce – from 28.52 ppm to 75.07 ppm, Th – from 1.7 ppm to 15.89 ppm. They belong to the singular magmatic process that confirms the ratio of Zr/Nb - Nb/Th. According to the ratio all this formations were melted from the source that is close to the source type EN [50] with the presence of a component of a depleted source. Series of the geochemical indicators (Nb/La, La/Ta, Ta/Th, et al.) show that magmas of the volcanites were also underwent by the contaminations of the crust material. According to the ratio Ce/Y (less than 4) and La/Nb (less than 3.5) the formation of the rocks of the researched complexes occurred at the expense of the mixture with the crust of the product of a partly melt of the spinel peridotite of the mantle [51].

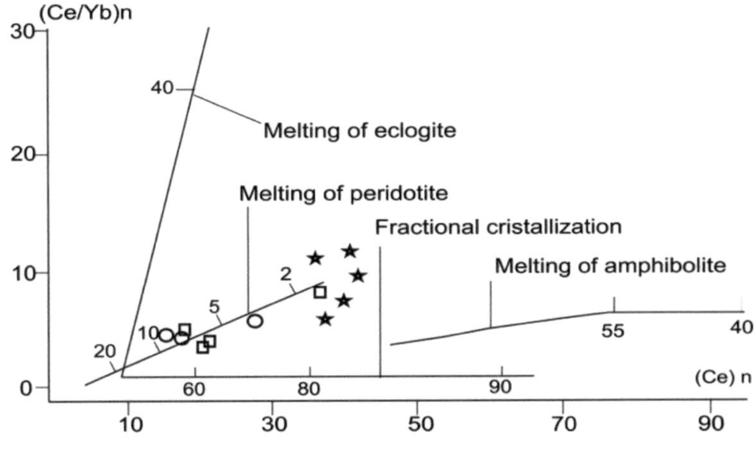

Figure 10. The location of the rocks of the andesite volcanic complexes on the diagram $(Ce/Yb)_n$-$(Ce)_n$ (conditional marks are listed on the fig. 8). The model trends of melting by [38].

Based on the comparison of the formation time, petro- and geochemical characteristics, belonging to the singular magmatic process of the rocks of the andesite formation of the northern flank of Bureya-Jziamysi superterrain of Poyarka, Burunda and Stanolir complexes, the followings can be stated:

1. By the ratio (Nb/La)n –(Nb/Th)n [52] the rocks of Burunda complex correspond with island arc lavas - the contaminated formations of the continental crust. The affection of the crust contamination is confirmed by the high values of the ratio $^{206}Pb/^{204}Pb$ (389.6) [23].

2. Formations of Burunda complex are inconsistently laying on the terrigene deposits of Ogodzha suite of Aptian – Albian age. The Ogodzha deposits are traced in the mining excavations under the formations of Burunda rock strata (Fig. 3). They are stated in singular tectonic blocks in southern direction. That points to a more wide development of the rocks as in primary variant. The coal-bearing deposits of the suite lay on the borders of Paleozoic and Early Mesozoic (Triassic) age of the superterrain transgressively. They form a flat monocline with the dip from 8 to 30°. Tuff material is present in the composition of the suite. It points at the parallel volcanic activity during the Ogodzha sediment accumulation. The following intensification of the volcanic activity (111-105 Ma) leads to the formation of Burunda volcanic complex. It can be proposed that Burunda volcanites are a fragment of an island arc that was formed on the edge of Bureya-Jziamysi superterrain and the deposits of Ogodzha suite are part of a sediment complex formed in a behind-the-arc basin.

3. The volcanic activity completes in the frames of the researched territory by the formation of the acid – alkali volcano-plutonic complex at 101 – 99 Ma. The formations of the complex are characterized as intraplate origin. Their geochemical signatures indicate of its source is close to an enriched mantle in the primary melt of the source region of melts [11].

4. Geodynamic situations of the formation of the Late Mesozoic volcanic complexes of Bureya-Jziamysi superterrain

The first introductions about possible geodynamical situations in the frames of the researched region were made L.P. Sonenshein with co-authors (1990), who thought that Mesozoic magmatic formations could be a product of activity of a subduction-related volcanism or a "hot spot" activity. Farther the same variants were elaborated by V.V., Yarmoluk and B.I. Kovalenko [53, 54], I.V. Gordienko [55], V.G. Moiseenko [56]. B.A. Natalin proposed both subductional and collisional situations [57]. Chinese geologists [58] researched a model of the formation of Mesozoic magmatism in the situation of a transformed continental edge. Analogical point of view was presented by A.I. Khanchuk and V.V. Ivanov [59]. The high precision values about the age and the volcanites geochemical composition of the rocks of the volcano-plutonic complexes of the region were absent during the composition of the geodynamical reconstructions listed above. The values are partly derived at the present time.

All the suggested geodymanical reconstructions include the interdependence of North-Asian and Sino-Korean cratons and platforms of the oceanic crust of the Pacific.

Northern border of Bureya-Jziamysi superterrain is a southern border of Mongol-Okhotsk orogenic belt. Along the belt the Late Mesozoic formations of the bimodal series are developed [12, 13]. The formation of the complexes occurred under conditions of collisional compression, agreed by the approaching of North-Asian and Sino-Korean cratons and a possible influence of plume on the area that is under conditions of the collisional compression. The bimodal complexes have a lineal separation in the frames of Mongol Okhotsk belt. But on the East their separation is framed by the structures of Bureya-Jziamysi superterrain. It might be proposed that the given theory had not suffered such processes. The Chinese geologists worked out a scheme of tectonic development for the territory of Bureya-Jziamysi superterrain on the results based on a seismic transect Manchzhuria – Suifankhe laid transversely the Songliao depression [58, 60]. It is stated that the extension, provoked by the changes of the Izanaga plate movement dominated in Late Jurassic – Early Cretaceous in the Songliao basin [58, 55]. According to the data [61] a sharp change of the direction (on 50°) and speed (from 5.3 to 30 cm per year) of the subduction of Izanaga plate under the eastern edge of Bureya-Jziamysi superterrain took place at about 135 Ma. This provoked the formation of series of the left displacements CB and C-CB extension and formation of the rift-like structures [58]. The structures were field with coal-bearing terrigenous sediments and volcanic formations of acid composition. During the period a complex of acid volcanites aged 136 Ma is forming in the frames of the studied territory.

Farther than 136 – 120 Ma the territory became as a passive continental edge. The temporal stage of the formation of Poyarka, Burunda and Stanolir volcanic complexes relates to the moment when the Izanaga plate changed its movement direction from northern to northwestern. With that, the angle of the turn of the plate was almost 30° [61]. During the period there was a flat subduction of the oceanic plate under the eastern edge of Asia with a speed more than 20 cm per year [61]. That's why the formation of the rocks on the continental crust under the conditions of the subduction seems to be possible.

Paleomagnetic data were obtained by U.S. Bretshtein and A.V. Klimov [6] for the main tectonic units of the southern part of the Far East of Russia (Fig. 11). According to the data, during the Jurassic the Bureya-Jziamysi superterrain was at large distance from the North-Korean plate. The distance was about a few thousands of kilometers.

140 Ma the superterrain (Bureya block dew to [62]) was located much more on the north from it's nowadays dislocation according to the data of geological and geophysical including GPS data [62].

Thus, it might be proposed, that during the period 120 – 105 Ma there was a volcanic activity on the territory of Bureya-Jziamysi superterrain which was controlled by the subduction processes. During the period the volcanic formations were loosing their typical subduction petro-chemical characters, for instance the composition of Sr decreased while the amount of Nb, Ta, Rb, K, Ti increased over time. Such values in the composition of the rocks show the attenuation of the active subduction processes. The temporal stage of the formation of the rocks of the three complexes correlates to the stage of the flat subduction of the oceanic plate Izanaga under the edge of Bureya-Jziamysi superterrain. The biggest magmatic activity took place during the period of the change in the movement direction of the oceanic plate from almost northern to north-western with the growth of speed till 23.5 cm per year [61].

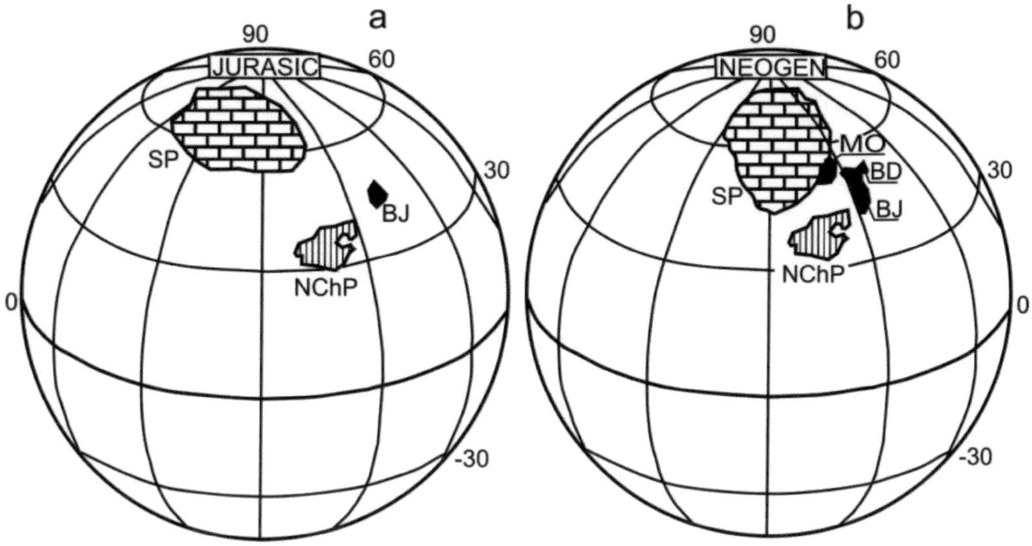

Figure 11. Palinspatic reconstruction of the location of the main tectonic units of the south of the Far East of Russia in Jurassic (a) and Neogene (b) by U.S. Bretshtein and A.V. Klimov [6]. SP – Siberian plate NChP – North-Chinese plate, BJ – Bureya-Jziamysi superterrain (by U.S. Bratshtein and A.B. Klimov – Khingan – Bureya plate), MO – Mongol-Okhotsk terrain, BD – Badzhal terrain

The magmatic processes decay completely in the interval of 105–101 Ma on the territory of Bureya-Jziamysi superterrain. The situation of the continental "riftogenesis" or the situation of a transforming continental edge begins to appear 101 Ma [59, 63], what was reflected on the formation of the acid – alkaline rocks of the intraplate volcano-plutonic complex. As the most possible tectonic scenario by the formation of the volcano-plutonic complex the author examines the collision of Bureya-Jziamysi and Badzhalsky terrains [11] which is confirmed by the paleomagnetic data (Fig. 10).

5. Conclusion

On the base of the confrontation of the formation time, petro- and geochemical characteristics, belonging to a singular magmatic focus of the rocks of andesite formation of Bureya-Jziamysi superterrain, the Poyarka, the Bureya and the Stanolir volcanic complexes, it might be stated that their formation happened more or less simultaneously (with a leading at the formation of Poyarka complex at the beginning). All the formations of the studied volcanic complexes have similar characteristics and are related to subductional volcanites of calc alkali series. The changes of the content of major- and minor element composition of the volcanic complexes may be explained by the mixture of the mantle source, fluids at the partial melt of the lower continental crust and a subducting plate at its contact with the mantle. The last fact is confirmed by the presence of "adakite component" – the shows of melt of the oceanic plate in the rocks of Poyarka and Burunda complexes: the presence of magnesial andesites and andesites, high concentrations of Sr and Ba, low concentrations of HREE with the high ratios of La/Yb and low ratios of K/La. Thus, it might be proposed that the existence of a simultaneous volcanic activity during 120 – 105 Ma on the territory of Bureya-Jziamysi superterrain, conditioned by the subductional processes. During the period, the volcanic formations loose their typical subductional signatures as reflected by the lower Sr concentration of the rocks, the increase in concentration of Nb, Ta, Rb, K, Ti, what is inferred to be connected to the decay of the active subduction processes.

The dislocation and the geochemical characteristics of the rocks of the complexes show the dislocation of the moving of the subducted oceanic plate. Its northern territory was pointing to the side of the ocean at that moment. It might be also proposed that Bureya-Jziamysi superterrain was not a component of Amur microcontinent during the period of the formation of the three complexes, but it was an independent geological object. Its annexation to the Amur microcontinent occurred much later than Albian.

Author details

I.M. Derbeko

Institute of Geology and Nature Management FEB RAS, Blagoveschensk, Russia

References

[1] Parfenov, L. M., Popeko, L. I., & Tomurtogoo, O. (1999). Problems of tectonic of Mongol-Okhotsk orogenic belt, Russian Journal of Pacific Geology, September- October 1999), ISSN-1819-7140., 18(5), 24-43.

[2] Sonenshein, L. P., Kuzmin, M. N., & Natapov, L. M. (1990). Tectonics of lithosphere plates on the territory of USSR. Moscow: Nedra, 328 p., 1

[3] Ashurkov, S. V., San'kov, A. I., Miroshnichenko, A. I., Lukhnev, A. V., Sorokin, A. P., Serov, M. A., & Byzov, L. M. (2011). GPS geodetic constraints on the kinematics of the Amurian plate, Geology & geophysics, February 2011), 0016-7886, 52(2), 299-311.

[4] Gatinsky, Yu. G., & Rundquist, D. V. (2004). Geodynamics of Eurasia- Plate tectonics and block tectonics, Geotectonics, January-February 2004), 0016-8521, 38(1), 3-20.

[5] Parfenov, L. M., Berezin, N. A., Khanchuk, A. I., Badarh, G., Belichenko, V. G., Bulgatov, A. N., Drill, S. I., Kirillova, G. L., Kuzmin, M. I., Nokleberg, U., Prokopiev, A. V., Timofeev, V. F., Tomurtogoo, O., & Yan', H. (2003). The model of the formation of the orogenic belts of Central and Northern-Eastern Asia, Russian Journal of Pacific Geology,Pacific geology, November- December 2003), ISSN-1819-7140., 22(6), 7-41.

[6] Khanchuk A.I. Geodynamics, magmatism and metallogeny of the Russian East, Vladivostok: Dalnauka,(2006). 580-4-40634-557-2p.

[7] Geology of BAM zone. Editor L.I. Krasny, Leningrad: Nedra,(1988). p

[8] Sharov, V. N., Fefelov, N. I., Jablonovsky, B. V., et al. (1992). Dating of the low Proterozoic stratificated formations of Patomsky plateau Pb/Pb method, Reports of the Earth Science, April 1992), 0102-8334X, 324(5), 1081-1084.

[9] Pavlov, Ju. A., & Parfenov, L. M. The abyssal structure of the Eastern Sayan and Southern Aldan frames of the Siberian plate, Novosibirsk: Nauka, (1973). p.

[10] Parfenov, L. M., Bulgatov, A. N., & Gordienko, I. V. (1996). Terrains and the formation of the orogenic belts of Transbaikal, Russian Journal of Pacific Geology, July-August 1996), ISSN-1819-7140., 15(4), 3-15.

[11] Derbeko, I.M(2012a). Later Mesozoic volcanism of Mongol-Okhotsk belt (eastern end and the southern framing of eastern member of the belt), Saarbruken: LFMBERT Academic Publishing GmbH&Co.KG, 97 p. 978-3-84734-060-7

[12] Bogatikov, O. A., & Kovalenko, V. I. (2006). Types of magma and their sources in the history of the Earth, Moscow: Institute of Geology of ore deposits Russian Academy of Science, 588-9-18013-428-0p.

[13] Derbeko, I. M. (2012). Bimodal volcano-plutonic complexes in the frames of Eastern member of Mongol-Okhotsk orogenic belt, as a proof of the time of final closure of Mongol-Okhotsk basin, In: Updates in volcanology- A Comprehensive Approach to Volcanological Problems. Chapter 5. InTech, 978-9-53307-434-4, 99-124.

[14] Geologycal map of Priamurie and neighbouring territories.Scale 1:2 500 000 ((1999). Explanatory note, Editors L.I. Krasny, A.S Volsky, I.A. Vasilev, Pen Yunbiao, Suy Yancyan, Van In, St. Petersburg- Blagoveschensk- Harbin: Ministry of nature resources of Russian Federation, Ministry of geology and mineral resources of China, 135 p.

[15] Sorokin, A. A., Sorokin, A. P., Ponomarchuk, V. A., Travin, A. V., Kotov, A. B., & Melnikova, O. V. (2008). Basaltic andesites of Amur-Zeja depression in Aptian: new geochemical and 40Ar/39Ar- geochronological data, Reports on Earth Science, April 2010), 0102-8334X, 421(4), 525-529.

[16] Derbeko, I. M., Sorokin, A. A., Ponomarchuk, V. A., & Sorokin, A. P. (2004). Timing of Mesozoic magmatism in Khingan-Okhotsk volcano-plutonic belt (Russian Far East), Geochim. et Cosmochim. Acta. S. 1, 0016-7037, 68, A226.

[17] Sorokin, A. A., Ponomarchuk, V. A., Derbeko, I. M., & Sorokin, A. P. (2004). New data on the geochronology of the magmatic associations of Khingan-Olonoy volcanic zone (Far East of Russia), Russian Journal of Pacific Geology, ISSN-0207-4028., 23(2), 52-62.

[18] Sorokin, A. A., Ponomarchuk, V. A., Derbeko, I. M., & Sorokin, A. P. (2005). Geochronology and geochemical peculiarities of Mesozoic associations of Khingan-Olonoisk

volcanic zone (the Far East of Russia), Stratigraphy and geological correlation, ISSN-0869-5938., 13(3), 63-78.

[19] Martinuk, M. V., Riamov, S. A., & Kondratieva, V. A. (1990). Explanatory report to the scheme of dismemberment and correlation of magmatic complexes of Khabarovsky region and Amur region, Khabarovsk, Russia: Industrial geological organization, 215 p.

[20] Le Bas, M., Le Maitre, R. W., Streckeisen, A., & Zanettin, B. (1986). A chemical classification of volcanic rocks based on the total-silica diagram, Journal of Petrology, 27, 0022-3530, 745-750.

[21] Irvine, T. N., & Baragar, W. R. (1971). A guide to the chemical classification of the common volcanic rocks, Canadian Journal Earth Science, , 8, 523-548.

[22] Resolutions of IV interdepartmental regional stratygraphic conference about Cambrian and Phanerozoic of the South of the Far East and East of Transbaikal, Scheme 35, (1993). Khabarovsk: Khabarovsk state mining-geological enterprise, 22 p.

[23] Agaphonenko, S. G. Explanatory note for state geological map of Russian Federation. Scale 1: 200 000 (Second edition). Tugur series. Page N-XXVI, St.- Petersburg: VSGEI, (2002). , 53.

[24] Derbeko, I. M., Agafonenko, S. G., Kozyrev, S. K., & Vyunov, D. L. (2010). The Umlekan-Ogodzha volcanic belt (the problem bodily separation), Lithosphere, May-June 2010), 1681-9004(3), 70-77.

[25] Rasskasov, S. V., Ivanov, A. V., Travin, A. V., Brandt, I. S., & Brandt, S. B. (2003). Ar-39Ar and K-Ar dating of the volcano rocks of Albian of Priamuria and Transbaikal, In: Isotopic geochronology in solving problems of geodynamics and ore genesis, St.-Petersburg: Center of information culture, 2003, , 410-413.

[26] Evtushenko, V. A. (1978). Stratigraphy and geochronology of the Cretaceous formations Small Khingan, In: Stratigraphy of the Far East, Vladivostok: FEGI, , 152-153.

[27] Gonevchuk, V. G. (2002). Tin-bearing systems of the Far East: magmatism and ore genesis, Vladivostok: Dalnauka, 580440251p.

[28] Kirjanova, V. V. (2000). New in stratigraphy of the Cretaceous South of the Amur region. In: Correlation of Mesozoic continental formations of the Far East and East of Trans-Baikal region, Chita: GGUP, , 49-52.

[29] Wang, P. J., Liu, W. Z., Wang, S. X., & Song, W. H. (2002). Ar/39Ar and K/Ar dating on the volcanic rocks in the Songliao basin, NE China; constraints on stratigraphy and basin dynamics, International Journal of Earth Science, 0167-4487X., 91, 331-340.

[30] Velicoslavinsky, S. D., & Glebovicky, V. A. (2005). New discrimination diagram for classification of the island arc and continental basalts on the base of petrochemical data, Reports of Akademii Nauk, March 2005), 0869-5652, 401(2), 213-216.

[31] Piskunov, B. I. (1987). Geologo-petrological specifics of the island arcs volcanism. Moscow: Nauka, 236 p.

[32] Churikova, T., Dorendorf, F., & Woerner, G. (2001). Sources and fluids in mantle wedge below Kamchatka, evidence from across-arc geochemical variation, Journal of Petrology, August 2001), 0022-3530, 42(8), 1567-1593.

[33] Kirillova, G. L. (2005). Late Mesozoic- Cenozoic sedimental basins of the continental edge of the south-eastern Russia: geodynamical evolution, coal- and oil-gas-bearing, Geotektonika, October-November 2005), 0001-6853X.(5), 62-82.

[34] Rollinson, H. R. (1995). Using Geohemical Data: Evalution, Presentation, Interpretation, London, 352 p.

[35] Tatsumi, Y., Hamilton, D. L., & Nesbitt, R. W. (1986). Chemical characteristics of fluid phase realeased from a subducted lithosphere and origin of are magmas: Evidence from high-pressure experiments and natural rocks, Journal of Volcanology and Geothermal Research, September 1986), 0377-0273, 29(1-4), 293-303.

[36] Volynec, A. O., Antipin, V. S., Perepelov, A. B., & Anoshin, G. N. (1990). Geochemistry of the volcanic series of the island arc system in application to geodynamics (Kamchatka), Geology & geophysics, 0016-7886(5), 3-13.

[37] Pusankov, Ju. M., Volinec, O. N., Seliverstov, V. A., et al. (1990). Geochemical typification of magmatic and metamorphic rocks of Kamchatka, Novosibirsk: IGG SB AS RF, 1990, 259 p.

[38] Gill, J. B. (1981). Orogenic andesites and plate tectonic, New York, 354-0-10666-939-0p.

[39] Wang, K., Plank, T., Walker, J. D., & Smith, E. I. (2002). A mantle melting profile the Basin and Range, SW USA, Journal of Geophysical Research, 0148-0227, 107(B1)

[40] Troshin, U. P., Grebenschikova, V. I., & Boiko, S. M. (1983). Geochemistry and petrology of the rare-earth plumaezite granites, Novosibirsk: Nauka, 183 p.

[41] Sun, S. S., & Mc Donough, W. F. (1989). Chemical and isotopic systematics of oceanic basalts: implications for mantle composition and processes, In: Magmatism in the ocean basins (Editors: Saunders A.D., Norry M.J.), Special Publications of the Geological Society, London, , 42, 313-345.

[42] Kogarko, L. I. (1973). Relation of Ni/Co- indicator of the mantle origin of magmas, Geokhimiya, October 1973), 0016-7525(10), 1446-1449.

[43] Balashov, Ju. A. (1976). Geokhimija of the rare elements, Moscow: Nauka, 267 p.

[44] Mc Kenzi, D. E., & Chappel, B. W. (1972). Shoshonitic and calc-alkaline lavas from the Highlands of Papua New Guinea, Contributions to mineralogy and petrology, 0010-7999, 35(1), 50-63.

[45] Whitford, D. J., Nicholls, J. A., & Taylor, S. R. (1979). Spatial variations in the geochemistry of Quaternary lavas across the Sunda arc in Java and Bali, Contributions to mineralogy and petrology, 0010-7999, 70(3), 341-356.

[46] Riou, R., Dupuy, C., & Dostal, J. (1981). Geochemistry of coexisting alkaline and calc-alkaline volcanic rocks from Northern Azerbaijan (NW Iran), Journal of Volcanology and Geothermal Research, ISSN-0377-0273., 11(2), 253-276.

[47] Allan, J. F., & Garmichael, J. S. E. (1984). Lamprophyric lavas in the Colima graben, SW Mexico, Contributions to mineralogy and petrolog, 0010-7999, 88(3), 203-216.

[48] Mitropoulos, P., Tarney, J., Saunders, A. D., & Marsh, N. G. (1987). Petrogenesis of Cenozoic volcanic rocks from the Aegean Island arc, Journal of Volcanology and Geothermal Research, 32, SSN-0377-0273.(1), 177-194.

[49] Saunders, A. D., Rogers, D., & Marriner, G. F. (1987). Geochemistry of Cenozoic volcanic rocks, Baja California, Mexico: implications for the petrogenesis of post-subduction magmas, Journal of Volcanology and Geothermal Research, ISSN-0377-0273., 32(1), 223-246.

[50] Genshaft, Yu. S., Grachev, A. F., Saltykovsky, A., & Ya, . (2006). Geochemistry of Cenozoic basalts of Mongolia: the problem of genesis of mantle sources, Geology & Geophysics, March 2006), 0016-7886, 47(3), 377-389.

[51] Hoffman, A. W. (1997). Mantle geochemistry: the message from oceanic volcanism, Nature, 1752-0894, 385, 219-229.

[52] Puchtel, I. S., Hofmann, A. W., Mezger, K., Jochum, K. P., Shchipansky, A. A., & Samsonov, A. V. (1998). Oceanic plateau model for continental crustal growth in the Archaean: A case study from the Kostomuksha greenstone belt, NW Baltic Shield, Earth and Planetary Science Letters, 1998, 155, 0001-2821X., 57-74.

[53] Yarmoluk, V. V., & Kovalenko, V. I. (1991). Riftogene magmatism of the active continental margins and their ore content. Moscow: Nauka, 263 p.

[54] Yarmoluk, V. V., Kovalenko, V. I., & Kuzmin, M. I. (2000). North-Asian superplume in Phanerozoic: magmatism and abyssal geodynamics, Geotectonics, September- November 2000), 0016-8521(5), 3-29.

[55] Gordienco, V. I., Klimuk, V. S., & Cuan, Khen. (2000). Upper Amur vulkano-plutonic belt of East Asia, Geology & Geophysics, 0016-7886, 41(12), 1655-1669.

[56] Moiseenco, V. G., & Sahno, V. G. (2000). Plum magmatism and mineralogy of Amur megastructure, Blagoveschensk: AmurKSRI, 160 p.

[57] Natalin, B. A. (1991). Mezozoic accretion and collision tectonics of the Far East sout of the USSR, Russian Journal of Pacific Geology, ISSN-1819-7140.(5), 3-23.

[58] Liu Zhaojun, Wang Xiaolin, Lui Wanghu, Xue Fang, Zhao(1994). Mapping Formational mechanism of the Songliao and Hailaer Mesozoic basins of Mongholui- Sui-

fenhe geoscience transekt region. In: M-SGT geological research group ed. Geological Research on Litosphere Structure and its Evolution of Mongholui- Suifenhe Geoscience Transect Region of China, Beijing: Seismic Publishing House. in Chinese)., 14-25.

[59] Khanchuk, A. I., & Ivanov, V. V. (1999). Mezo-Cenozoic geodynamical situations and a golden ore formation of the Far East, Geology & Geophysics, November 1999), 0016-7886, 40(11), 1635-1645.

[60] Yang Baojun, Liu Cai, Zhou Yang, Liang Tiecheng, Tang Dayi.Study of the crust structure in the Anda-Zhaozhou-Harbin transect region using deep reflecthion method. In: M-SGT geophusical research group ed. Research on geophusical field and deep structural characteristics of Manzhouli- Suifenhe geoscience transekt region of China. Beijing: Seismic Publishing House, (1995). In Chinese with Engl. Abstr.)., 100-113.

[61] Maruyama, S., & Seno, T. (1986). Orogeny and relative plate motions: example of the Japanese Islands, Tectonophysics, 127, 3-4, 1, (August 1986), 0040-1951, 305-329.

[62] Pisarevsky, S. A. (2005). New edition of the Global Paleomagnetic Database. EOS Transactions American Geophysical Union, 0096-3941, 86(17), 170.

[63] Khanchuk, A. I. (2001). Pre-Neogene tectonics of the Sea-of-Japan region: A view from the Russian, Earth Science, 1674, 55(5), 275-291.

Section 2

Volcanic Hazards

Chapter 3

Application of the Bayesian Approach to Incorporate Helium Isotope Ratios in Long-Term Probabilistic Volcanic Hazard Assessments in Tohoku, Japan

Andrew James Martin, Koji Umeda and
Tsuneari Ishimaru

Additional information is available at the end of the chapter

http://dx.doi.org/10.5772/51859

1. Introduction

Geological hazard assessments based on established statistical techniques are now commonly used as a basis to make decisions that may affect society over the long-term (0.1 – 1 Ma). Volcanic risk essentially consists of:

(1) The probability of a 'volcanic event' occurring such as a dike intrusion or a new strato- or caldera volcano forming e.g. [1- 7]

(2) The consequences of the volcanic event e.g. [8 - 9].

A challenge with the long-term probabilistic assessment of future volcanism in relation to the siting of, for example geological repositories is that because new volcano formation is rare, uncertainties in models are inherently large [10]. Sites for nuclear facilities in particular must be located in areas of very low geologic risk [11]. Recent studies have been carried out looking at the hazard posed by volcanoes to nuclear power plants in Armenia [e.g. 9, 12] and Java, Indonesia [e.g. 13]. Here the focus was more on the consequences of an eruption at an existing volcano on the safety of an operating nuclear power plant. In the case of a geological repository for high and/or low level radioactive waste, the emphasis is on the consequences of new igneous activity such as a dike that may intrude the repository [e.g. 14] and transport the waste to the surface. In this case, the probability of a new volcano forming in the first place is very low (typically < 10^{-7}/a) since by definition such facilities should be located away from existing Quaternary volcanoes. However the lack of volcano 'data' implies that addition information on the processes that control future long-term spatio-temporal distribution of volcanism are

INTECH
open science | open minds

needed. This has motivated several investigators to incorporate datasets in addition to the distribution and timing of past volcanic activity in volcanic probabilistic analyses [e.g. 15]. Bayesian inference has been used to combine geophysical datasets to probability distributions constructed from known historic volcano locations in order to estimate the location of future volcanism over a regional scale [1]. More recently, [16] used Bayesian inference to merge prior information and past data to construct a probability map of vent opening at the Campi Flegrei caldera in Italy.

Here we revisit the Bayesian approach developed by [1] where seismic tomographs and geothermal gradients were incorporated into probabilistic assessments by Bayesian inference in Tohoku. We apply the same Bayesian technique in the same study area to incorporate recently acquired helium isotopes into probabilistic hazard assessments; such noble gases have been shown to be excellent natural tracers for mantle-crust interaction owing to their inert chemical properties which means they are not altered by complex chemical processes. Moreover helium isotopes provide evidence for the presence of mantle derived materials in the crust, owing to the distinct isotopic compositions between the crust and the upper mantle [e.g. 17, 18]. We examine the link between volcanism and $^3He/^4He$ ratios that may infer possible regions of magma generation and hence volcano formation. Such links between magmatism and elevated $^3He/^4He$ ratios have been proposed [e.g. 19, 20], but the link has not been examined quantitatively in probabilistic based models. Finally we discuss the Bayesian method in developed by [1] in the context of recent approaches to incorporate multiple datasets [e.g. 21, 22].

2. Japan and the Tohoku region

Japan is one of the most tectonically active regions in the world. Due to the dynamics of four plates, Quaternary volcanoes have formed along distinct volcanic fronts in east and west Japan (Figure 1).

The Tohoku region (Figure 2) is arguably one of the most extensively studied volcanic arcs in the world, particularly regarding the relationship between volcanism and tectonics. Moreover there have been numerous geological and geophysical investigations yielding high-quality datasets e.g., [23 - 29].

Tohoku is a mature double volcanic arc with a back-arc marginal sea basin located on a convergent plate boundary of the subducting Pacific plate and the North American plate (Figure 1). The location and orientation of the volcanic front (grey line in Figure 2) has been linked to the opening of the Sea of Japan and subduction angle of the Pacific plate [e.g., 26, 33]. From 60 Ma up until about 10 Ma the volcanic front migrated east and west several times, however, it has been relatively static during the last 8 Ma [26].

Presently there are 15 known historically active volcanoes in the Tohoku region and a total 170 volcanoes that formed during the Quaternary [30]. Volcanism has gradually become more clustered and localized over a period from 14 Ma to present [34], thus volcano clustering is a characteristic feature in Tohoku.

http://dx.doi.org/10.5772/51859

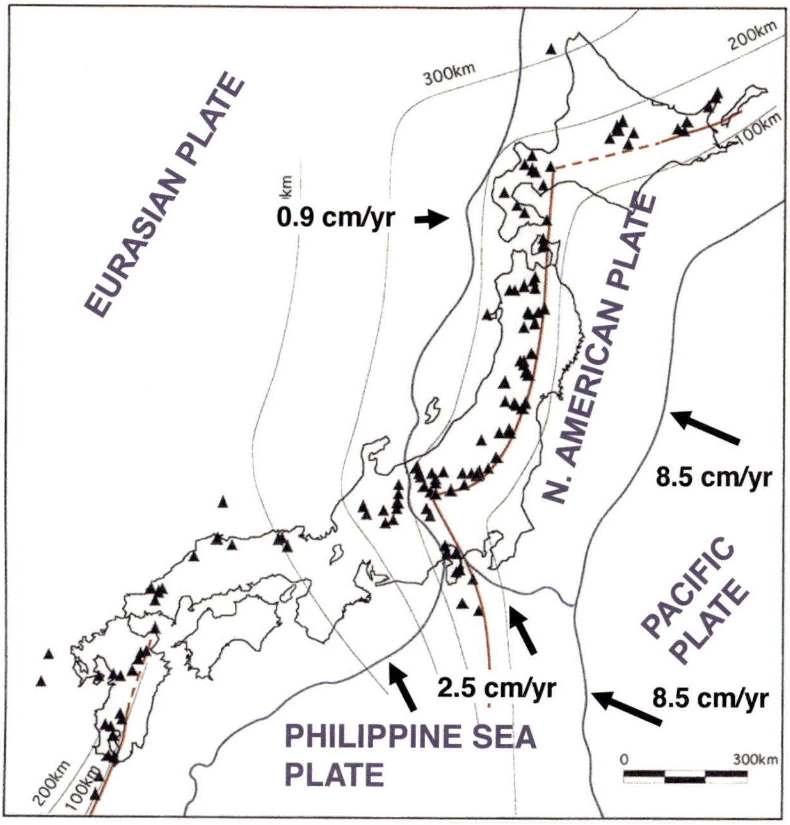

Figure 1. Map showing the tectonic setting of Japan (Figure modified from [1]). The four main islands that make up Japan are located on or near the boundaries of four plates. Black triangles denote Quaternary volcanoes and the red lines depict the main volcanic fronts. The thin contour lines denote the depth of the subducting Pacific Plate beneath Japan. The velocities and arrows indicate the subduction rates and directions respectively of the Pacific, Philippine Sea and Eurasian plates relative to the North American plate.

3. Defining the volcanic event

What do we mean here by 'volcano'? In developing probabilistic based models, one of the most difficult and challenging tasks is defining the 'volcanic event'. This is because the volcanic event defined has to be simple and consistent enough for the probabilistic based models to handle. To a certain extent the degree of consistency that can be realistically included in a model is largely constrained by the size of the study area and by the amount and quality of geological, geochemical and/or geophysical data available. The volcanic event could range from a single eruption to a series of eruptions. It could be defined as the existence of a relatively young cinder cone, spatter mound, maar, tuff ring, tuff cone, pyroclastic fall, lava flow or even a large composite volcano. On the other hand older edifices may have been eroded and/or covered by sedimentary deposits such as alluvium and thus be more difficult to locate and/or

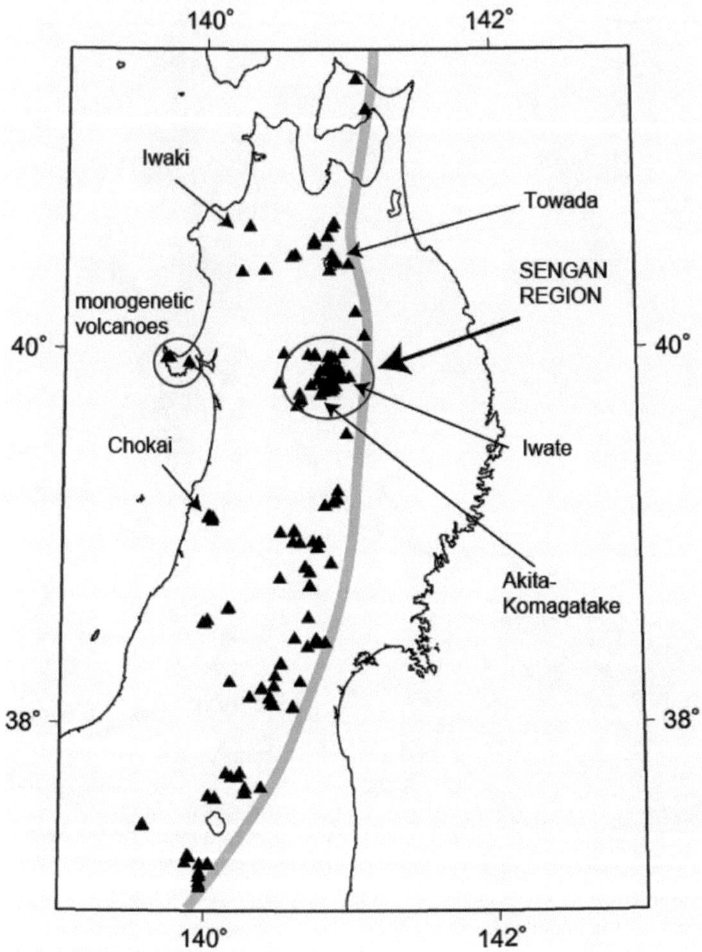

Figure 2. The volcanic arc in Tohoku consists of approximately 170 Quaternary edifices [28, 30] (modified from [1]). The highest density of volcanoes in Tohoku is the cluster in the Sengan region. Other notable volcanoes are the Towada volcano which has been the site of late Quaternary large-volume felsic eruptions resulting in large caldera formation [e.g., 27, 31], and the Iwaki and Chokai [32] volcanoes which are active andesitic volcanoes on the back-arc side of Tohoku. The grey line denotes the present day volcanic front.

are easily overlooked. Results of magnetic and gravity data have been used as evidence for locating such hidden volcanic events which in turn had an impact on resulting probabilities at given locations [e.g. 15].

If we were carrying out a hazard assessment on a single volcano we may be interested in defining the event as a series of pyroclastic flows or surges or eruptions that generate lava flows that exceed a certain volume [e.g. 9]. This is particular relevant to volcanic hazard assessments carried out at volcanoes near densely urbanized areas such as the Campi Flegrei caldera in southern Italy [16].

http://dx.doi.org/10.5772/51859

Several aligned edifices with the same eruption age may also be considered as a single volcanic event. Such vent alignments typically developed simultaneously as a result of magma supply from a single dike. For example the vent alignments in the Higashi-Izu monogenetic volcanic group [e.g. 35], could well be classified as a single volcanic event temporally but spatially are multiple. Where age data has been limited, some authors have implemented a condition whereby a cone or cones can only be defined as a volcanic event if they are associated with a single linear dike or a dike system with more complex geometry [e.g. 36].

Many of the advances made on modelling future spatial or spatio-temporal patterns of volcanism where carried out in monogenetic volcano fields due to the apparent relative ease of defining such volcanoes as point processes [e.g., 37, 38]. However as composite or established polygenetic volcanoes represent multiple eruptions from the same conduit occurring over several tens to hundreds of thousands of years, defining the volcanic event is not so easy if the focus is on single eruption episodes as the type of eruption can evolve significantly during the lifetime of the composite volcano. In fact the temporal definition of a monogenetic volcano appears to be not so straightforward either as this can range from several days to a few weeks or longer. For example the Ukinrek maar in Alaska formed in about eight days [39] and the 1913 eruption forming the Ambrym Volcano, Vanuatu in the south west pacific in just a few days [40]. Moreover, [41] argued that monogenetic volcanoes can be both spatially and temporarily more complex than a single eruptive event. In other words so called 'monogenetic' volcanoes can also be 'polygenetic' albeit smaller scaled than large volume complex strato or caldera volcanoes. Based on this there could be a case to look again at the volcanic event definition used in earlier probabilistic assessments carried out in monogenetic volcanic fields [e.g. 42, 43].

In Tohoku, new volcanoes forming at new locations typically evolve into large complex strato and/or caldera volcanoes containing multiple vents e.g. Akitakomagatake volcano [44]. Such large polygenetic volcanoes in Tohoku have been sub-grouped into unstable types where the eruptive centre has migrated more than 1.5 km within 10 ka and stable types were the vents are more concentrated around the geographic centre of the volcano [44].

The volcanic event definition requires information on both the temporal and spatial aspects; the temporal definition relates to the recurrence rate, λ_t(number of volcanic events per unit time), and spatial definition to the intensity or spatial recurrence rate $\lambda_{x,y}$ (number of volcanic events per unit area). $\lambda_t$and $\lambda_{x,y}$ can also be combined as a spatio-temporal recurrence rate $\lambda_{x,y,t}$ (number of volcanic events per unit area per unit time) [42].

The temporal definition of a volcanic event could range from a single eruption occurring in one day or less, to an eruption cycle in which active periods of eruptions occur between dormant periods. The time scale of an active period may vary from several years to thousands of years. In previous volcanic hazard analyses carried out on complex, large-volume strato and/or caldera volcanoes, volcanologists have typically defined volcanic events as single eruptions or several eruptions within some defined time period separated by periods in which there is no activity e.g. [4]. This is because the focus at such established volcanoes is not on the probability of a new volcano forming in the vicinity of the existing volcano but rather on the probability of the next eruption or eruption phase.

3.1. Tohoku volcanic event definition

In the context of siting of a geological repository, the main concern is the formation of a new volcano in a region where volcanoes do not already exist. Thus the distinction between monogenetic (simple or complex) and polygenetic (complex strato and/or caldera) volcanism is not relevant for the definition of volcanic event here. Table 1 is a compilation of all Quaternary volcanoes in the Tohoku volcanic arc modified from the Catalog of Quaternary Volcanoes in Japan [1, 30]. Volcano complexes refer to magma systems that have evolved over the long-term (order of 0.1 Ma) which appear as regional scale clusters. In this chapter we use the same definition of volcanic event as [1] taking into account eruption volumes. This is depicted as a white triangle in Figure 3 and is the average geographic location of the vents (white dots). The eruption products released from the vents are represented by the dark grey regions in Figure 3. The lighter grey areas in Figure 3a are the eruption products of a separate volcanic event. Each volcanic event typically has a time gap of more than 10 ka, and/or is differentiated from other volcanic events according to geochemistry.

Volcano Complex	Volcanic event	Location		Age (Ma)			Dating Method	Eruptive volume (km^3, DRE)
		Latitude	Longitude					
Mutsuhiuchi-dake	Older Mutsuhiuchi-dake	41.437	141.057		ca.0.73		K-Ar	5.9
Mutsuhiuchi-dake	Younger Mutsuhiuchi-dake	41.437	141.057	0.45		0.2	Strat.	3.6
Osorezan	Kamabuse-yama	41.277	141.123		ca.0.8		K-Ar	11.4
Hakkoda	Hakkoda P.F.1st.	40.667	140.897		0.65		K-Ar	17.8
Hakkoda	South-Hakkoda	40.600	140.850	0.65	–	0.4	K-Ar	52.4
Hakkoda	Hakkoda P.F.2nd.	40.667	140.897		0.4		K-Ar	17.3
Hakkoda	North-Hakkoda	40.650	140.883	0.16	–	0	K-Ar	30.4
Okiura	Aoni F. Aonigawa P.F.	40.573	140.763		ca.1.7		K-Ar	17.6
Okiura	Aoni F. Other P.F.	40.573	140.763	1.7	–	0.9	K-Ar	3.7
Okiura	Okogawasawa lava	40.579	140.759	0.9	-	0.65	Strat.	0.9
Okiura	Okiura dacite	40.557	140.755	0.9	-	0.7	K-Ar	2.1
Ikarigaseki	Nijikai Tuff	40.500	140.625		ca.2.0		K-Ar	20.2
Ikarigaseki	Ajarayama	40.490	140.600	1.91	–	1.89	K-Ar	2.1
Towada	Herai-dake	40.450	141.000					5.1
Towada	Ohanabe-yama	40.500	140.883	0.4	–	0.05	K-Ar	8.9
Towada	Hakka	40.417	140.867					1.4
Towada	Towada Okuse	40.468	140.888		0.055		14C	4.8
Towada	Towada Ofudo	40.468	140.888		0.025		14C	22.1

Volcano Complex	Volcanic event	Location		Age (Ma)			Dating Method	Eruptive volume
		Latitude	Longitude					(km³, DRE)
Towada	Towada Hachinohe	40.468	140.888		0.013		14C	26.9
Towada	Post-caldera cones	40.457	140.913	0.013	–	0	Strat.	14.4
Nanashigure	Nanashigure	40.068	141.112	1.06	–	0.72	K-Ar	55.5
Moriyoshi	Moriyoshi	39.973	140.547	1.07	–	0.78	K-Ar	18.1
Bunamori	Bunamori	39.967	140.717		1.2		K-Ar	0.1
Akita-Yakeyama	Akita-Yakeyama	39.963	140.763	0.5	–	0	K-Ar	9.9
Nishimori/Maemori	NIshimori/Maemori	39.973	140.962	0.5	–	0.3	K-Ar	2.6
Hachimantai/Chausu	Hachimantai	39.953	140.857	1	–	0.7	K-Ar	5.5
Hachimantai/Chausu	Chausu-dake	39.948	140.902	0.85	–	0.75	K-Ar	13.7
Hachimantai/Chausu	Fukenoyu	39.953	140.857		ca.0.7		Strat.	0.2
Hachimantai/Chausu	Gentamri	39.956	140.878					0.2
Yasemori/Magarisaki-yama	Magarisaki-yama	39.878	140.803	1.9	–	1.52	K-Ar	0.3
Yasemori/Magarisaki-yama	Yasemori	39.883	140.828		1.8		K-Ar	0.9
Kensomori/Morobidake	Kensomori	39.897	140.871		ca.0.8		Strat.	0.8
Kensomori/Morobidake	Morobi-dake	39.919	140.862	1	–	0.8	Strat.	2.5
Kensomori/Morobidake	1470m Mt. lava	39.909	140.872					0.1
Kensomori/Morobidake	Mokko-dake	39.953	140.857		ca.1.0		Strat.	0.5
Tamagawa Welded Tuff	Tamagawa Welded Tuffs R4	39.963	140.763		ca.2.0		K-Ar	83.2
Tamagawa Welded Tuff	Tamagawa Welded Tuffs D	39.963	140.763		ca.1.0		K-Ar	32.0
Nakakura/Shimokura	Obuka-dake	39.878	140.883	0.8	–	0.7	K-Ar	2.9
Nakakura/Shimokura	Shimokura-yama	39.889	140.933					0.4
Nakakura/Shimokura	Nakakura-yama	39.888	140.910					0.4
Matsukawa	Matsukawa andesite	39.850	140.900	2.6	–	1.29	K-Ar	11.6
Iwate/Amihari	Iwate	39.847	141.004	0.2	–	0	K-Ar	25.1
Iwate/Amihari	Amihari	39.842	140.958	0.3	–	0.1	K-Ar	10.6
Iwate/Amihari	Omatsukura-yama	39.841	140.919	0.7	–	0.6	K-Ar	3.3
Iwate/Amihari	Kurikigahara	39.849	140.882					0.2
Iwate/Amihari	Mitsuishi-yama	39.848	140.900		0.46		K-Ar	0.6
Shizukuishi/Takakura	Marumori	39.775	140.877	0.4	–	0.3	K-Ar	2.4
Shizukuishi/Takakura	Shizukuishi-Takakura-yama	39.783	140.893	0.5	–	0.4	Strat.	5.2
Shizukuishi/Takakura	Older Kotakakura-yama	39.800	140.900		1.4		K-Ar	2.7
Shizukuishi/Takakura	North Mikado-yama	39.800	140.875					0.3
Shizukuishi/Takakura	Kotakakura-yama	39.797	140.907	0.6	–	0.5	K-Ar	1.8
Shizukuishi/Takakura	Mikado-yama	39.788	140.870		ca.0.3			0.2

Volcano Complex	Volcanic event	Location		Age (Ma)			Dating Method	Eruptive volume
		Latitude	Longitude					(km³, DRE)
Shizukuishi/Takakura	Tairagakura-yama	39.808	140.878		ca.0.3			0.1
Nyuto/Zarumori	Tashirotai	39.812	140.827	0.3	–	0.2	K-Ar	0.6
Nyuto/Zarumori	Sasamori-yama	39.770	140.820	0.23	–	0.1	K-Ar	0.4
Nyuto/Zarumori	Yunomori-yama	39.772	140.827		ca.0.3			0.5
Nyuto/Zarumori	Zarumori-yama	39.788	140.850		0.56		K-Ar	0.9
Nyuto/Zarumori	Nyutozan	39.802	140.843	0.58	–	0.5	K-Ar	5.0
Nyuto/Zarumori	Nyuto-kita	39.817	140.855		ca.0.4		K-Ar	0.1
Akita-Komagatake	Akita-Komagatake	39.754	140.802	0.1	–	0	K-Ar	2.9
Kayo	Kayo	39.803	140.735	2.2	–	1.17	K-Ar	5.9
Kayo	KoJiromori	39.828	140.787		0.94		K-Ar	0.3
Kayo	Akita-Ojiromori	39.839	140.788	1.7	1.7	1.7	Strat.	0.3
Innai/Takahachi	Takahachi-yama	39.755	140.655	1.7	1.7	1.7	K-Ar	0.0
Innai/Takahachi	Innai	39.692	140.638	2	–	1.6	K-Ar	0.5
Kuzumaru	Aonokimori andesites	39.543	140.983		2.06		K-Ar	0.3
Yakeishi	Yakeishidake	39.161	140.832	0.7	–	0.6	K-Ar	9.5
Yakeishi	Komagatake	39.193	140.924		ca.1.0		K-Ar	7.6
Yakeishi	Kyozukayama	39.178	140.892	0.6	–	0.4	K-Ar	5.7
Yakeishi	Usagimoriyama	39.239	140.924	0.07	–	0.04	K-Ar	2.3
Kobinai	Kobinai	39.018	140.523	1	–	0.57	FT, K-Ar	2.3
Takamatsu/Kabutoyama	Kabutoyama Welded Tuff	39.025	140.618		1.16		TL	3.2
Takamatsu/Kabutoyama	Kiji-yama Welded Tuffs	39.025	140.618		0.30		K-Ar	5.1
Takamatsu	Takamatsu	38.965	140.610	0.3	–	0.27	K-Ar	3.8
Takamatsu	Futsutsuki-dake	38.961	140.661		ca.0.3			0.8
Kurikoma	Tsurugi-dake	38.963	140.792	0.1	–	0	K-Ar	0.2
Kurikoma	Magusa-dake	38.968	140.751	0.32	–	0.1	K-Ar	1.5
Kurikoma	Kurikoma	38.963	140.792	0.4	–	0.1	K-Ar	0.9
Kurikoma	South volcanoes	38.852	140.875		ca.0.5		K-Ar	0.3
Kurikoma	Older Higashi Kurikoma	38.934	140.779		ca.0.5		K-Ar	2.2
Kurikoma	Younger Higashi Kurikoma	38.934	140.779	0.4	–	0.1	K-Ar	0.7
Mukaimachi	Mukaimachi	38.770	140.520		ca.0.8		K-Ar	12.0
Onikobe	Shimoyamasato tuff	38.830	140.695	0.21	0.21	0.21	FT	1.0
Onikobe	Onikobe Centeral cones	38.805	140.727		ca.0.2		TL	1.1
Onikobe	Ikezuki tuff	38.830	140.695	0.3	–	0.2	FT	17.3
Naruko	Naruko Central cones	38.730	140.727		ca.0.045		14C	0.1
Naruko	Yanagizawa tuff	38.730	140.727		ca.0.045		FT	4.8

Volcano Complex	Volcanic event	Location		Age (Ma)			Dating Method	Eruptive volume
		Latitude	Longitude					(km^3, DRE)
Naruko	Nizaka tuff	38.730	140.727		ca.0.073		FT	4.8
Funagata	Izumigatake	38.408	140.712	1.45	–	1.14	K-Ar	2.3
Funagata	Funagatayama	38.453	140.623	0.85	–	0.56	K-Ar	19.0
Yakuraisan	Yakuraisan	38.563	140.717	1.65	–	1.04	K-Ar	0.2
Nanatsumori	Nanatsumori lava	38.430	140.835	2.3	–	2	K-Ar	0.5
Nanatsumori	Miyatoko Tuffs	38.428	140.793		ca.2.5		Strat.	6.1
Nanatsumori	Akakuzure-yama lava	38.433	140.768	1.6	–	1.5	Strat.	1.5
Nanatsumori	Kamikadajin lava	38.447	140.772	1.6	–	1.5	K-Ar	0.8
Shirataka	Shirataka	38.220	140.177	1	–	0.8	K-Ar	3.8
Adachi	Adachi	38.218	140.662		ca.0.08		FT	0.9
Gantosan	Gantosan	38.195	140.480	0.4	–	0.3	K-Ar	4.6
Kamuro-dake	Kamuro-dake	38.253	140.488		ca.1.67		K-Ar	5.7
Daito-dake	Daito-dake	38.316	140.527					5.7
Ryuzan	Ryuzan	38.181	140.397	1.1	–	0.9	K-Ar	4.6
Zao	Central Zao 1st.	38.133	140.453	1.46	–	0.79	K-Ar	0.8
Zao	Central Zao 2nd.	38.133	140.453	0.32	–	0.12	K-Ar	15.2
Zao	Central Zao 3rd.	38.133	140.453	0.03	–	0	K-Ar	0.0
Zao	Sugigamine	38.103	140.462		1		K-Ar	9.9
Zao	Fubosan/byobudake	38.093	140.478	0.31	–	0.17	K-Ar	15.2
Aoso-yama	Gairinzan	38.082	140.610	0.7	–	0.4	K-Ar	6.1
Aoso-yama	Central Cone	38.082	140.610	0.4	–	0.38	K-Ar	3.0
Azuma	Azuma Kitei lava	37.733	140.247	1.3	–	1	K-Ar	24.7
Azuma	Higashi Azumasan	37.710	140.233	0.7	–	0	K-Ar	22.8
Azuma	Nishi Azumasan	37.730	140.150	0.6	–	0.4	K-Ar	7.2
Azuma	Naka Azumasan	37.713	140.188	0.4	–	0.3	K-Ar	4.6
Nishikarasugawa andesite	Nishikarasugawa andesite	37.650	140.283		ca.1.5		K-Ar	1.9
Adatara	Adatara Stage 1	37.625	140.280	0.55	–	0.44	K-Ar	0.3
Adatara	Adatara Stage 2	37.625	140.280		ca.0.35		K-Ar	0.4
Adatara	Adatara Stage 3a	37.625	140.280		ca.0.20		K-Ar	2.0
Adatara	Adatara Stage 3b	37.625	140.280	0.12	–	0.0024	K-Ar	0.3
Sasamori-yama	Sasamari-yama andesite	37.655	140.391	2.5	–	2	K-Ar	0.4
Bandai	Pre-Bandai	37.598	140.075		ca.0.7		K-Ar	0.1
Bandai	Bandai	37.598	140.075	0.3	–	0	Strat.	14.0
Nekoma	Old Nekoma	37.608	140.030	1	–	0.7	K-Ar	11.4
Nekoma	New Nekoma	37.608	140.030	0.5	–	0.4	K-Ar	0.9

Volcano Complex	Volcanic event	Location		Age (Ma)			Dating Method	Eruptive volume (km^3, DRE)
		Latitude	Longitude					
Kasshi/Oshiromori	Kasshi	37.184	139.973					0.1
Kasshi/Oshiromori	Oshiromori	37.199	139.970					0.7
Kasshi/Oshiromori	Matami-yama	37.292	139.886					0.3
Kasshi/Oshiromori	Naka-yama	37.282	139.899					0.0
Shirakawa	Kumado P.F.	37.242	140.032		1.31		K-Ar	19.2
Shirakawa	Tokaichi A.F. tuffs	37.242	140.032	1.31	–	1.24	Strat.	12.0
Shirakawa	Ashino P.F.	37.242	140.032		1.2		FT	19.2
Shirakawa	Nn3 P.F.	37.242	140.032	1.2	–	1.17	Strat.	0.0
Shirakawa	Kinshoji A.F. tuffs	37.242	140.032	1.2	–	1.18	Strat.	9.0
Shirakawa	Nishigo P.F.	37.252	139.869		1.11		FT	28.8
Shirakawa	Tenei P.F.	37.242	140.032		1.06		Strat.	7.7
Nasu	Futamata-yama	37.244	139.971		0.14		K-Ar	3.2
Nasu	Kasshiasahi-dake	37.177	139.963	0.6	–	0.4	K-Ar	12.3
Nasu	Sanbonyari-dake	37.147	139.965	0.4	–	0.25	K-Ar	5.5
Nasu	Minamigassan	37.123	139.967	0.2	–	0.05	K-Ar	8.7
Nasu	Asahi-dake	37.134	139.971	0.2	–	0.05	K-Ar	4.6
Nasu	Chausu-dake	37.122	139.966	0.04	–	0	K-Ar	0.3
Chokai	Shinsan Lava flow	39.097	140.053	0.02	–	0	Strat.	
Chokai	Higashi Chokai	39.097	140.053	0.02	–	0.02	K-Ar	4.3
Chokai	Nishi Chokai	39.097	140.020	0.09	–	0.02	Stra.	0.8
Chokai	Nishi ChokaiII	39.097	140.020	0.13	–	0.01	K-Ar	21.0
Chokai	Old Chokai	39.103	140.030	0.16	–	0.55	K-Ar	67.0
Chokai	Uguisugawa Basalt	39.103	140.030	0.55	–	0.6	K-Ar	1.0
Chokai	Tengumari volcanics	39.103	140.031	0.55	–	0.6	K-Ar	11.0
Gassan	Ubagatake	38.533	140.005	0.400	–	0.300	K-Ar	3.5
Gassan	Yudonosan lavas/ pyroclastics	38.534	139.988	0.800	–	0.700	K-Ar	7.5
Gassan	Gassan	38.550	140.020	0.500	–	0.400	K-Ar	18.0
Numazawa	Sozan lava domes	37.452	139.577					
Numazawa	Mizunuma pyroclastic dep.	37.452	139.577		ca.0.05		FT	2.0
Numazawa	Numazawa pyroclastics	37.452	139.577	0.005	–	0.005	14C	2.5
Numazawa	Mukuresawa lava	37.452	139.577		–			
Ryuzan	Ryuzan	38.181	140.397	1.130	–	0.940	K-Ar	6.0
Sankichi-Hayama	Sankichi-Hayama	38.137	140.315	2.400	–	2.300	K-Ar	2.9
Daitodake	Daitodake	38.316	140.527		ca.1		Strat.	7.5

Volcano Complex	Volcanic event	Location		Age (Ma)			Dating Method	Eruptive volume
		Latitude	Longitude					(km^3, DRE)
Hijiori	Hijiori Pyroclastic flow	38.610	140.159		ca.0.01		Strat.	1.0
Hijiori	Komatsubuchi lava dome	38.613	140.171		ca.0.01		Strat.	0.0
Tazawa	Tazawa (lake)	39.723	140.667					
Daibutsu	Daibutsu	39.817	140.517	2.340	–	2.160	K-Ar	3.2
Kampu	Kampu	39.928	139.877	0.030	–	0.000	Strat.	0.6
Toga	Toga	39.950	139.718		ca.0.42		FT/K-Ar	2.0
Megata	Megata	39.952	139.742	0.030	–	0.020	Strat.	0.1
Inaniwa	Inaniwa	40.195	141.050	7.000	–	2.700	K-Ar	14.0
Taira-Komagatake	Taira-Komagatake	40.410	140.254	0.200	–	0.170	Strat.	3.0
Tashiro	Tashiro	40.425	140.413	0.600	–	0.470	K-Ar	9.0
Tashiro	Hirataki nueeardente deps.	40.420	140.413	0.020	–	0.020	Strat.	0.7
Iwaki	Iwaki	40.653	140.307	0.330	–	0.000	K-Ar	49.0

Table 1. Tohoku volcanic arc [30, 28]. Dense-rock equivalent (DRE) of eruptive volumes is the product of volume and density of the respective volcanic deposits.

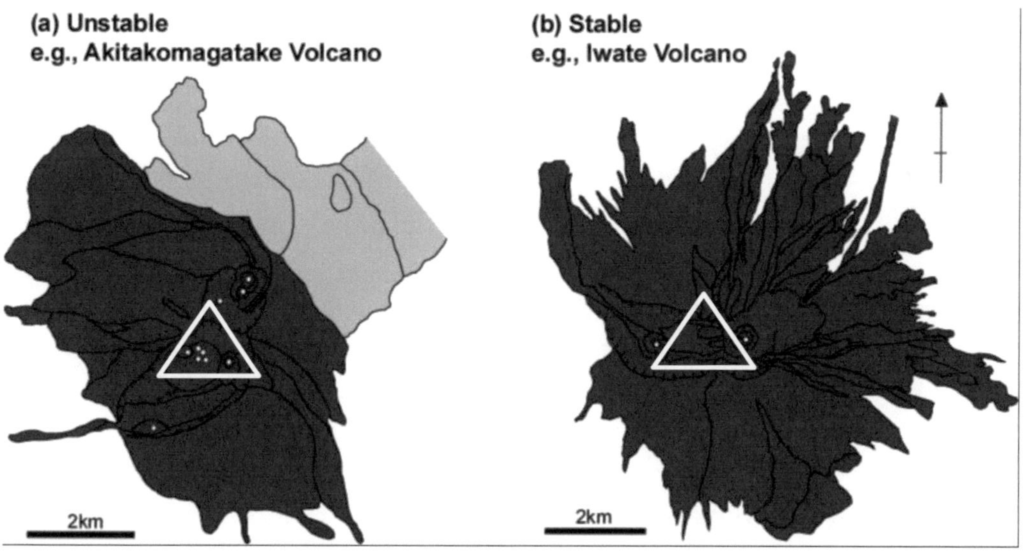

Figure 3. Two volcano types in Tohoku classified according to migration distance from eruption centre [44]: (a) an unstable type with vent (white dots) migration exceeding 1.5 km in 10 ka resulting in a summit with multiple peaks; and (b) a stable type commonly with a narrow saw-tooth or pointed appearance. For consistency in volcanic event definition over the Tohoku region, both types are treated as a single volcanic event (white triangle) in the probability analysis. However they are also optionally weighted with the corresponding eruption products (dark grey regions). (Figure modified from [1]).

4. Bayesian model

The following is a slightly shorter description of the Bayesian methodology published in [1]. A two-dimensional surface distribution is set-up showing the continuous probability of one or more volcanic event(s) forming within a region of interest, in an arbitrarily time frame of the order of 0.1 – 1 Ma. The volcanic event definition defined above means that we are estimating with known uncertainty, the probability of a new volcano forming at a given location (x, y). [1] noted that a challenge with estimating the long-term future spatial distribution of volcanism is the fact that we are trying to model something that we cannot sample directly; namely the locations of future volcanoes. In this chapter we incorporate $^3He/^4He$ ratios, as these may be indicative of conduits in the earth's crust through which magma may rise through resulting in future volcano formation [19, 20].

Information, no matter how obtained, can be described by a probability density function (PDF) [e.g. 45, 46]. Once the dataset is expressed as a PDF, it is possible to combine with our initial PDF created based on *a priori* assumptions on volcanism. Bayesian inference is a powerful tool that allows us to construct an a *posteriori* PDF given *a priori* assumptions and the PDF generated by our new dataset.

Essentially, two stages are performed yielding the *a posteriori* PDF. The first is to make a long-term future prediction based solely on the distribution and ages of past volcanic events, creating an *a priori* PDF. The *a priori* assumption is that the past and the present provide information about the future; in other words the locations of past and present volcanism are used as an initial guide to estimating future long-term spatial patterns of volcanism. The basic logic behind the *a priori* assumption is that a new volcano doesn't form far from existing volcanoes. The *a priori* assumption can be quite vague in the first step as it is simply the starting point. The next stage is to update or modify the *a priori* assumptions by incorporating information that is likely to be indicative of the locations of future volcanism and/or we have increased our understanding of the process that controls the location of volcanism. This new information and/or knowledge, obtained from chemical and/or geophysical data, is used to modify the *a priori* PDF to form an *a posteriori* PDF that is expected to better reflect the location of future volcanism. The cycle can be repeated any number of times for other datasets by treating the *a posteriori* PDF as the new *a priori* PDF in the first step above.

4.1. Bayesian inference and Bayes' theorem

Bayes' theorem [e.g. 47] is used to setup a model providing a joint probability distribution for the location known volcanic events (*a priori* PDF) and current R/R_A contoured datasets recast as a PDF (likelihood function). The joint probability density function or *a posteriori* PDF can be written as the product of two PDFs; the *a priori* PDF and the sampling or likelihood PDF

$$\mathbf{P}(x,y \mid \theta) = \frac{\mathbf{P}(x,y)\mathbf{L}(\theta \mid x,y)}{\int_A \mathbf{P}(x,y)\mathbf{L}(\theta \mid x,y)dA} \tag{1}$$

where x and y represent grid point locations within the volcanic field A, $\theta$is additional dataset, $\boldsymbol{P}(x, y)$is the *a priori* PDF, $\boldsymbol{L}(\theta \mid x, y)$the likelihood function generated by conditioning additional data on the locations of volcanic events, and $\boldsymbol{P}(x, y \mid \theta)$ the resulting *a posteriori* PDF [1]. The *a posteriori* PDF is normalized to unity by integrating over the entire Tohoku volcanic field; hence total cumulative probability will not change but the shape of the 2-D surface distribution will be modified according to the likelihood function.

4.2. A priori PDF

We assume that past and present volcanic events can be used to estimate future locations of volcanoes over the long-term, as well as constraining upper bound recurrence rates in the volcanic field. The spatial distribution of volcanoes in volcanic arcs like Tohoku are random [48] hence by treating volcanism in Tohoku volcanic arc as a low frequency, random event, it is assumed that the underlying process could be approximated to a Poisson process [1]. Moreover, by treating the location of volcanic events as random points within some set, the spatial distribution of volcanism can be modeled as a spatial point process [1] where a spatial point process is a stochastic model that can be described as the process controlling the spatial locations of the events$\boldsymbol{s}_1,\ldots,\boldsymbol{s}_1$ in some arbitrary set $\boldsymbol{S}$ [49]. In applying point process models to volcanism, [42] eloquently defined$\boldsymbol{s}_1,\ldots,\boldsymbol{s}_n$ as volcanic events and $\boldsymbol{S}$ as the volcanic field.

The Poisson process is 'homogeneous' if the spatial distribution of point events are completely random [49]. However, as with many volcanic fields, spatial patterns of volcanism in the Tohoku volcanic arc are clustered [34, 50], hence the distribution of volcanoes are not completely random and therefore non-homogeneous (also referred to as in-homogenous). Applying the Clark-Evans nearest-neighbour test [51], [1] showed that the distribution of the volcanic events defined above is clustered with greater than 95% confidence. A non-homogeneous Poisson process is the simplest alternative for modeling such clustered events. Moreover, point process models based on non-homogeneous Poisson processes have been extensively used in modeling the spatial and spatio-temporal characteristics of several volcano fields (e.g. the Springerville volcanic fields in Arizona [38] and the Higashi-Izu monogenetic volcano group, Shizuoka Prefecture, Japan [43]. In these models the local spatial density of volcanic events $\lambda_{x,y}$ is calculated using a kernel function [37, 52]. The kernel function itself is a density function used to obtain the intensity of volcanic events at a sampling pointx_p, y_p, calculated as a function of the distance to nearby volcanoes and a smoothing constant h (Figure 4).

As noted by [42] the choice of kernel function with appropriate values of h has some consequence for the parameter estimation because it controls how $\lambda_{x,y}$ varies with distance from existing volcanoes. The Gaussian kernel has been used a lot in probabilistic assessments carried out in monogenetic volcanic fields, [e.g. 15, 43] since it was assumed that the next volcano to form would not be far from an existing volcanoes. In order to include extreme volcanic events further afield however, [1] modelled spatial patterns using the Cauchy kernel which has thicker tails than the Gaussian kernel. [1] also showed that the spatial distribution of volcanic events in the Tohoku volcanic arc fit a Cauchy distribution whereas monogenetic fields such as the Higashi-Izu Monogenetic Volcano Group [43] tend to be Gaussian. We therefore also use a two-dimensional Cauchy kernel here to calculate the spatial recurrence rate $\lambda_{x,y}$ at grid point x_p, y_p where:

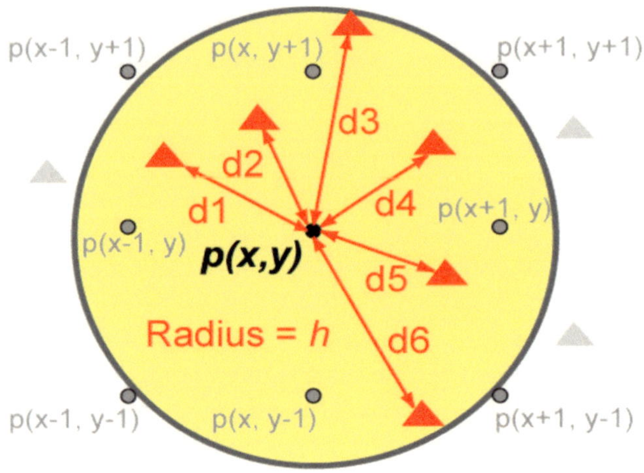

Figure 4. Local volcano intensity $\lambda_{x,y}$ at each grid point (*x*, *y*) is computed using a Cauchy kernel function. $\lambda_{x,y}$ is a function of volcano distance from grid point (*x*, *y*) for *N* = 6 volcanoes (modified from [1]).

$$\lambda_{x,y} = \frac{1}{\pi h^2 N} \sum_{i=1}^{N} \left\{ \frac{l_{vi}}{1 + \left[\left(\frac{x_p - x_{vi}}{h} \right)^2 + \left(\frac{y_p - y_{vi}}{h} \right)^2 \right]} \right\} \tag{2}$$

x_{vi}, y_{vi} are Cartesian coordinates of the *i*th volcanic event, *N* the number of volcanic events used in the calculation and l_{vi} is a factor for weighting eruption volume of the corresponding *i*th volcanic event. l_{vi} is set to unity when eruption volume is excluded. The calculation is repeated on a 10 km mesh in the study area 139 to 143 longitude and 37 to 41.6 latitude and the resulting PDF is normalized to unity. The 10 km grid spacing was selected taking into account the resolution of available geophysical or geochemical datasets across the entire Tohoku volcanic arc.

4.3. Estimating an optimum smoothing coefficient h for the volcanoes in Tohoku

The choice of the smoothing coefficient depends on a combination of the size of the volcanic field, size and degree of clustering and the amount of robustness and conservatism required at specific points within or nearby the volcanic fields in question. In order to estimate the most likely optimum value of smoothing coefficient, [1] plotted cumulative probability density functions with varying values of smoothing are compared with the fraction of volcanic vents and nearest-neighbour volcanic event distances in Tohoku (Figure 5).

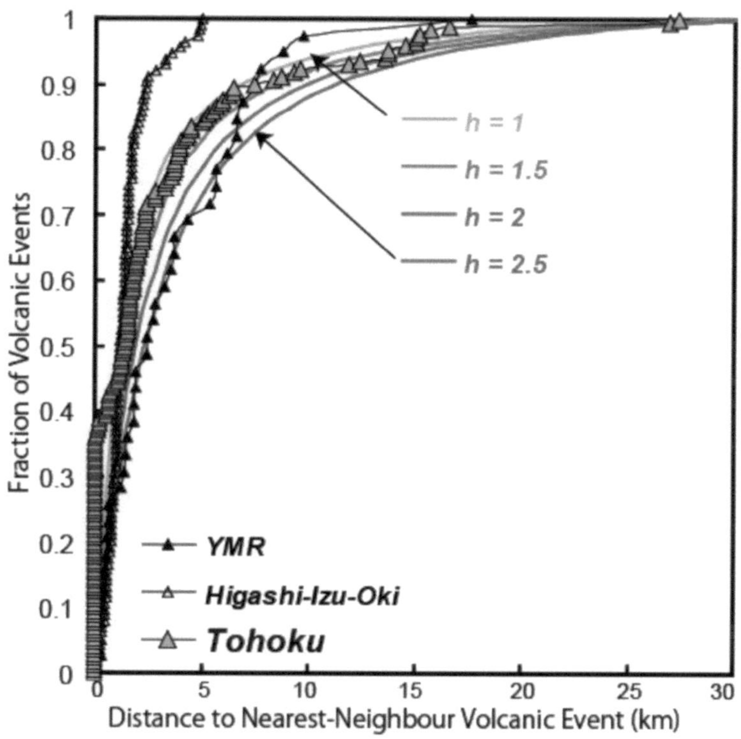

Figure 5. Suitable values of smoothing coefficient *h* are estimated by plotting cumulative distances to nearest neighbour volcanic events and cumulative probability distribution with differing values of smoothing coefficient. From this plot, suitable values of smoothing coefficient for known volcanic events in Tohoku volcanic are estimated to be 1-1.5 km for the Cauchy kernel. (Figure modified from [1]). As a comparison, the monogenetic volcanoes in the Yucca Mountain Region (YMR) and the Higashi-Izu-Oki monogenetic volcano group are also plotted.

The cumulative plots in Figure 5 suggest that the spatial distribution of volcanic events in the Tohoku volcanic arc fit a Cauchy distribution with smoothing coefficients of $h = 1–1.5$ km.

4.4. A priori probabilities

Probability estimates for each grid point x_p, y_p are computed by using a Poisson distribution where $\lambda_{x,y}$ represents the intensity parameter computed using equation (2) :

$$P_{x,y}\left\{N(t) \geq 1\right\} = 1 - \exp\left(-t\lambda_t\lambda_{x,y}\Delta x\Delta y\right) \tag{3}$$

where, $N(t)$ represents the number of future volcanic vents that occur within time t and area $\Delta x \Delta y$ (10 km x 10 km). The parameter $\lambda_{x,y}$ is normalized to unity across the Tohoku, so, equation (3) represents the probability of one or more volcanic event(s) forming in an area $\Delta x \Delta y$centred on point x_p, y_p given the formation of a new volcanic event in Tohoku. This calculation is repeated on a grid throughout Tohoku. The resolution is such that the spatial recurrence rate $\lambda_{x,y}$does not vary within each cell. For the regional recurrence rateλ_t an average of 120 volcanic events per million years is used, effectively taking average Quaternary activity [1].

Using smoothing coefficients of 1 - 1.5 km for the Cauchy kernel, as well as weighting eruption volumes, probability plots were constructed using equation (3). A probability contour plot for one case is shown in Figure 6.

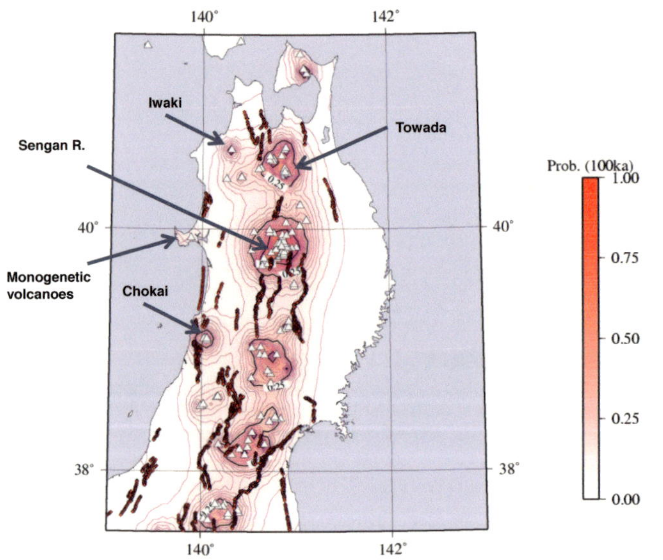

Figure 6. Probabilities of one or more volcanic events occurring in the next 100 ka based on *a priori* PDF (Cauchy (h = 1.5 km). White triangles denote the volcanic events used in the calculation and black lines are active faults [35]

The highest probabilities are located in the Sengan region (10^{-6} - 10^{-5} / a) which has the highest density of volcanic events in the Tohoku volcanic arc. By testing the two volcanic event subdefinitions (weighted with and without eruption volume), [1] found that the probabilities in the vicinity of monogenetic volcanoes on the back-arc region were higher when volcanic events were not weighted with eruption volumes (1 - 4 x 10^{-7}/a, weighted; 1 - 4 x 10^{-6}/a, unweighted), whereas the probabilities around established centers such as Iwaki, Towada, Sengan and Chokai were reduced slightly. This is expected as volcanoes with large eruption volumes are the sites of highest magma production. However if the focus of the assessment is on new volcano event formation, irrelevant of whether the new volcano evolves into are large complex stratovolcano and/or caldera or not, then selecting the volcanic event definition that is not weighted with eruption volume would seem more appropriate.

4.5. The likelihood function

Here the *a priori* PDF is conditioned $^3He/^4He$ ratios. This is done by normalizing additional data into a likelihood function according to how such information is judged by the expert and/or indicated by experimental result to relate to the distribution of volcanism [1]. Helium isotopes have been shown to provide evidence for the presence of mantle derived materials in the crust, and hence potential volcanism based on distinct $^3He/^4He$ ratios (Figure 7) [17, 18]. [1] looked at seismic tomographs and geothermal gradients. This is because *P* velocity perturbations ($\Delta V / V$) in particular at 40 km depth [29] is a good estimate of the minimum depth of partial melting in the mantle for most of the volcanoes in Tohoku. Geothermal gradients on the other hand were used by [1] as an additional aid to *P* velocity perturbations since it is not possible to differentiate heat from *P* wave velocity alone.

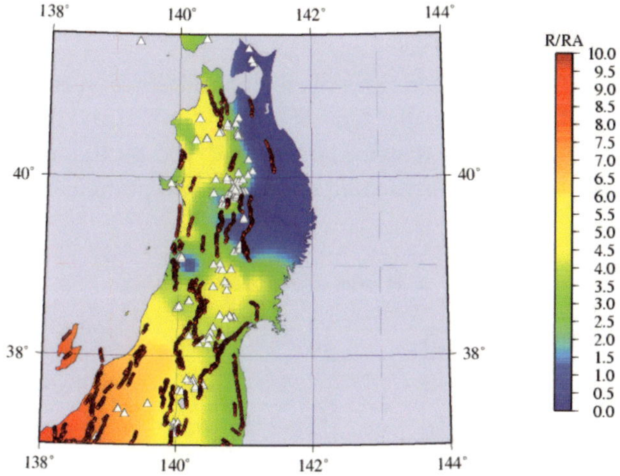

Figure 7. Distribution of R/R_A data (Ra denotes the atmospheric $^3He/^4He$ ratio of 1.4×10^{-6}) taken from boreholes and hot springs [20, 54, 55]

In order to compare the R/R_A ratios, cumulative plots of values around all volcanic events and values of 10 km^2 bins over all of Tohoku are plotted. Figure 8 shows R/R_A ratios below all volcanic events (8a) and volcanic events less than 100 ka (8b). In both cases approximately 90% of all volcanic events are distributed in regions with R/R_A ratios greater than 3. In other words 90% volcanoes are located in regions where $^3He/^4He$ is elevated.

The R/R_A ratios are interpolated to represent a continuous, differentiable surface and then the spatial data are mapped into a likelihood function based on the percentage of recent volcanic events that lie within the binned R/R_A ratios in Figure 8. For low P velocity perturbation, [1] assumed an inverse linear relationship; based on the interpretation that low *P* velocity perturbation corresponds to partial melting (and hence increased probability of volcanism). In this case, 10% of volcanoes less than 100 ka located in regions where $\Delta V / V$ ranged from -6% to -5% etc. For geothermal gradients [1] used a linear relationship for recasting the data values as a PDF.

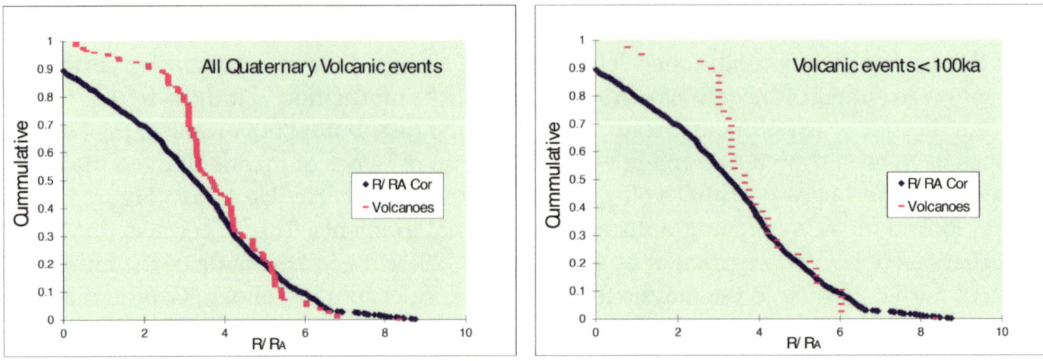

Figure 8. umulative plots of R/R_A values below volcanic events and the whole of Tohoku (10km² grid spacing).

4.6. A posteriori probabilities

Finally the *a posteriori* PDF is calculated from the likelihood function and the *a priori* PDF using equation (1). The integral across the entire field of both the *a priori* and the *a posteriori* PDFs is set to unity; however the shape of the distribution is modified by the likelihood function. The probability of a new volcanic event is calculated for each grid point using equation (3).

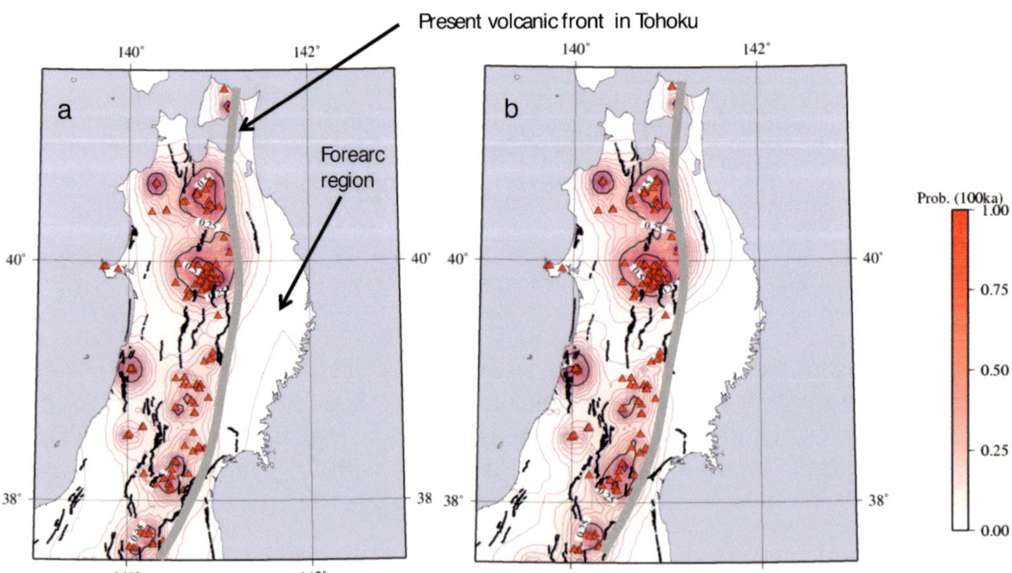

Figure 9. Comparison of *a priori* (a) and *a posteriori* probability plots (b) calculated with a Cauchy kernel (h = 1.5 km) conditioned on R/R_A ratios. Both PDFs a weighted by eruption volume. Black lines are active faults [53].

Using equations (1) to (3) above, two dimensional probability plots are subsequently constructed showing the probability of one or more future volcanic event(s) forming during the long-term, given that a volcanic event will occur in the Tohoku volcanic arc during 100 ka. Figure 9 shows a comparison of the *a priori* probability (9a) and a posteriori probability (9b) conditioned on R/R_A ratios of one or more volcanic events forming in 100 ka.

The probability of new volcanic event formation in the forearc region to the east of the volcanic front is reduced slightly in the *a posterior* probability calculation. This is more evident when we repeat the calculation for new volcanic events forming in the next 1Ma (Figure 10).

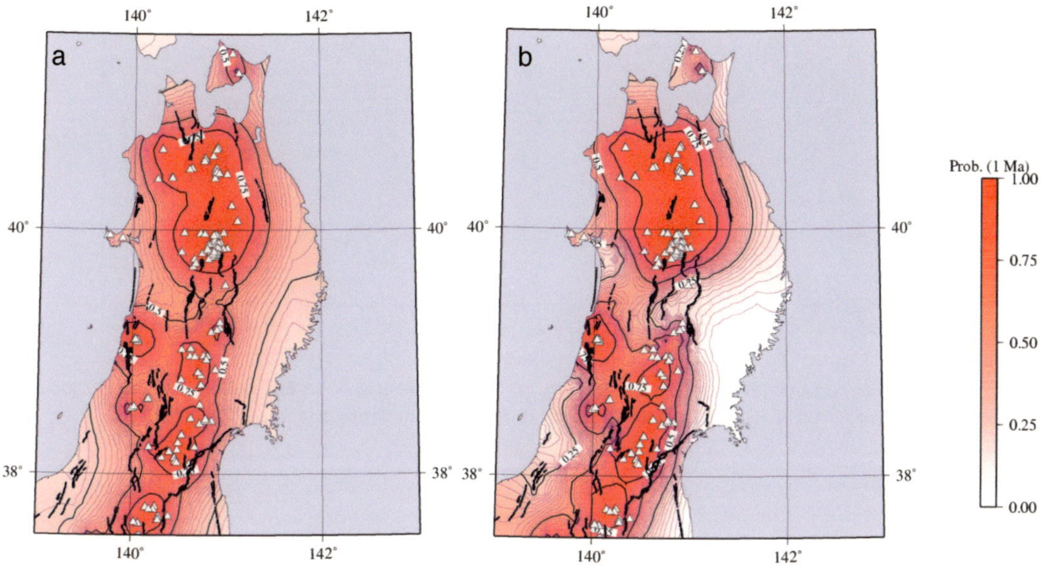

Figure 10. Probability of the formation of a new volcanic event over the next 1Ma; *a priori* (a) and *a posteriori* (b) probability plots calculated with a Cauchy kernel (*h* = 1.5 km) conditioned on R/R_A ratios.

The R/R_A analyses are compared with the probability calculations conditioned on *P* velocity perturbations (10 or 40 km) (Figure 11) and geothermal gradients (Figure 12) [1]. The *a posteriori* probability below Iwaki volcanic event is particularly low when conditioned on 40 km depth P velocity perturbation datasets but that there was no significant change beneath Chokai, another andesitic volcano on the back-arc side of Tohoku. With the *a posteriori* calculation conditioned on the R/R_A analyses, no decrease is seen in the probabilities below Iwaki volcano when compared to *a priori* plots. Similar results can be seen when probabilities are conditioned on 10km depth P velocity perturbations (Figure 11a) or geothermal gradients (Figure 12) [1].

[1] found that *a posteriori* probabilities are not reduced when compared to *a priori* probabilities in the northern regions when conditioning on shallower (10 km) *P* velocity perturbations or on geothermal gradients. This seems reasonable as seismic velocity structure [57] and the depth of Curie isotherms [58] in this part of Tohoku reveal high-temperature-like geophysical anomalies at depths of up to 10 km below Iwaki volcano which may be indicative of the shallower depths (ca. 10km) of magma chambers.

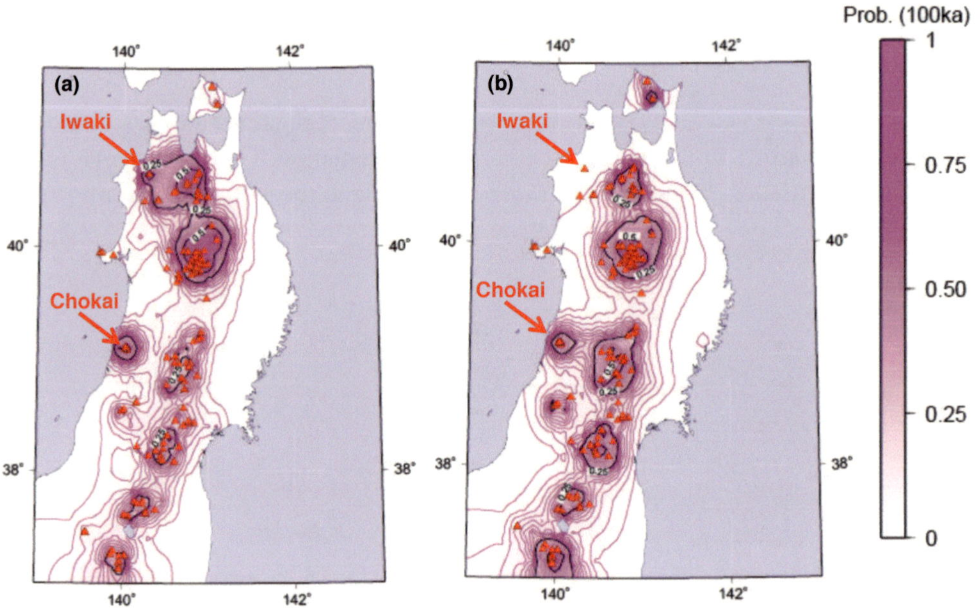

Figure 11. *A posteriori* probability plots calculated with: (a) Cauchy kernel (h = 1.5 km) conditioned on ΔV / V at 10 km depth; (b) Cauchy kernel (h = 1.5 km) conditioned on ΔV / V at 40 km depth (modified from [1]).

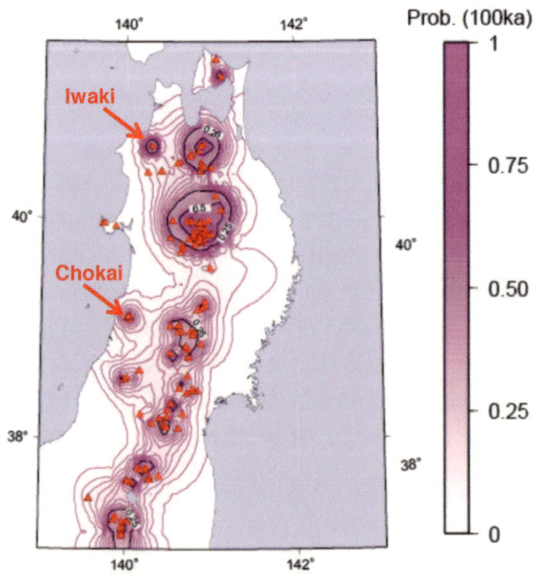

Figure 12. *A posteriori* probability plots calculated with Cauchy kernel (h = 1.5 km) conditioned on geothermal gradients (modified from [1])

5. Discussion

The main advantage of probabilistic based models over deterministic models is that the probability of new volcano event formation is never zero. [1] showed that Bayesian inference is well-suited for formally combining observations relevant to the imaging of the magma source region (e.g. seismic tomography) with quantitative methods for estimation of volcano intensity. Moreover, the strength of Bayesian inference is that probabilistic assessments can be improved with increased understanding of the physical processes governing magmatism and/ or data that may be indicative of future volcanism such as the helium isotope ratios presented here. Nevertheless it is worth examining the logic behind what we perceive to be 'data' and what we mean by *a priori* information and knowledge.

5.1. Which datasets are a priori information?

[1] used the volcano geographical datasets themselves as a starting point in their analysis. The same approach was applied in this chapter. In the first step a Cauchy kernel was used to calculate$\lambda_{x,y}$. This means that the probability new volcanic event formation decreases with increasing distance from existing volcanic events. In the case of selecting a location for a geological repository, there may be a need to have a conservative estimate and accept that extreme events may occur. In this case, selection of the Cauchy as the *a priori* PDF would be most appropriate due to the thickness of the tails. This is especially the case if we have to make probability calculations for periods for 1Ma where the tectonic setting can change, and we may have a shift in the location of the volcanic front. The probabilities in distal regions would only be reduced in the *a posterior* probability calculation if newly obtained evidence in such regions shows that volcano formation is zero or close to zero. Since R/R_A ratios vary due to the heterogeneous release of mantle helium and elevated ratios and are likely to indicate the presence of partial melting [e.g. 20] datasets may give some indication on the future location of volcanism even in non-volcanic regions. Seismic tomography on the other hand offers a direct view of the mantle that can be interpreted in terms of degree of partial melting [e.g. 58, 59].

It could be equally argued, however that the logic of [1] should be reversed in that the models based on seismic tomography or elevated helium isotope ratios are in fact *a priori* information or knowledge, and the location of volcanic events the 'data'. The philosophy here is that we assume new volcanic events will form in regions where partial melting is likely to be occurring now and that the distribution of known volcanic events are the datasets updating our model and/or knowledge. However, this may be true for the very recent volcanism up to about 1,000 years say, but how relevant are volcanic events that formed over 100,000 years ago or more to the present day geophysical snap-shot of the Earth's crust or upper mantle? This question is difficult to answer as there is very little information on the temporal behaviour of partial melting in the mantle. This is also evident when we try to evaluate our forecasts below. A problem here is that we are always trying to predict the formation of future volcanic events which may or may not be related to historic volcanic events. Our closest 'data' to such future events are thus present day geophysical snap shots of the current conditions in the crust or upper mantle and/or newly formed or forming volcanic events. This has been the motivation for [1] to use such geophysical data or models as the basis of the likelihood function.

On the other hand there are also practical aspects to be considered particularly when starting a hazard analysis in a region where there have not been many studies. In such a case, the only data available to begin with might be just the geographical location of volcanoes. Information from more complicated and expensive surface based investigations might not come until later.

5.2. Model evaluation

Since it is not possible to infer directly the location of future volcanic events that will form in the next 0.1 to 1 Ma from now, models can instead by evaluated by calculating the probability of the new volcanic events that formed after some time in the past, using all volcanic events that formed before that time [1, 38]. Since we calculate the probability of future volcanism in the next 100 ka in most of the analyses described here, 100 ka is selected as the timeframe in the verification calculations. In Tohoku, as there are a large number of dated volcanic events it is possible to verify the Bayesian models developed to a certain extent by using all volcanic events that formed before 100 ka to predict the location of volcanic events that formed between 100 ka and the present day. Since the 'new' volcanic events are still in the past, it is possible to compare probability plots with the locations of volcanic events we are attempting to forecast. Figure 13 shows probability plots for the Cauchy PDF (h=1.5 km) and the *a posterior* probability conditioned on R/R_A ratios. All volcanic events that formed before 100 ka (white triangles) during the Quaternary were used to make a forecast for the period from 100 ka ago to the present day. All subsequent volcanic events that formed during the forecast period are shown in red. Probability calculations are then compared with the locations of volcanoes that formed during the forecast period.

In both cases, all subsequent volcanoes formed in regions where the probability was at least 10%. Approximately 50% of newly formed volcanic events formed in regions where the probability was at least 25%. There was approximately 10% increase in probabilities in the locations were volcanoes formed in the *a posteriori* probability calculations.

Probability calculations above were made using single inferences on one set of data. However, Bayes' theorem allows beliefs to be updated as additional information becomes available. [1] attempted this by combining geothermal and seismic tomography datasets (Figure 14).

By conditioning on *P* velocity perturbations at 40 km depth, the model assigned a low probability for the Iwaki volcano which formed in region where probability was calculated to be low ($< 10^{-9}$/a). This could be improved upon by including both P velocity perturbations at 10 km depth and geothermal gradients [1].

5.3. Varying the temporal recurrence rate

The temporal recurrence rates in Tohoku have been steady state from 0.5 Ma to present [28]. This implies that recurrence rates are likely to remain steady state for at least the next 0.1 Ma. However if we need to assess volcanism over a much longer time frame such as 1.0 Ma more

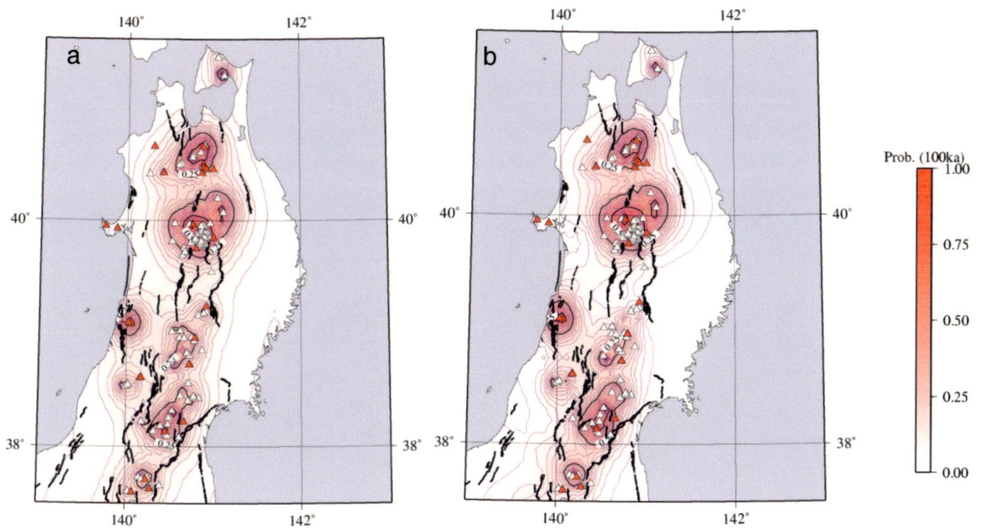

Figure 13. Verification probability plots calculated using all volcanic events before 100 ka (white triangles) in order to predict the subsequent distribution of volcanic events that formed from 100 ka to present (red triangles) for (a) the *a priori* probability (Cauchy, h=1.5km, eruption volume weighting included) and the *a posteriori* probability (b) conditioned on R/R_A ratios.

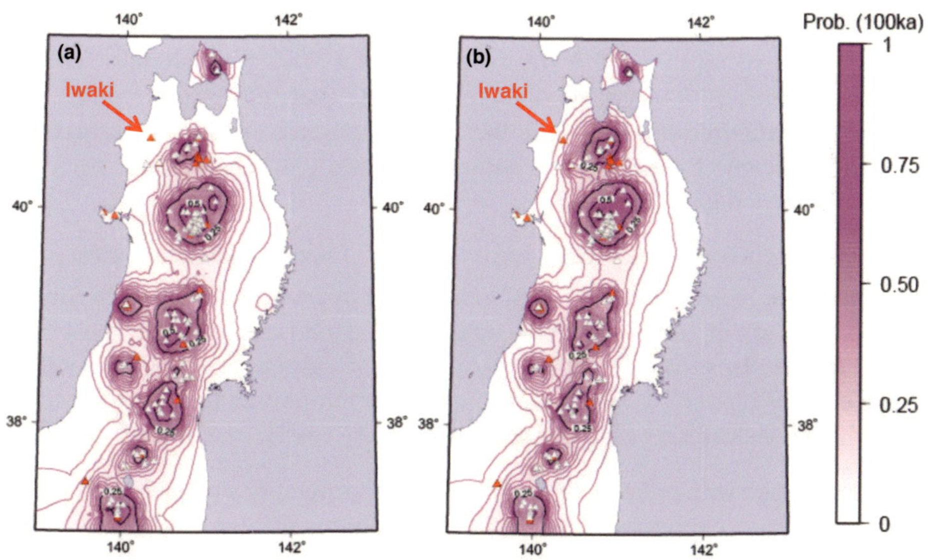

Figure 14. Verification probability plots calculated using all volcanic events before 100 ka (white triangles) in order to predict the subsequent distribution of volcanic events that formed from 100 ka to present (red triangles): (a) Cauchy kernel (h = 1.5 km) conditioned on ΔV / V at 40 km depth, (b) Cauchy kernel (h = 1.5 km) conditioned on geothermal gradient and ΔV / V at 10 and 40 km depths. (Figure modified from [1]).

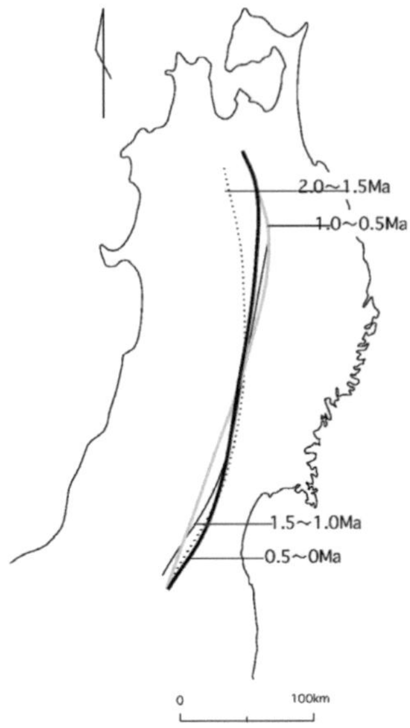

Figure 15. Shift in the volcanic front in Tohoku (compiled from [60])

care is needed. In addition to temporal recurrence rates, the type of volcanism can also change over extended timeframes. For example, [28] used eruptive volumes of volcanic products along the volcanic front in Tohoku to identify three sub-stages with distinct types of volcanism and volumetric changes in the last 2.0 Ma. From 2.0 to 1.2 Ma large-scale felsic eruptions were predominant; during 1.2 to 0.5 Ma, the crustal stress changed to compression yielding the formation of strato-volcanoes all along the Tohoku volcanic arc. Finally, from 0.5 Ma to the present day, volcanically active areas became localized [34]. The volcanic front also shifted over a 2.0 Ma period [60] (Figure 15)

It can thus be argued that for periods beyond 0.1Ma, it is unreasonable to treat $\lambda_t$in equation (3) as constant or steady state. One option might be to assign say a Weibull function where recurrence rates can increase or decrease with time [61] if there is sufficient age data to indicate temporal trends statistically. Alternatively one could assume that the temporal recurrence rates are entirely random with a tendency to cluster temporally [e.g. 22, 62]. Moreover, [22] showed that time clustering can have an impact on the spatial intensity of volcanoes.

A challenge though with utilizing temporal data are the quantity and quality of the age datasets and being consistent enough with the temporal definitions since eruptions may last for several days, weeks, months, years even longer. Having a consistent temporal definition is especially challenging when handling volcanic datasets on the regional scale described in this chapter. As highlighted in section 3, even for monogenetic volcanoes, the temporal definition is not so straightforward [41]. It was for this reason [42] argued that a drawback with nearest-neighbour models which are a function of both spatial and temporal parameters is that they require the ages of every single volcanic event within the volcanic field in question. Nevertheless in certain cases such as tectonically controlled basaltic fields, eruptions can be time predictable, [63] hence there is potential to improve on the Bayesian model presented here by taking into account time clustering in the temporal rate parameter.

6. Conclusions

Bayes' thereom is a powerful statistical tool for incorporating additional datasets. In this chapter R/R_A ratios were used in probabilistic volcanic hazard assessments applying the methodology developed by [1]. These were compared with earlier assessments in Tohoku incorporating low P perturbations at 10km and 40km depth and geothermal gradients. Probabilities of one or more volcanic event(s) forming in Tohoku for both analyses were found to be similar ranging from $10^{-10} - 10^{-9}$ /a between clusters and 10^{-5} /a within clusters. The Cauchy kernel, combined with multiple datasets successfully captures all subsequent volcanic events, including extreme events. This is particularly important when making calculations over 1Ma when the tectonic setting is likely to change resulting in a potential shift of the volcanic front. Although the Cauchy kernel appears to be over conservative for regions east of the volcanic front, where probabilities are expected to be negligible, values are reduced when R/R_A ratios are included.

Acknowledgements

Diagrams of the probability plots were made using Generic Mapping Tools (GMT) [64]. The authors thank the constructive comments made by two anonymous reviewers which improved the manuscript.

Author details

Andrew James Martin[1], Koji Umeda[2] and Tsuneari Ishimaru[2]

1 National Cooperative for the Disposal of Radioactive Waste (NAGRA), Wettingen, Switzerland

2 Japan Atomic Energy Agency (JAEA), Toki, Japan

References

[1] Martin, A. J., Umeda, K., Connor, C. B., Weller, J. N., Zhao, D. and Takahashi, M. (2004), Modeling long-term volcanic hazards through Bayesian inference: An exam-

ple from the Tohoku volcanic arc, Japan. *J. Geophys. Res., 109,* B10208, doi: 10.1029/2004JB003201.

[2] Wickman, F. E. (1966), Repose period patterns of volcanoes, *Ark. Mineral. Geol., 4,* 291-301.

[3] Wadge, G. (1982), Steady state volcanism: Evidence from eruption histories of polygenetic volcanoes, *J. Geophys. Res. 87,* 4035-4049.

[4] Klein, F. W. (1984), Eruption forecasting at Kilauea Volcano, Hawaii, *J. Geophys. Res., 89,* 3059-3073.

[5] Sornette, A., J. Dubois, J. L. Cheminee, and D. Sornette (1991), Are sequences of volcanic eruptions deterministically chaotic? *J. Geophys. Res., 96,* 11,931-11,945.

[6] Dubois, J., and J. L. Cheminee (1991), Fractal analysis of eruptive activity of some basaltic volcanoes, *J. Volcanol. Geotherm. Res., 45,* 197-208.

[7] Pyle, D. M. (1998), Forecasting sizes and repose times of future extreme volcanic events, *Geology, 26,* 367-370.

[8] Wadge, G., P. A. V. Young, and I. J. McKendrick (1994), Mapping lava flow hazards using computer simulation. *J. Geophys. Res. 99,* 489-504.

[9] Connor L. J., C. B. Connor, K. Meliksetian and I. Savov (2012). Probabistic approach to modeling lava flow inundation: a lava flow hazard assessment for a nuclear facility in Armenia. *J. App. Volc. 1,* 3-19.

[10] Crowe, B. M., M. E. Johnson, and R. J. Beckman (1982), Calculation of the probability of volcanic disruption of a high-level radioactive waste repository within southern Nevada, USA. *Radioact. Waste Manage. Nucl, Fuel Cycle, 3,* 167-190.

[11] International Atomic Energy Agency (1997), Volcanoes and associated topics in relation to nuclear power plant siting, a safety guide, *Provisional Safety Stand. Ser.* 1, 49 pp., Vienna, Austria.

[12] Weller, J. N., A. J. Martin, C. B. Connor, and L. Connor (2006), Modelling the spatial distribution of volcanoes: An example from Armenia, in *Statistics in Volcanology*, edited by Mader, H. M., Coles, S. G., Connor, C. B. and Connor, L. J, pp. 296, Geol. Soc. Lon. on behalf of IAVCEI.

[13] McBirney, A., L. Serva, M. Guerra and C. B. Connor (2003), Volcanic and seismic hazards at a proposed nuclear power site in central Java, *J. Volcanol. Geotherm. Res., 126,* 11-30.

[14] Woods A. W., Sparks S., Bokhove, O., LeJeune A. M., Connor C. B. and Hill B. E. (2002), Modeling magma-drift interaction at the proposed high-level radioactive waste repository at Yucca Mountain, Nevada, USA. Geophys. Res. Lett. 29 (13), 1641, doi:10.1029/2002GL014665

[15] Connor, C. B., J. A. Stamatakos, D. A. Ferrill, B. E. Hill, I. Goodluck,, F. Ofoegbu, M. Conway, S. Budhi, and J. Trapp (2000), Geologic factors controlling patterns of small-volume basaltic volcanism: Application to a volcanic hazards assessment at Yucca Mountain, Nevada, *J. Geophys. Res., 105,* 417-432.

[16] Selva J., Orsi G., Di Vito M. A., Marzocchi W., Sandri L. (2012) Probability hazard map for future vent opening at the Campi Flegrei caldera, Italy. *Bull. Volc.* 74: 497-510

[17] Ozima, M., and F. A. Podosek (2002), *Noble Gas Geochemistry*, 2nd ed., 286 pp., Cambridge Univ. Press, New York.

[18] Hilton, D. R. (2007), Geochemistry - The leaking mantle, *Science, 318,* 1389-1390.

[19] Sano, Y., and H. Wakita, (1985), Geographical distribution of $^3He/^4He$ ratios in Japan: Implications for arc tectonics and incipient magmatism, *J. Geophys. Res., 90,* 8729-8741.

[20] Umeda, K., K. Asamori, A. Ninomiya, S. Kanazawa, and T. Oikawa (2007), Multiple lines of evidence for crustal magma storage beneath the Mesozoic crystalline Iide Mountains, northeast Japan. *J. Geophys. Res., 112,* B05207, doi:10.1029/2006JB004590.

[21] Marti J., and Felpeto A. (2010), Methodology for the computation of volcanic susceptibility An example for mafic and felsic eruptions on Tenerife (Canary Islands). *J. Volcanol. Geotherm. Res.,* 195, 69-77.

[22] Jaquet, O, C. Lantuejoul, and J. Goto (2012), Probabilistic estimation of long-term volcanic hazard with assimilation of geophysics and tectonic data. *J. Volcanol. Geotherm. Res.* 235-236: 29-36, doi:10.1016/j.jvolgeores.2012.05.003.

[23] Hasegawa, A., N. Umino, and A. Takagi (1978), Double-planed structure of the deep seismic zone in the northeastern Japan arc, *Tectonophysics, 47,* 43-58.

[24] Nakagawa, M., H. Shimotori, and T. Yoshida (1986), Aoso-Osore volcanic zone- The volcanic front of the northeast Honshu arc, Japan, *Journal of the Japan Association of Mineralogy and Economic Geology, 81,* 471-478.

[25] Hasegawa, A., S. Horiuchi, and N. Umino (1994), Seismic structure of the northeastern Japan convergent margin: A synthesis, *J. Geophys. Res., 99,* 22,295-22,311.

[26] Yoshida, T., T. Oguchi, and T. Abe (1995), Structure and evolution of source area of the Cenozoic volcanic rocks in Northeast Honshu arc, Japan, *Memoirs of the Geological Society of Japan, 44,* 263-308.

[27] Takahashi, M. (1995), Large-volume felsic volcanism and crustal strain rate, *Bulletin of the Volcanological Society of Japan, 40,* 33-42.

[28] Umeda, K., S. Hayashi, M. Ban, M. Sasaki, T. Oba, and K. Akaishi (1999), Sequence of volcanism and tectonics during the last 2.0 million years along the volcanic front in Tohoku district, NE Japan, *Bulletin of the Volcanological Society of Japan, 44,* 233-249.

[29] Zhao, D., F. Ochi, A. Hasegawa, and A. Yamamoto (2000), Evidence for the location and cause of large crustal earthquakes in Japan, *J. Geophys. Res., 105,* 13,579-13,594.

[30] Committee for Catalog of Quaternary Volcanoes in Japan (1999), eds. Catalog of Quaternary volcanoes in Japan [CD-ROM], *The Volcanological Society of Japan.*

[31] Hayakawa, Y. (1985), Pyroclastic geology of Towada volcano, *Bulletin of the Earthquake Research Institute, University of Tokyo, 60,* 507-592.

[32] Ban, M., S. Hayashi and N. Takaoka, K-Ar dating of the Chokai volcano, northeast Japan arc: A compound volcano composed of continuously established three stratovolcanoes, 46, 317 – 333.

[33] Oki, J., N. Watanabe, K. Shuto, and T. Itaya (1993), Shifting of the volcanic fronts during Early to Late Miocene age in the Northeast Japan arc, *The Island Arc, 2,* 87-93.

[34] Kondo, H., K. Kaneko, and K. Tanaka (1998), Characterization of spatial and temporal distribution of volcanoes since 14 Ma in the northeast Japan arc, *Bulletin of the Volcanological Society of Japan, 43,* 173-180.

[35] Koyama, M., Y. Hayakawa, and F. Arai (1995), Eruptive history of the Higashi-Izu Monogenetic Volcano field 2: Mainly on volcanoes older than 32,000 years ago, *Kazan, 40,* 191-209.

[36] Sheridan, M. F. (1992), A Monte Carlo technique to estimate the probability of volcanic dikes, paper presented at Third International Conference on High-Level Radioactive Waste Management, Am. Nucl. Soc., La Grange Park, Ill., 2033-2038.

[37] Lutz, T. M., and J. T. Gutmann (1995), An improved method for determining and characterizing alignments of point like features and its implications for the Pinacate volcanic field, Sonora, Mexico, *J. Geophys. Res. 100,* 17,659-17,670.

[38] Condit, C. D., and C. B. Connor (1996), Recurrence rates of volcanism in basaltic volcanic fields: An example from the Springerville volcanic field, Arizona, *Geol. Soc. Am. Bull., 108,* 1225-1241.

[39] Kienle J., P. R. Kyle, S. Self, R. J. Motyka and V. Lorenz, (1980), Ukinrek Maars, Alaska: I, April 1977 eruption sequence, petrology and tectonic setting: *J. Volcanol. Geotherm. Res., 7,* 11-37.

[40] Nemeth K. and S. J. Cronin (2011), Drivers of explosivity and elevated hazard in basaltic fissure eruptions: The 1913 eruption of Ambrym Volcano, Vanuatu (SW-Pacific). *Jour. Volcanol. Geoth. Res., 201,* 194-209, doi: 10.1016/j.jvolgeores.2010.12.007.

[41] Kereszturi, G., K. Nemeth, G. Csillag, K. Balogh and J. Kovacs (2011). The role of external environmental factors in changing eruption styles of monogenetic volcanoes in a Mio/Pleistocene continental volcanic field in western Hungary *J. Volcanol. Geotherm. Res.,* 201, 227-240. doi: 10.1016/j.jvolgeores.2010.08.018.

[42] Connor, C. B., and B. E. Hill (1995), Three nonhomogenous Poission models for the probability of basaltic volcanism: Application to the Yucca Mountain region, Nevada. *J. Geophys. Res., 100,* 10,107-10,125.

[43] Martin, A. J., M. Takahashi, K. Umeda, and Y. Yusa (2003), Probabilistic methods for estimating the long-term spatial characteristics of monogenetic volcanoes in Japan, *Acta. Geophys., Pol.,51,* 271-291.

[44] Takahashi, M. (1994), Structure of polygenetic volcano and its relation to crustal stress field: 1. Stable and unstable vent types, *Bulletin of the Volcanological Society of Japan, 39,* 191-206.

[45] Tarantola, A. (1990), Probabilistic Foundations of Inverse Theory, in *Oceanographic and Geophysical Tomography,* edited by Y. Desaubies, A. Tarantola and J. Zinn-Justin, pp. 1-27, Elsivier Science Publishers B. V.

[46] Debski, W. (2004), Application of Monte Carlo techniques for solving selected seismological inverse problems, *Publications of the Institute of Geophysics, Polish Academy of Sciences, B-34 (367),* 207 pp., Warszawa, Poland.

[47] Gelman, A., J. B. Carlin, H. S. Stern, and D. B. Rubin (1995), *Bayesian Data Analysis,* 526 pp., Chapman and Hall/CRC, Boca Raton.

[48] de Bremond d'Ars, J., C. Jaupart, and R. S. J. Sparks (1995), Distribution of volcanoes in active margins, *J. Geophys. Res., 100,* 20,421-20,432.

[49] Cressie, N. A. C. (1991), *Statistics for Spatial Data,* 900 pp., John Wiley, New York.

[50] Hayashi, S., K. Umeda, M. Ban, M. Sasaki, M. Yamamoto, T. Oba, K. Akaishi, and T. Oguchi (1996), Temporal and spatial distribution of Quaternary volcanoes in northeastern Japan (1) - Spreading of volcanic area toward back-arc side – *Program and Abstract, Volcanological Society of Japan,* no. 2, 71-71.

[51] Clark, P. J., and F. C. Evans (1954), Distance to nearest neighbour as a measure of spatial relationships in populations. *Ecology, 35,* 445-453.

[52] Diggle, P. J. (1985), A kernel method for smoothing point process data, *Appl. Statist., 34,* 138-147.

[53] AIST (2009), Active fault database of Japan, version 2009. Research Information Database DB095, National Institute of Advanced Industrial Science and Technology. http://riodb02.ibase.aist.go.jp/activefault/index_e.html

[54] Wakita, H., and Y. Sano (1983), $^{3}He/^{4}He$ ratios in CH_4-rich natural gases suggest magmatic origin, *Nature, 305,* 792-794.

[55] Sano, Y., Y. Nakamura, and H. Wakita (1985), Areal distribution of $^{3}He/^{4}He$ ratios in the Tohoku district, Northeastern Japan, *Chem. Geol., 52,* 1-8.

[56] Nakajima, J., T. Matsuzawa, A. Hasegawa and D. Zhao (2001), Three-dimensional structure of Vp, Vs and Vp/Vs beneath northeastern Japan: Implications for arc magmatism and fluids, J. Geophys. Res., 106, 21,843 – 21,857, doi:10.1029/2000JB000008

[57] Tamanyu S., K. Sakaguchi, T. Sato and M. Kato (2008), Integration of geological and geophysical data for extraction of subsurface thermal and hydrothermal anomaly areas – Examples in Tohoku and Chugoku/Shikoku districts, Japan, *Bulletin of the Geological Survey of Japan, 59,* 7-26.

[58] Zhao, D., A. Hasegawa, and S. Horiuchi (1992), Tomographic imaging of P and S wave velocity structure beneath northeastern Japan, *J. Geophys. Res., 97,* 19,909-19,928.

[59] Zhao, D. (2001), Seismological structure of subduction zones and its implications for arc magmatism and dynamics, *Phys. Earth Planet. Int., 127,* 197-214.

[60] Umeda. K., H. Osawa, T. Nohara, E. Sasao, O. Fujiwara, K. Asamori, and N. Nakatsuka (2005), Current status of the geoscientific research for long-term stability of the geological environment in JNC's R& D programme, *Journal of Nuclear Fuel Cycle and Environment,* 11, 97-112.

[61] Ho, C-H. (1991), Nonhomogenous Poisson model for volcanic eruptions, *Mathematical Geology, 23,* 167-173.

[62] Jaquet, O., S. Low, B. Martinelli, V. Dietrich, and D. Gilby (2000), Estimation of volcanic hazards based on Cox stochastic processes, *Physics and Chemistry of the Earth, Part A: Solid Earth and Geodesy, 25,* 571-579.

[63] Valentine, G. A. and F. Perry (2007). Tectonically controlled, time-predictable basaltic volcanism from a lithospheric mantle source (central Basin and Range Province, USA). Earth Planet. Sci. Lett., 261, 201-216, doi: 10.1016/j.epsl.2007.06.029.

[64] Wessel, G. P. L., and W. H. F. Smith (1998), New improved version of the generic mapping tools released, *Eos, 79,* 579.

Section 3

Volcanic Resources and Geoconservation

Chapter 4

Volcanic Rock-Hosted Natural Hydrocarbon Resources: A Review

Jiaqi Liu, Pujun Wang, Yan Zhang, Weihua Bian,
Yulong Huang, Huafeng Tang and Xiaoyu Chen

Additional information is available at the end of the chapter

http://dx.doi.org/10.5772/54587

1. Introduction

Evolution and the hydrocarbon bearing capacity of basins are closely related to volcanic activity, and not only source rock maturity, but also hydrocarbon trapping are influenced by volcanism within a basin. Volcanic rocks act as important basin filling material in different types of basins, for instance, rift basins, epicontinental basins, basins in a trench-arc system, back-arc foreland basins, etc. [1]. Volcanic accumulation of oil and gas is a new global field of hydrocarbon exploration and has been proved in more than 300 basins in 20 countries and regions [2]. The Cenozoic volcanic rocks, especially Jurassic, Cretaceous and Tertiary, contribute about 70% of the total preservation globally [3-7].

Derivations of hydrocarbon in volcanic accumulation have organic as well as inorganic sources [8-10]. Volcanic rocks could act as a reservoir or cover within hydrocarbon traps, whose thermal effects could accelerate the maturity of source rocks or destroy preserved hydrocarbon [11-13]. Primary hydrocarbon accumulations could be reformed or destroyed during tectonic and volcanic processes, the preserved hydrocarbon remobilized to other traps or the ground surface [14]. Effective reservoirs have been found in most lithology [15-17]. Lithofacies, including deposits and rocks formed by explosive, effusive, extrusive and subvolcanic processes, could bear hydrocarbon, and the facies combination close to a volcanic conduit shows better porosity and permeability due to an increased number of fractures and reservoir spaces, or an elevated volume of coarse-grained fragmented rocks [7, 18]. Reservoir spaces within volcanic rocks are composed of primary pores, secondary pores and fissures with significant heterogeneity [19]. Tectonism, weathering and fluid saturation and/or movement could modify reservoir space [6, 20-23]. Upward cover and lateral seal of volcanic rocks could form hydrocarbon traps [24]. Lateral distribution of volcanic rocks can be mapped based on aeromagnetic and gravity data [25]. The reflective seismic features of volcanic rocks are summarized [26] and visualized [12] based on qualified seismic data. Volcanic edifice is identified by trend surface analysis and spectrum imaging methods [27]. A volcanic reservoir has been predicted based on seismic wave impedance [28].

Over the last half a century the Songliao Basin (Figure 1) has been the most productive basin in China for hydrocarbon and the Xujiaweizi fault depression has been proved as a typical volcanic accumulation. Based mainly on the achievements of hydrocarbon exploration in volcanic rocks in the Songliao Basin, the authors reviewed the hydrocarbon-related volcanic impacts, volcanic lithofacies and key geophysical techniques for volcanic accumulation exploration.

2. Volcanism impacts on the formation of oil and gas accumulation

2.1. Volcanic activity provides a catalyst for the evolution of organic matter

During the transformation from organic matter to hydrocarbon, the role of volcanic is mainly to supply a catalyst and thermal energy. Volcanogenic zeolite and olivine can be a catalyst in turning organic matter into hydrocarbon [29]. Hydrothermal liquid contains many transition metals, such as Ni, Co, Cu, Mn, Zn, Ti, V etc. [30]. The transition metals are catalysts for organic matter thermal degradation [31]. Studies have shown that some volcanic minerals undergo catalysis and hydrogenation which can produce more oil and gas source rocks at lower temperature and pressure. Jin [32] performed a catalysis and hydrogenation experiment on volcanic minerals and source rocks. He used zeolite as a catalyst collected from volcanic rocks, olivine as intermediates of accelerating hydrogen generation and type II and type III organic matter as source rocks. The experimental results show that the hydrogen production rate increased after olivine addition, while when adding zeolite and olivine hydrogen, the production rate still improved. This is due to olivine alteration occuring and reacting with water to produce hydrogen in the organic matter into hydrocarbon conversion process. The reaction is as follows:

$$6(Mg_{1.5}Fe_{0.5})SiO_4 + 13H_2O \rightarrow 3Mg_3SiO_2O_5(OH)_4 + Fe_3O_4 + 7H_2$$

The results show that after the source rocks interact with zeolite and olivine, the production rate of methane improved 2 to 3 times, which is related to the hydrogen increasing. The results also show that the better the organic matter or kerogen types, the higher the production rate of hydrogen and methane. However, the catalytic minerals are not only the clay minerals such as zeolite, but also pyrite.

Pyrite can be found in kerogen commonly, whose mass fraction is closely related to the type of kerogen. The data analysis of the Songliao Basin from Zhang et al. [33] shows that pyrite mass fraction is up to 38% ~ 76% in type I kerogen, which is 10% to 30% in type II kerogen and 2.5% to 3.5% in type III kerogen, i.e., the better the kerogen type, the higher the pyrite mass fraction. Electron microscopy reveals that type I and type II kerogen often closely coexist with pyrite and kerogen, exhibiting obvious zoning around pyrite [34]. Kerogen forms have some

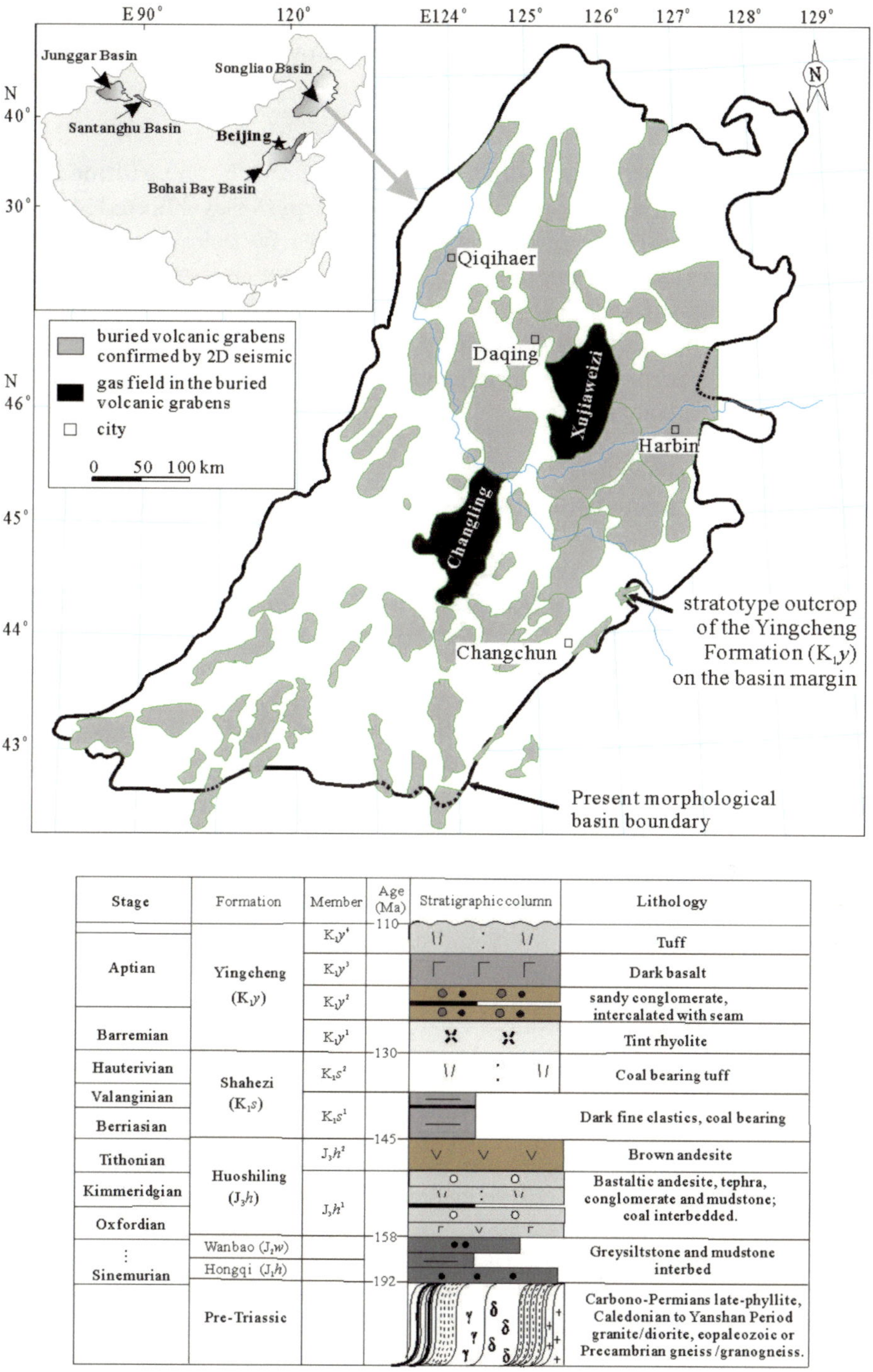

Figure 1. Geological structure and stratigraphic framework of the fault-depression sequence of the Songliao Basin, China

connection with pyrite forms, different kerogen forms and different pyrite forms: type III kerogen is a mainly contour shape without microcrystal pyrite inclusion; type III kerogen is mainly amorphous with rich microcrystal pyrite inclusion; type II kerogen form is a mixed type whose relative proportion of components change greatly.

Pyrite is the most widely distributed sulphide in the crust, which can be formed in a variety of different geological conditions [35]. Copper containing a pyrite layer hosted in volcanic rock series is the largest pyrite mass fraction deposit, formed by volcanic sedimentation and hydrothermal processes [36]. Although most pyrite in the kerogen is usually of organic origin, pyrite mass fraction in kerogen will undoubtedly be affected by volcanism. Sulphur-rich material provided by the volcanism can increase the sulphur content in an aqueous medium participating in the formation of pyrite related to kerogen, while volcanic rocks or pyroclastic rocks can form pyrite directly [37].

2.2. Volcanic activity provide thermal energy for the evolution of organic matter

The thermal effect of volcanic activity has a dual function on organic matter hydrocarbon generation, which can accelerate the maturation of immature source rocks and hydrocarbon generation [38], and also can make mature source rocks over-mature or destroy oil and gas reservoirs formed earlier [39]. The temperature of magma can be more than 1,000 degrees Celsius [40-41], which of hydrothermal fluid can be up to 300-400 Celsius degree, making it a carrier with large amounts of heat energy. The heat will accelerate the maturity of organic matter [42-44]. Studies at the Illinois Basin by Schimmelmann et al. [45] have shown that R_o values increased from 0.62% to 5.03% within 5 metres at the coal contact to large intrusion, while R_o values increased from 0.63% to 3.71% within 1 metre at the coal contact to small intrusion. George [46] found that intrusion made R_o rise from 0.55% to 6.55% when he investigated the maturity of the Scottish Midland Valley oil shale. Raymond and Murchison [47] found that vitrinite reflectance was significantly higher around bedrock in the carboniferous strata, Midland Valley, Scotland.

The research data show that the igneous body's effect on organic matter maturity incidence is relevant to the igneous body's size. Carslaw and Jaeger [48] thought that the sphere of influence of the intrusion is in the range of 1-1.5 times rock mass thickness. Through the vitrinite reflectance analysis of rocks around intrusion, Dow [49] concluded that influence scope can be up to twice the thickness of the intrusive body. According to the study of sill and dike on the east Siberia platform, Galushkin [50] considered the scope of sill and dike to be in general within 30-50% sill or dike thickness, rarely more than the thickness. Chen [51] thought that influence range of a sill to organic matter is from less than the sill thickness to more than double thickness, even reaching four times the thickness of the sill. Galushkin [50] reached the conclusion that the intensity of intrusion alternation was in the range of 50-90% sill or dike thickness through many analysed examples. Zhu et al. [52] thought that only within a 15m scope, was organic matter obviously affected by sill. When Raymond and Murchison [47] studied the sediments in in carboniferous strata, Midland Valley, he found that vitrinite reflectance of the organic matter in the tuff close to the volcanic neck is significantly higher than that in the sedimentary rocks. So the magmatic intrusion effect on the evolution extent of organic matter requires further study.

3. Biogenic and abiogenic hydrocarbon related to volcanic reservoirs

3.1. Organic hydrocarbon generation

The origin of oil and gas has been a long debated theoretical issue. There are two opposing points of view: 1) the organic origin theory and 2) the inorganic origin theory. Organic origin theory considers oil and gas to come from biological processes. Inorganic origin theory explains the origin of oil and gas through inorganic synthesis and mantle degassing. The earliest organic origin theory was proposed by Lomonosov in 1763 [53]. He thought that fertile substances underground, such as oil shale, carbon, asphalt, petroleum and amber, originated in plants. The hydrocarbon formation theory of kerogen thermal degradation proposed by Tissot and Welte [54] and Hunt [55] are the representatives of the organic hydrocarbon generation theory.

The hydrocarbon formation theory of kerogen thermal degradation is based on the diagenesis of organic matter resulting from biopolymers into geopolymers, then kerogen. Kerogen is the main precursor material of oil compounds during the process of hydrocarbon generation, when thermal degradation plays a major role [54]. For sufficient hydrocarbon class and commercial oil gathering, sedimentary rocks must experience the hydrocarbon generation and temperature threshold. Mass hydrocarbons are formed at temperatures from 60 to 150°C by heated organic matter [55]. According to this theoretical model, the sedimentary organic matter maturity, especially for kerogen, becomes the key factor for evaluating hydrocarbon potential. When the threshold burial depth reaches, kerogen will be changed from immature to mature. Oil and gas generates by series of thermal degradation.

Thus the organic origin theory of petroleum has been completely established - it is consistent with the object geological facts, especially the basic law of sedimentary organic matter evolution. The theory has been accepted gradually by the majority of petroleum geologists and plays a major role in oil and gas exploration [56].

3.2. Inorganic hydrocarbon generation

Although organic origin theory has been the guiding theory of modern oil exploration, with foundation of immature oil and ultra-deep liquid hydrocarbon, inorganic origin theory has aroused much attention among geologists. Take the Xujiaweizi area where the Daqing oilfield as an example (Figure 1). Here many reservoirs have been found to contain a lot of alkane gas and non-hydrocarbon gas with inorganic origin, such as CH_4 and CO_2 [57-60]. Carbon isotope of carbon dioxide ($\delta^{13}C_{CO2}$) is an important indicator to identify the carbon dioxide origin, and many domestic and foreign scholars have undertaken research on this [58, 61-64]. Dai [65] pointed out that the $\delta^{13}C_{CO2}$ value is from +7 ‰ to - 39 ‰ in China, in which the organic origin

$\delta^{13}C_{CO2}$ value is from - 10 ‰ to - 39.14 ‰, with the main frequency scope of -12 ‰ ~ -17 ‰; the inorganic origin $\delta^{13}C_{CO2}$ value is from +10 ‰ to -8 ‰ with the main frequency scope of -3 ‰ ~ -8 ‰ (Figure 2). Inorganic origin CO_2 can be divided into mantle-derived, carbonate pyrolysis, magma degassing and so on. The $\delta^{13}C$ value of mantle-derived CO_2 is around - 6‰, which of carbonate pyrolysis is from +3‰ to -3‰. CO_2 volume fraction of Well FS9 in the Xujiaweizi area, Songliao Basin, is 89.73%, and the $\delta^{13}C$ value is from -4.06‰ to -5.46‰. The $\delta^{13}C_{CO2}$ value is -6.61‰ of Well FS6, which confirms that CO_2 of the Xujiaweizi region belongs to the mantle-derived category.

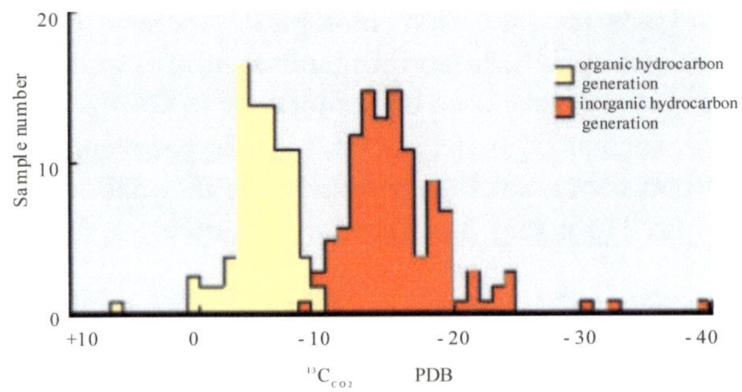

Figure 2. Organic and inorganic origin$\delta^{13}C_{CO2}$ frequency (Dai [62, 65])

4. Volcanic lithofacies related to oil and gas accumulation

Current classification of volcanic lithofacies is mainly based on the style of volcanism or eruptive and/or pyroclast/volcaniclast transport, and corresponds to a modern volcano architecture that forms volcanic rocks [66-70]. The diagenetic process significantly influences the porosity and permeability of volcanic deposits turning to volcanic rocks during burial in a sedimentary basin. According to the characteristics of volcanic rocks and the hydrocarbon exploration situation in the Songliao Basin (Figure 1), Wang et al. [71] introduced a classification system on volcanic lithofacies (Table 1). Volcanic lithofacies are classified as "*5 facies and 15 sub-facies*", and distinguished with representative features. These features include the transportation mechanism and origin, diagenesis style, representative lithology and structure, facies sequence and rhythm, and potential reservoir spaces. The classification emphasizes the relationship between lithofacies and reservoir capability, and will facilitate the recognition of volcanic lithofacies in various scales of outcrop section, well core, well cutting and thin section. Lithofacies can be identified with seismic and logging data by proper correlation. The reservoir capability of volcanic rocks can be primarily evaluated based on their facies and sub-facies.

Facies	Sub Facies	Material and Transport	Diagenesis	Lithology	Structure	Texture	Sequence	Reservoir Space
V Volcanogenic Sedimentary Facies	V3 Coal-bearing tuffaceous sediment	Tuffaceous pyroclast, plant-enriched turf	Compaction and consolidation	Interbedded tuff and coal seam	Pyroclastic/ clastic structure	Rhythmic bedding, horizontal bedding	Swamp, close to volcanic dome	Intergranular pore, primary and secondary pore, fissure, porosity and permeability similar to sedimentary rock
	V_2 Reworked volcanogenic sediment	Pyroclast reworked by fluid		Layered pyroclastic rocks/ tuff	Pyroclastic structure with rounded gravel, no epi-clast.	Cross bedding, trench bedding, graded bedding, massive	Depression between volcanic domes, conduit-close facies of large volcanic edifice	
	V_1 Epiclast-bearing volcanogenic sediment	Pyroclast dominating with epiclast		Epiclast-bearing tuff (tuffaceous sandy conglomerate)	Pyroclastic / clastic structure with rounded gravel. Few epi-clast		Depression between volcanic domes	
IV Extrusive Facies (late stage of a cycle)	IV3 Outer extrusive sub-facies	Lava front condense, deform, scrap and wrap new and old rocks. The mixture wriggles under internal force	Condensing lava weld new and old rock fragments	Breccia lave with deformed fluidal structure	Welded breccia and welded tuff structure	Deformed fluidal structure	Outer part of extrusive facie, transition to effusive facies	Inter-breccia fissure, micro-fissure, fissure between fluidal structure
	IV2 Middle extrusive sub-facies	Lava with high viscosity flows under internal force, domed near crater.	Lava condense and consolidate (quench)	Massive pearlite and cryptocrystalline rhyolite	Vitric structure, perlitic structure, mortar structure	Massive, layered, lenticular and wrapping	Middle part of extrusive facie	Primary micro-fissure, tectonic fissure, Inter-crystal pore
	IV 1 Inner extrusive sub-facies			Pillow or sphericity shaped perlite		Spherical, pillow, dome	Core of extrusive facies	Inter-perlite sphere space, pores within loosely packing perlite, micro-fissure, Inter-crystal pore
III Effusive Facies (middle stage of a cycle)	III3 Upper effusive sub-facies	Crystals and syn-eruption breccia bearing lava flows on surface under gravity and propelling of subsequent lava	Lava condense and consolidate	Vesicular rhyolite	Spherulitic structure, cryptocrystallin e structure, microcrystallin e structure	Vesicular, amygdaloidal, lithophysae	Upper part of flow unit	Vesicular, inner space of lithophysae, inner space of amygdaloidal
	III 2 Middle effusive sub-facies			Rhyolite with fluidal structure	cryptocrystallin e structure, microcrystallin e structure, porphyritic structure	Fluidal structure, few Vesicular-amygdaloidal structure	Middle part of flow unit	fissure between fluidal structure, vesicular, tectonic fissure
	III 1 Lower effusive sub-facies			Cryptocrystalline rhyolite, syn-genetic breccia bearing rhyolite	Vitric structure, cryptocrystallin e structure, porphyritic structure, breccia structure	Massive, dashed deformed fluidal structure	Lower part of flow unit	Slaty and wedge joint, tectonic fissure

Facies	Sub Facies	Material and Transport	Diagenesis	Lithology	Structure	Texture	Sequence	Reservoir Space
II Explosive Facies (early stage of a cycle)	II3 Pyroclastic flow deposti	Volatile bearing hot mixture of pyroclast and magma slurry flows on surface under gravity and propelling of subsequent lava	Lava condensation and consolidation as well as compaction	Crystal fragment, vitric fragment and magma slurry and lithic fragment bearing welded tuff (lava); lava cemented polymictic conglomerate	Welded tuff structure, pyroclastic structure	Massive, normal grading, inversed grading, orientated vesicular and vitric, matrix support	Early stage of cycle, Upper part of explosive facies, transition to effusive facies	Intergranular pore, vesicular, loose deposit of condensing unit
	II 2 Base surge deposit	Air ejecting multiphase turbidity of gas-solid-liquid flows rapidly under gravity on surface (maximum velocity: 240km/ hour)	Compaction	Crystal fragment, vitric fragment and magma slurry bearing tuff	Pyroclastic structure (crystal fragment bearing tuff structure dominating)	Parallel bedding, cross bedding, regressive sand wave bedding	Middle-lower part of explosive facies, interbedded with air fall deposit, normal grading, thinning bedding upward, cover the palaeo-slope	Loose deposit within volcanic body, intergranular pore, inter-breccia fissure
	II 1 Air fall deposit	Free fall of air ejecting solid and plastic material (under influence of wind)	Compaction	Bomb and pumice bearing agglomerate, breccia, crystal fragment bearing tuff	Agglomerate structure, breccia structure, tuff structure	Granular support, normal grading, trajectory falling blocks	Lower part of explosive facies, normal grading, thinning bedding upward, intercalated bedding	
I Conduit Facies (lower part of a edifice)	I3 Crypto-explosive breccia	Volatile-enriched magma intrudes and explodes within surrounding rocks underground. Surrounding rocks were in-situ broken and cemented by magma. Explosion and cementation function simultaneously.	Cement by condensation of hydrothermal or fine pyroclast	Cyrpto-explosive breccia	Crypto-explosive breccia structure, self-mortar structure, cataclastic structure	Column, layered, dike, branch, fissure filling	Near crater, top of sub-volcanic rock, fingered in surrounding rock	Inter-breccia pore, filled primary micro fissure
	I2 Dikes and sills	Syn or post magma intrusion	Lava condensation and crystallization	Sub-volcanic rock, porphyrite and porphyry	Porphyritic structure, holocrystalline structure.	Chilled border, flow plane, flow line, columnar and tabular, xenolith	Several to fifteen hundreds of meters beneath volcanic edifice, fingered with other lithofacies and surrounding rocks	Slaty and columnar joint, fissure between dike and surrounding rock
	I1 diatreme	Detained lave in conduit, Collapse of crater	Lava condensation, lava welding pyroclast, compaction	Lava, welded breccia/tuff lave, tuff/breccia	Porphyritic structure, welded structure, breccia/tuff structure	Packing structure, ring or radial joint, belted lithology	Diameter of hundreds of meters, vertical occurrence, penetrate other facies	Inter-breccia pore, ring and radial fissure

Table 1. Classification of volcanic facies and corresponding characteristics for each sub-facies

Sequences of facies and sub-facies assemblage follow certain principles. In the Songliao Basin, the felsic sequences are explosive facies → effusive facies/extrusive facies (probability: 50%±), conduit facies → effusive facies/extrusive facies (probability: 30%±) and explosive facies → conduit facies →extrusive facies/explosive facies (probability: 20%±). The intermediate – basic sequences are effusive facies → explosive facies (probability: 50%±), effusive facies → volcanogenic sedimentary facies (probability: 30%±), effusive facies → explosive facies → volcanogenic sedimentary facies (probability: 20%±). The sequences of felsic rocks inter-bedded with intermediate – basic rocks are more complex and mainly include effusive facies → explosive facies → volcanogenic sedimentary facies (probability: 30%±), explosive facies → volcanogenic sedimentary facies (probability: 20%±), explosive facies → effusive facies → volcanogenic sedimentary facies (probability: 20%±) and effusive facies → conduit facies → extrusive facies (probability: 10%±). Sequences of facies are the basis of volcanic lithofacies modelling, geological interpretation of seismic data and prediction of volcanic reservoir.

According to facies assemblage in drill cores and outcrops, lava of explosive facies and effusive facies can be directly linked to the pyroclastic rocks of volcanogenic sedimentary facies, especially in the proximity of a volcanic conduit, while most volcanogenic sedimentary facies form along volcanic edifice flanks. In general, felsic eruptive sequences start with explosive facies, while in conduit-close regions, they starts with conduit facies (Figure 3).

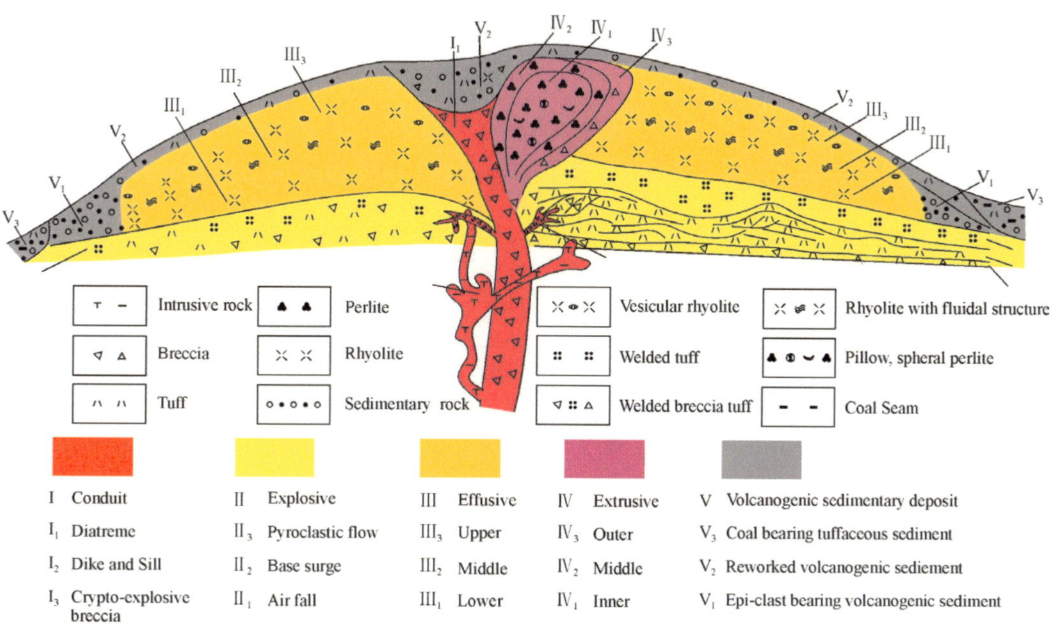

Figure 3. Model of the facies of Mesozoic acidic volcanic rocks in the Songliao Basin, China

Since diagenesis of volcanic lava is condensing consolidation-dominated, its porosity-change influenced by the burial is less pronounced than for sedimentary rocks, thus, volcanic rock will contribute more reservoirs when burying depth exceeds a threshold value. In the Songliao Basin, the threshold burying depth is about 3,500 m. Beneath this depth, sandstone is densely compacted and loses reservoir capability, the reservoir is volcanic rock-dominated. Reservoir spaces within volcanic rocks show complex structures and strongly heterogeneous distribution. Based on observation of well core, cutting and analysis of micro-structure, reservoir spaces of volcanic rocks in the Songliao Basin are classified as primary pores, secondary pores and fissures which include 13 types of elemental components (Figure 4). In general, the reservoir space of volcanic rocks has a dual-component of pore and fissure.

Primary vesicular pores and tectonic joints well develop within upper sub-facies of effusive facies, effectively connected intergranular pores are found within loose deposits between sub-facies of explosive facies, primary fissures and intergranular pores well develop within inner sub-facies of extrusive facies. Exploration of hydrocarbon accumulation in volcanic rocks should target these sub-facies in case of the existence of source rocks and traps.

5. Identification of volcanic reservoirs by well-loggings

Identification of volcanic lithology and lithofacies by logging is primarily based on calibration with drilling cores and cuttings, then logging parameters are used to make cross-plots and frequency distribution histograms. In addtion, logging facies' analysis and FMI image interpretation is used so as to discriminate volcanic rocks as well as their textures and structures.

5.1. Discrimination analysis of volcanic lithology and lithofacies by cross-plots of logging parameters

Cross-plots of logging parameters are simple and effective methods which are generally used to discriminate volcanic lithology and lithofacies in drilled wells [72-73]. Primary logging parameters includ natural gamma (GR), natural gamma-ray spectral logging (U, Th and K), electrical resistivity (RT), NPHI porosity, RHOB density, acoustic log (DT), photoelectric absorption coefficient (Pe) as well as compound parameters M and N. Two of theses parameters are plotted in a X and Y coordinate system, different regions are divided by the concentration of data points, then will be assigned with corresponding geological information. Generally, this method is used firstly on well sections with known lithology and lithofacies, so as to make master plates which are then applied to the other unknown sections in the same area. Applications in the Songliao Basin show that GR-Th, Pe-Th and M-N cross-plots are the most effective methods for discriminations of volcanic lithologies (Figure 5). Moreover, logging facies' analysis and FMI image interpretation are used to identify the textures and structures of volcanic rocks, and then finally determine the discrimination of volcanic lithology and lithofacies in detail.

http://dx.doi.org/10.5772/54587

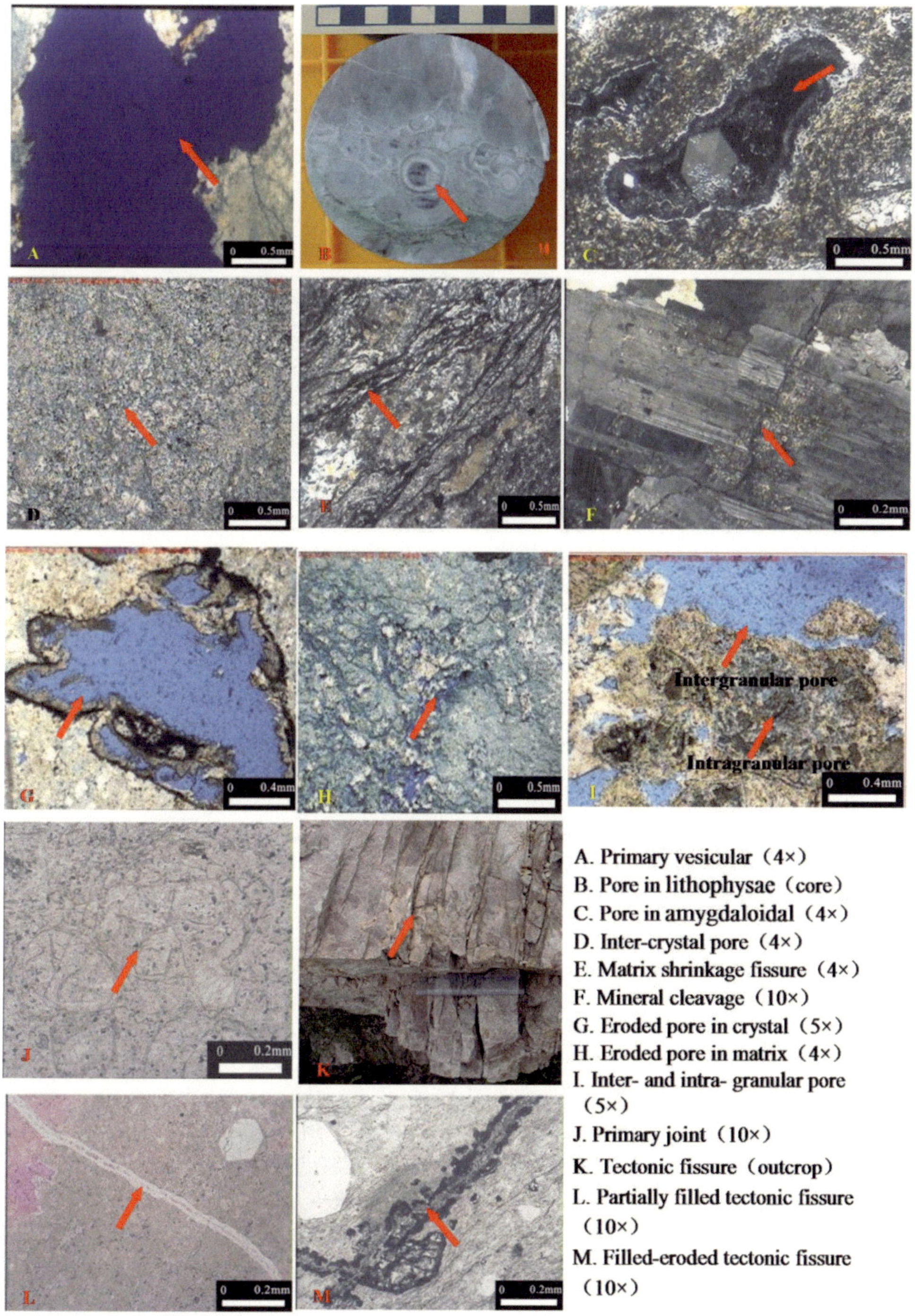

Figure 4. Volcanic reservoir space types

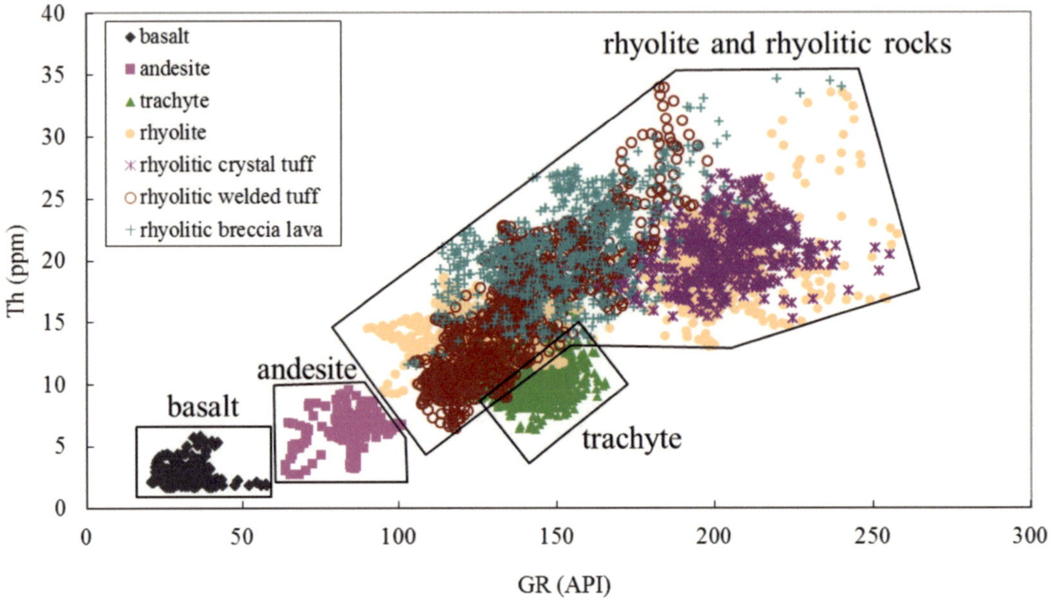

Figure 5. Cross-plot of GR versus Th for lithological identification of volcanic rocks of the Lower Cretaceous Yingcheng Formation in the Xujiaweizi depression, Songliao Basin, China

5.2. Logging facies' analysis of volcanic rocks

Comparative analysis between the volcanic facies and logging facies of drilling core sections is aimed at revealing and summarizing the relationship between geologic properties and logging responses, so as to solve the multiplicity of interpretation by logging parameters, and then set up identification standards of logging facies in the study area. Identification of logging facies is by means of configuration analysis of logging curves including SP, GR, RT, ML, RHOB, as well as dip logging interpretation. Moreover, the standard logging facies could be interpreted as lithofacies on the basis of geologic data.

Electrical conductivity of volcanic reservoirs is mainly influenced by lithology, porosity and permeability, saturation, content of metal elements and also burial depth. Occurrence of hydrocarbons will greatly increase the resistivity, while it will obviously decrease with water. The shape of logging curves and their assemblages are related closely to volcanic lithologies as well as their textures and structures which have become good markers for discrimination of volcanic lithofacies. For massive volcanic rocks, the framework is the main medium of conduction. Under this circumstance, lithology, lithofacies and burial depth are the main controlling factors to the conduction of rocks and changes of logging curve shapes. For example, intermediate-felsic volcanic rocks of vent facies are characterized with high-GR and mid-RT, and their logging curves appear as a high amplitude dentiform and peak shape. While basalts of volcanic vent facies show low-GR and the tuff displays low-RT.

The Mesozoic volcanic rocks are the most important gas reservoirs in the northern Songliao Basin. Five lithofacies and 15 sub-facies have been recognized in the volcanic rocks. The best reservoirs were generally found in three of the 15 sub-facies including pyroclastic bearing lava flow, upper effusive and inner extrusive sub-facies. The corresponding logging characteristics are as follows. The pyroclastic rock-bearing lava flow sub-facies show high-GR values with high amplitude dentiform and medium to mid-high RT with low frequency, low amplitude dentiform. The upper effusive sub-facies show high GR with high amplitude dentiform and mid-high RT with finger and peak shapes. The inner extrusive sub-facies show high GR with medium amplitude dentiform and mid-high to high RT with medium amplitude dentiform. In addition, crypto-explosive and outer extrusive sub-facies may also be good reservoirs. The occurrence of hydrocarbons will cause a remarkable increase of resistivity, while water does the contrary. The changing of resistivity without influence of fluids from low to high are respectively followed as volcanogenic sedimentary facies, extrusive facies, explosive facies, volcanic conduit facies and effusive facies [74].

5.3. Identification of volcanic textures and structures by FMI image interpretation

With the characteristics of high resolution, total borehole coverage and visibility, FMI image interpretation may reveal continuous geologic information such as lithology, textures and structures, as well as pores and fractures by means of calibrations with drilling core sections. Sizes and shapes of volcanic breccia and conglomerates, as well as features of rock structures and beddings, can give much geologic information on volcanic lithofacies and pore spaces, especially for well sections lacking drilled cores [75-77].

Features displayed by FMI images of volcanic rocks are the synthesized effects of logging response units including volcanic fragments, framework, fractures and pores. On the FMI images, bright tone corresponds to high resistivity, while dark tone relates to low resistivity, and warm colours, such as yellow and orange, indicate medium resistivity (Table 2).

Image type	Tone	Resistivity	Geological interpretation
static	bright (white)	high	massive structure
	dark (brown, black)	low	fractures and pores
	mottle	heterogeneous	heterogeneous rock mass
dynamic	bright (white)	high	volcanic breccias, rock fragments, crystal fragments, magma fragments, amygdala
	yellow, orange	middle	rock matrix or framework
	dark (brown, black)	low	fractures and pores

Table 2. Image interpretation of volcanic imaging logging (Li et al. [72])

Comparatively, rock fragments generally display bright tones due to high resistivity, while matrix shows dark as a result of low resistivity, and these featuresare common in pyroclastic lava rocks and pyroclastic rocks. The transformation of bright and dark zones on a FMI image not only indicates resistivity changes, but also reflects the contact relations among different parts of rocks. Descriptions of drilled cores reveal that there is great difference between volcanic fragments (breccia, conglomerates, rock fragments, crystal fragments and magma fragments) and their surrounding matrix (lava framework or volcanic ash) due to distinguishing colour, content and shape, which may result in colour diversities of the FMI images according to resistivity changes. Standard interpretation models of volcanic textures and structures which are used to identify lithofacies have been summarized through calibrations of FMI images with geologic information, for example, lava texture, welded texture, tuff texture, breccia texture and massive structure, vesicular-amygdaloidal structure, flow structure (Figure 6).

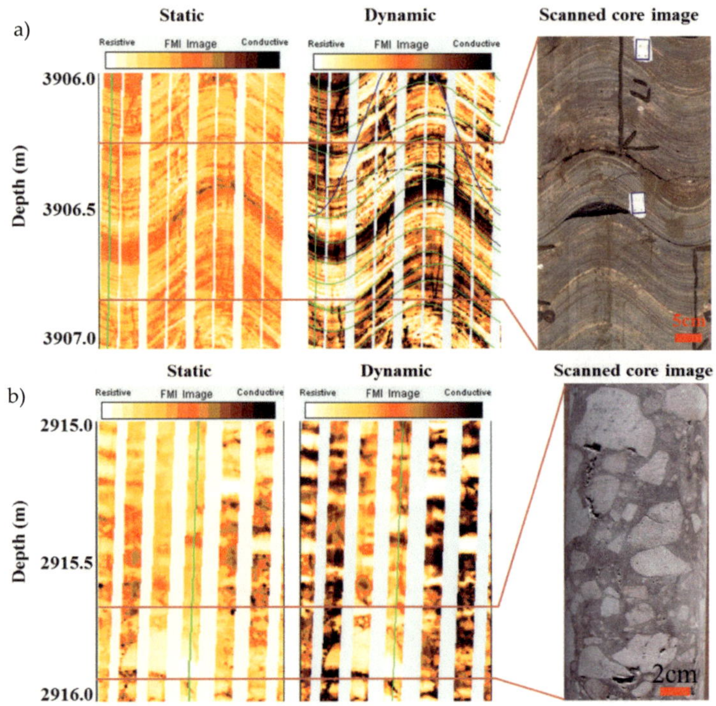

Figure 6. Typical lithological structures on FMI logging image. a) Flow bandings of extrusive rhyolite; b) Volcanic breccia structure of explosive facies.

6. Identification of volcanic reservoirs by seismic data

At the exploration stage, the volcanic facies mapping relies mainly on the artificial seismic facies' analysis and is under the control of well facies or according to an experience template,

http://dx.doi.org/10.5772/54587

converted into a volcanic facies map. At the development stage, the volcanic facies planar prediction relies mainly on the waveform classification method for obtaining the seismic facies map. The volcanic facies is interpreted under the control of facies of wells and the volcanic edifice belts [78-79]. This method can identify the volcanic facies and its combination. Now it is widely used in the volcanic exploration of the Songliao Basin. The waveform classification method of volcanic facies' identifying is explained in this paper.

The actual seismic facies were calculated by combining different amplitude, frequency, phase and time intervals. A seismic facies map can be obtained by the waveform classification calculation. Volcanic facies is predicted through observing the combination and distribution characteristics of the model trace in the seismic facies map. The number of waveform classifications (model trace) can be ensured by the seismic characteristics of volcanic facies. After that, we conduct the waveform classification experiment by selecting multiple waveform classifications or by using different thickness of time intervals. Lastly, the stability of the calculation results need be checked. The optimal time interval of the waveform classification calculation is between half a wavelength to two wavelengths.

Taking the volcanic rocks of the upper Yingcheng Formation in the Changling rift YYT work area of the southern Songliao Basin as an example (Figure 1), we introduce the prediction method of the volcanic facies plane. The volcanic facies is predicted by selecting 7 or 15, respectively, as the number of waveform classification. The prediction results of volcanic facies show good consistency. Volcanic facies is predicted by number waveform classification being set to 7. Firstly, overlap the volcanic seismic facies maps with the structure maps. The results show that, in some high tectonic areas, the waveforms characteristics of the seismic facies have unorganized distribution, but in the relatively flat tectonic area, the waveforms characteristics of seismic facies show continuous distribution. Next, the waveform classification characteristics of the seismic facies can be calibrated with well facies, the waveform characteristics of seismic facies in the central region of volcanic edifices are multi-waveform clutter distribution, while the waveform characteristics of the far-source area far away from the centre of volcanic edifices are continuous distribution (Figure 7). In this way, the centre's facies belt (volcanic conduit / effusive facies) distribution of volcanic edifices can be predicted. The different waveforms' seismic facies are calibrated by volcanic facies revealed by the well. A waveform classification planar map should be converted to the volcanic facies map. The seismic characteristics of volcanic conduit facies and its combination are rounded, massive, banded and messy reflection, in the edge is ring banded. The seismic characteristics of explosive facies and their combination are banded, messy, good continuous reflection. The seismic characteristics of effusive facies and their combination are mottled massive, middle-bad continuum reflection. According to the seismic characteristics, volcanic facies planar distribution is identified in the YYT area. The effusive facies distribution is dominating, and the explosive facies distribution is less. The effusive facies distribute mainly on both sides of the central fault. Explosive facies distribute mainly in the southeast far away the central fault. There are two volcanic facies sequences, one is the volcanic conduit facies-effusive facies, the other is the volcanic conduit facies-explosive facie/volcanic sedimentary facies.

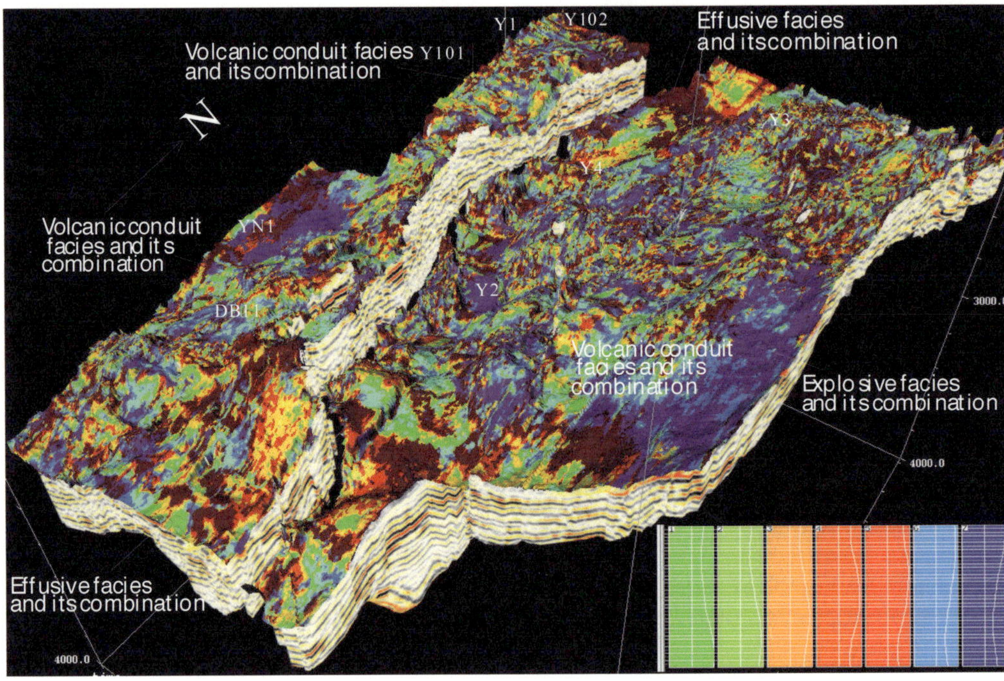

Figure 7. Waveform classification (seismic facies) map of volcanic rocks of the upper Yingcheng Formation in the YYT work area

7. Mechanism and geological occurrence of oil and gas accumulation

Volcanic oil and gas reservoirs are mainly not only accumulations in volcanic rocks, but also those hydrocarbon reservoirs with volcanic rocks as seals or forming traps. The formation and distribution of hydrocarbon accumulations in volcanic rocks is different from non-volcanic (silici-)clastic rocks. Since volcanic rocks cannot produce hydrocarbons, neighbouring source rocks are essential to the formation of oil and gas accumulations in volcanic rocks, thus it will be more favourable for better matching relationships between volcanic reservoirs and source rocks [80-81].

7.1. Volcanic rocks as oil and gas reservoirs

Since the porosity and permeability of volcanic rocks do not decrease remarkably according to the increase of burial depth, they are more favourable for hydrocarbon accumulations compared to sedimentary rocks in the deep part of the basin. So far, hundreds of volcanic reservoirs, such as the Niigata Basin in the Honshu Island of Japan [82], Austral and Neuquen Basins of Argentina [7] and the Bohai Bay Basin, Songliao Basin and Junggar Basin in China have been found [5, 24, 71, 81]. As a whole, the volcanic reservoirs are mostly Cenozoic and Mesozoic, especially Jurassic, Cretaceous and Tertiary, which may be related to the frequent global volcanism in these epochs. As revealed by petroleum explorations, occurrences of hydrocarbons have been found in almost all types of volcanic rocks, and in detail, basalts have the largest proportion while the rest follow as andesite, rhyolite and pyroclastic rocks [83]. Prolific volcanic reservoirs have been found in explosive, extrusive and volcanic-sedimentary facies, while considering inside a volcanic edifice, they have generally the best reservoir properties in volcanic vents and near vent facies [71, 84]. Vertically, favourable reservoirs are developed in the upper part of volcanic sequences due to post-eruption weathering, leaching and dissolving [85].

7.2. Formation of oil traps related to volcanic activities

Besides being reservoirs, volcanic rocks can also be good cap rocks. After volcanic ash falls into water, it will inflate and form layers of bentonite or mudstone with bentonite which may become excellent cap rocks [86]. While mudstone lacks sealing abilities in the mid-shallow part of basins, unaltered massive basalts are generally rather better cap rocks, taking the Scott Reef oil field in the Browse Basin of Australia and Eastern Sag in the Liaohe Basin of China as examples [2, 24]. Some layered intrusive rocks may also be good cap rocks, for example, the Lin 8 oil trap in the Huimin depression of the Bohai Bay Basin, north China [87]. In the deep part of the Xujiaweizi depression of the Songliao Basin, two types of volcanic rocks have been found to be cap rocks including the pyroclastic type and lava type. Due to better sealing abilities, the pyroclastic type cap rocks control the regional accumulation and distribution of gas in volcanic reservoirs, while the lava type only control the local accumulation and distribution of gas in volcanic rock bodies [88]. Besides, diverse oil and gas traps can be formed by the local structures of intrusions as well as their matching with sedimentary layers, commonly forming arched uplifts and lateral barriers [24, 89].

All aspects of the common hydrocarbon accumulating conditions and their favourable matching relationships are also necessary to the volcanic oil and gas reservoirs. So far, most of the discovered volcanic oil and gas reservoirs are structural-lithologic and stratigraphic. Near-source accumulations are formed when volcanic rocks emplace close to source rocks, developing concentration zones of volcanic oil and gas reservoirs. While there is a long distance between source rocks and volcanic rocks, certain accumulations may also form due to communications of faults and unconformities. There are two accumulation patterns divided by volcanic reservoir forming conditions such as near-source play and distal play [81]. Near-source plays are mostly discovered in eastern Chinese depressions, for example, the Paleogene mafic volcanic rocks in the Bohai Bay Basin and the Lower Cretaceous felsic volcanic rocks which developed prolific oil and gas accumulations emplaced right on the top of high-quality source rocks. Both near-source and distal plays are found in western Chinese basins, for instance, volcanic rocks and source rocks have developed together in carboniferous-Permian of the Junggar and Santanghu Basins which formed near-source plays, and the source rocks mainly developed in the Lower Paleozoic while the reservoir volcanic rocks emplaced in the Permian, thus forming distal plays in the Sichuan and Tarim Basins (Figure 8).

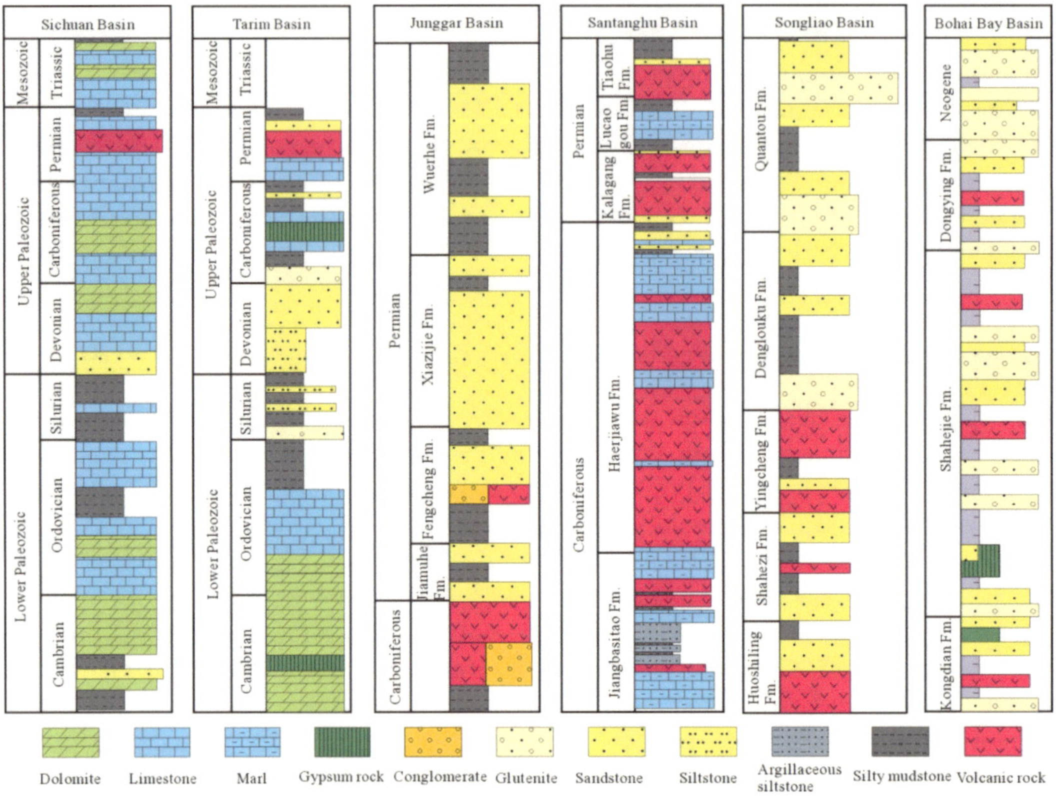

Figure 8. Source-reservoir-cap assemblages of the volcanic oil and gas accumulations in main petroliferous basins of China (Zou et al. [81])

8. Exploration of volcanic accumulation of oil and gas

8.1. The characteristics of volcanic oil and gas accumulation

The characteristics of volcanic oil and gas accumulation mainly include volcanic reservoirs and reservoir forming elements. Taking the characteristics of volcanic oil and gas accumulation in the Songliao Basin as an example, volcanic gas accumulation can be classified into acid type and intermediate-basic type by lithology. By the characteristics of volcanic edifices, these two types mentioned above can be furthe rclassified into six sub-types, including acid pyroclastic sub-type, lava sub-type, complex sub-type and intermediate-basic pyroclastic sub-type, lava sub-type, complex sub-type. Great differences have been discovered in developmental degrees among the types in the volcanic gas accumulation (Figure 9). The acid and intermediate-basic lava sub-types account for 72% of volcanic gas accumulation of the Yingcheng Formation in the north of the Songliao Basin, and the contribution degree of acid lava sub-type can reach 50%. The acid type accounts for 92% of volcanic gas accumulation in the south of the Songliao

http://dx.doi.org/10.5772/54587

Basin, while only the intermediate-basic lava sub-type gains industrial gas. The highest deliverability in a single well is gained in the acid complex sub-type; the deliverability of the intermediate-basic type is lower than the acid type. There are few differences among the intermediate-basic pyroclastic sub-type, lava sub-type and complex sub-type. However, there are great differences among the acid pyroclastic sub-type, lava sub-type and complex sub-type.

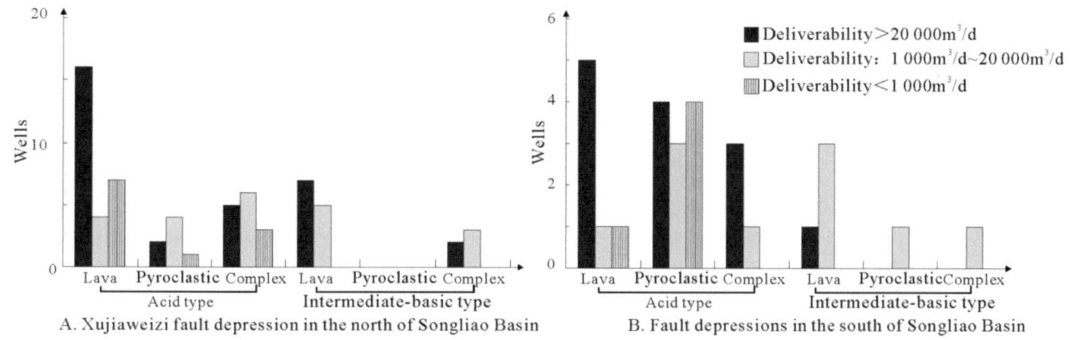

Figure 9. The relationship between hydrocarbon accumulations and volcanic edifices in the Songliao basin

By analysing the relationship between reservoirs and reservoir forming structures, most of the volcanic gas accumulations are structural-lithologic gas accumulations. The gas mainly originated from the mud and coal-bearing strata of the early Cretaceous Shahezi Formation and Huoshiling Formation. The fluid transforming system is made up by the faults, joints and high porosity-permeability transforming zones. The distribution range of gas layers is not absolutely controlled by the structural trap. When above the water-gas contact (WGC), the high porosity-permeability zone forms a gas layer, the medium-low porosity-permeability zone forms a poor gas layer and the zone which has fewer pores and fractures forms a dense layer or baffle layer. The proportion of poor gas layers increases gradually from the acid complex type to the acid pyroclastic sub-type to intermediate-basic lava sub-type (Figure 10). WGC is an uneven contact surface caused by the peculiarity of the stratigraphic construction of volcanic edifices. There are great differences in shape among the different volcanic edifices. The gas thickness of acid pyroclastic sub-type and acid complex sub-type changes little, forming a tabular and sill-like shape. There are great changes in the gas thickness of intermediate-basic lava sub-type, the maximum thickness is 2~3 times thicker than the minimum thickness and forms a mound or wedge shape. Moreover, in the same gas layer, the deliverability is also different in the different locations.

8.2. The reservoir forming elements of volcanic oil and gas accumulation

By the comparison between industrial gas wells and other wells, the advantages of forming high production gas include reservoir space diversity, high porosity, good source rocks, anticlinal /faulted anticline traps and the vertical migration pathway. The reservoir forming effects will be poor once one of the above conditions is absent.

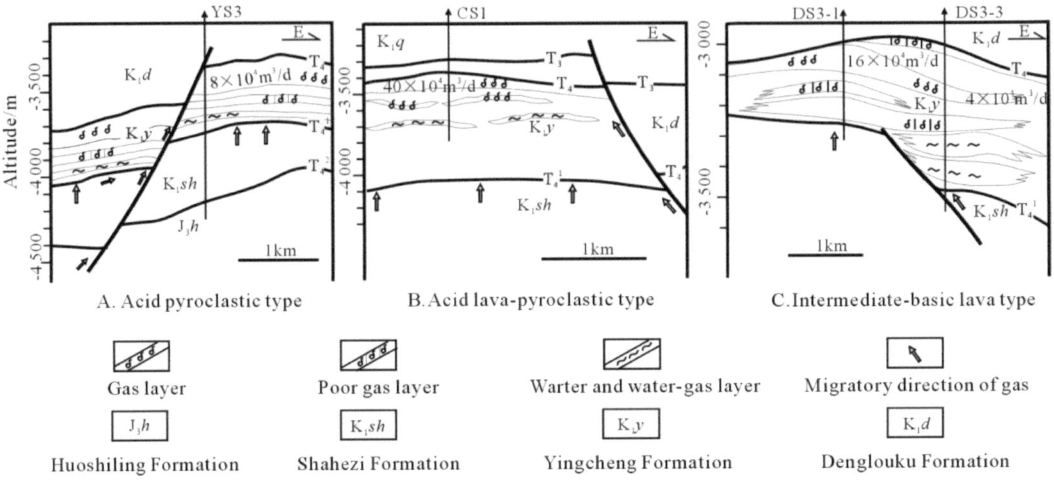

Figure 10. Forming pattern of gas pools of volcanic edifices in the Songliao Basin

For example, although having good source rocks, a DB10 well only gains low producing gas because of unitary reservoir spaces and poor porosity-permeability. A YN1 well does not even gain industrial gas because of poor source rocks and porosity-permeability despite having diversiform reservoir spaces. An SS1 well has good porosity-permeability and diversiform reservoir spaces, while its source rock is poor, it does not gain industrial gas. The wells show that overlying strata on the volcanic rocks of the Yingcheng Formation can be regional caps in the Songliao Basin and there are a wealth of high angle joints and faults in the volcanic rocks of the Yingcheng Formation. So the main reservoir forming elements of volcanic gas accumulation include effective source rocks, faults connecting to the source rocks and reservoir porosity-permeability.

9. Summary

Hydrocarbon exploration in volcanic rocks is a relatively new and important topic today. As a typical example in China, the Songliao Basin has been introduced here.

Natural transformation from organic matter to hydrocarbon is a slow process. This slow process can be accelerated by volcanic heat that is the thermal effect of volcanic activity. In addition, it can also be catalyzed by volcanogenic minerals, such as zeolite and olivine, and transition metals in hydrothermal liquid, such as Ni, Co, Cu, Mn, Zn, Ti, V, etc.

The origin of natural gas hosted in volcanic reservoirs can be both biogenic and abiogenic. In the Songliao Basin of China, most of hydrocarbon has been proved as having organic derivation, a few alkane gas and non-hydrocarbon gas have inorganic derivation. The origin of hydrocarbon can be distinguished with isotopes such as C and He.

Volcanic gas accumulation type		Primary pore		Secondary pore		Trap type	Migration type	Relations among source, reservoir and cap	Deliverability $10^4m^3/d$	Typical well (Formation)
Type	Sub-type	Pore	Fracture	Pore	Fracture					
Acid	Pyroclastic	Intergranular pore, Intercrystalline pore, Intercentric pore (breccia)	Crash fractures, width : 0.1 mm , filling degree: 95%	/	High-angle fractures, width : 1~2 mm, upper part with no fill, lower part filled by calcite filling degree: 100%	faulted anticline trap	vertical	self generation and self preservation, lower generation and upper preservation	8	YS3 (k_1y)
		Interbreccial pore, Corrosion pore	Intercrystalline crash fractures	Dissolved pore	High-angle fracture	fault nose trap	lateral, vertical	self generation and self preservation, lower generation and upper preservation	gas show	SS1 (k_1y)
	Lava	Intercentric pore	Crash fractures	/	Joints and high-angle fracturesfilling degree: 80%	fault nose trap	lateral	self generation and self preservation	gas show	YN1 (k_1y)
		Vesicle, Intercentric pore	Intercrystalline crash fractures	/	High-angle fractures and oblique crossing fractures, reticular partly, width: 0.5~2 mm	anticline trap	lateral vertical	self generation and self preservation, lower generation and upper preservation	30	YS2 (k_1y)
	Complex	Vesicle, Interbreccial pore, Corrosion pore	Intercrystalline crash fractures	Dissolved pore	high-angle fractures and Joints	anticline trap	lateral vertical	self generation and self preservation, lower generation and upper preservation	40	YS1 (k_1y)
		/	Crash fractures, filled by calcite	/	Branched fractures, width:1 cm, filled by red magma, filling degree: 30%~100%	fault nose trap	lateral vertical	self generation and self preservation, lower generation and upper preservation	gas show	YS4 (k_1y)
intermediate-basic	Pyroclastic	Interbreccial pore, vesicle (breccia)	Crash fractures	Dissolved pore	High-angle structural fractures	anticline trap	vertical	lower generation and upper preservation	5.6	DS3 (k_1y)
	Lava	Micro-vesicle	/	/	High-angle fractures, width: 2~30 mm, filled by calcite, filling degree: 10%~100%	fault nose trap	lateral vertical	self generation and self preservation, lower generation and upper preservation	0.4	DB11 (k_1y)
		Amygdale, vesicle, Interbreccial pore	Crash fractures	/	Reticular fractures, width:2~50 mm, filled by calcite, filling degree: 10%~100%	anticline trap	lateral vertical	lower generation and upper preservation	4.2	DX5,DS3-1 (k_1y)
	Complex	Amygdale, vesicle,	/	/	Reticular fractures, width:2~5 cm, filled by calcite, filling degree 10%~100%	anticline trap	lateral vertical	lower generation and upper preservation	5.0	DS4 (k_1y)

Notes:"/"shows not discovered by the cores/debris, k_1y is Yingcheng Formation ofLower Cretaceous

Table 3. The characteristics of gas pool forming of volcanic edifices of faulted sequences in the Songliao Basin

Volcanic lithofacies can be classified into "*5 facies and 15 sub-facies*" with respect to the lithification mechanism and reservoir significance. In general, a felsic sequence starts with explosive facies, and an intermediate – basic sequence starts with effusive facies. The reservoir spaces of volcanic rocks are composed of primary pores, secondary pores and fissures. The upper part of effusive facies, loose intercalation in between explosive facies, and inner sub-facies of extrusive facies are the main targets for hydrocarbon exploration in the Songliao Basin because of their good porosity and permeability.

Volcanic lithology and lithofacies in drilled wells are effectively discriminated with cross-plots of different logging parameters. The most effective methods are GR-Th, Pe-Th and M-N cross-plots for lithological discrimination. Lithofacies are characterized by GR and RT with respect to amplitude, outer shape and frequency. The texture and structure information of volcanic rocks can be depicted with FMI images.

The spatial distribution of lithological and lithofacies associations can be characterized by seismic parameters, such as amplitude, frequency, phase and time intervals. Seismic facies are mapped with waveform classification between half a wavelength to two wavelengths. Geological facies are interpreted from seismic facies coupled with core section description and well-log information.

Volcanic rocks mainly act as reservoir or seal rocks in hydrocarbon accumulations. Most of the discovered volcanic oil and gas reservoirs are structural-lithologic and stratigraphic. Source rocks are essential to the formation of oil and gas. Although plays of proximal facies are predominant, distal facies have also been discovered to be productive in the Songliao Basin.

Porosity and permeability in volcanic rocks are more heterogeneous than sedimentary rocks. High resolution data are necessary for hydrocarbon exploration in volcanic rocks. Furthermore, diagenesis of volcanic rocks is one of the most important topics in the future because it is the controlling factor on the porosity and permeability of volcanic reservoirs.

Acknowledgements

This research was supported by a grant from the Major State Basic Research Development Programme of China (no. 2009CB219300), Key laboratory of Evolution of Past Life and Environment in Northeast Asia (Jilin University), Ministry of Education, China.

Author details

Jiaqi Liu[1], Pujun Wang[1*], Yan Zhang[1], Weihua Bian[1], Yulong Huang[1], Huafeng Tang[1] and Xiaoyu Chen[2]

*Address all correspondence to: WangPJ@jlu.edu.cn

1 College of Earth Sciences, Jilin University, Changchun, China, Institute of Geology and Geophysics, Chinese Academy of Sciences, Beijing, China

2 Institute of Geology and Geophysics, Chinese Academy of Sciences, Beijing, China

References

[1] Einsele, G. Sedimentary Basins: Evolution, Facies, and Sediment Budget. Berlin: Springer (2000).

[2] Schutter, S. R. Occurrences of Hydrocarbons in and around Igneous Rocks. Geological Society, London, Special Publications (2003). doi:GSL.SP.2003.214.01.03, 214, 35-68.

[3] Benyamin, B. Facies distribution approach from log and seismic to identification hydrocarbon distribution in volcanic fracture. In: 9th SPWLA Japan Formation Evaluation SYMP (2007)., 25-26.

[4] Dutkiewicz, A, Volk, H, Ridley, J, & George, S. C. Geochemistry of Oil in Fluid Inclusions in a Middle Proterozoic Igneous Intrusion: Implications for the Source of Hydrocarbons in Crystalline Rocks. Organic Geochemistry (2004)., 35(8), 937-957.

[5] Feng, Z. Q. Volcanic Rocks as Prolific Gas Reservoir: A Case Study from the Qingshen Gas Field in the Songliao Basin, NE China. Marine and Petroleum Geology (2008).

[6] Kawamoto, T. Distribution and Alteration of the Volcanic Reservoir in the Minami-Nagaoka Gas Field. Journal of the Japanese Association for Petroleum (2001)., 66(1), 46-55.

[7] Sruoga, P, & Rubinstein, N. Processes Controlling Porosity and Permeability in Volcanic Reservoirs from the Austral and Neuquen Basins, Argentina. AAPG Bulletin (2007)., 91(1), 115-129.

[8] Beeskow, B, Treloar, P. J, Rankin, A. H, et al. A Reassessment of Models for Hydrocarbon Generation in the Khibiny Nepheline Syenite Complex, Kola Peninsula, Russia. Lithos (2006).

[9] Dai, J. X, Shi, X, & Wei, Y. Z. Summary of the Abiogenic Origin Theory and the Abiogenic Gas Pools (Fields). Acta Petrolei Sinica (2001). in Chinese with English abstract).

[10] Liu, W. H, Chen, M. J, Guan, P, et al. Ternary Geochemical-Tracing System in Natural Gas Accumulation. Science in China (Series D) (2007a)., 50(10), 1494-1503.

[11] Othman, R, Arouri, K. R, Ward, C. R, et al. Oil Generation by Igneous Intrusions in the Northern Gunnedah Basin, Australia. Organic Geochemistry (2001)., 32(10), 1219-1232.

[12] Thomson, K. Volcanic Features of the North Rockall Trough: Application of Visualisation Techniques on 3D Seismic Reflection Data. Bulletin of Volcanology (2005)., 67(2), 116-128.

[13] Wang, X, Lerche, I, & Walter, C. Effect of Igneous Intrusive Bodies on Sedimentary Thermal Maturity. AAPG Bulletin (1989)., 73(9), 1177-1178.

[14] Jin, Q, Xu, L, Wan, C, et al. Interactions between Basalts and Oil Source Rocks in Rift Basins: CO_2 Generation. Chinese Journal of Geochemistry (2007)., 26(1), 58-65.

[15] Gu, L. X, Ren, Z. W, Wu, C. Z, et al. Hydrocarbon Reservoirs in a Trachyte Porphyry Intrusion in the Eastern Depression of the Liaohe Basin, Northeast China. AAPG Bulletin (2002)., 86(10), 1821-1832.

[16] Levin, L. E. Volcanogenic and Volcaniclastic Reservoir Rocks in Mesozoic- Cenozoic Island Arcs: Examples from the Caucasus and the Nw Pacific. Journal of Petroleum Geology (1995)., 18(3), 267-288.

[17] Vernik, L. A New Type of Reservoir Rock in Volcaniclastic Sequences. AAPG Bulletin (1990)., 74(6), 830-836.

[18] Wang, P. J, Chen, S. M, Liu, W. Z, et al. Relationship between Volcanic Facies and Volcanic Reservoirs in Songliao Basin. Oil & Gas Geology (2003). in Chinese with English abstract)

[19] Bergman, S. C, Talbot, J. P, & Thompson, P. R. The Kora Miocene submarine andesite stratovolcano hydrocarbon reservoir, northern Taranaki Basin, New Zealand. In: New Zealand Oil Exploration Conference, Wellington, New Zealand, (1992)., 178-206.

[20] Liu, W. Z, Pang, Y. M, Wu, H. Y, et al. Relationship between Pore-Space Evolution and Deeply Buried Alteration during Diagenetic Stage of the Volcanic Fragments in Reservoir Greywacke from the Songliao Basin, NE China. Journal of Jilin University (Earth Science Edition) (2007). in Chinese with English abstract).

[21] Luo, J. L, Zhang, C. L, & Qu, Z. H. Volcanic Reservoir Rocks: A Case Study of the Cretaceous Fenghuadian Suite, Huanghua Basin, Eastern China. Journal of Petroleum Geology (1999)., 22(4), 397-416.

[22] Rogers, K. L, Neuhoff, P. S, Pedersen, A. K, & Bird, D. K. CO_2 Metasomatism in a Basalt-Hosted Petroleum Reservoir, Nuussuaq, West Greenland. Lithos (2006).

[23] Zhao, H. L, Liu, Z. W, Li, J, et al. Petrologic Characteristics of Igneous Rock Reservoirs and their Research Orientation. Oil & Gas Geology (2004). in Chinese with English abstract).

[24] Chen, Z. Y, Huo, Y, Li, J. S, et al. Relationship Between Tertiary Volcanic Rocks and Hydrocarbons in the Liaohe Basin, People's Republic of China. AAPG Bulletin (1999)., 83(6), 1004-1014.

[25] Yang, H, Zhang, Y, Zou, C. N, et al. Volcanic Rock Distribution and Gas Abundance Regularity in Xujiaweizi Faulted Depression, Songliao Basin. Chinese Journal of Geophysics, 49(4): 1136-1143 (in Chinese with English abstract).

[26] Stewart, S. A, & Allen, P. J. D Seismic Reflection Mapping of the Silverpit Multi-Ringed Crater, North Sea. Geological Society of America Bulletin (2005).

[27] Jiang, C. J, Feng, X. Y, Zhan, Y. J, et al. New Methodology to Explore Gasbearing Volcanic Reservoir in Xujiaweizi Fault Depression of the Northern Songliao Basin. Petroleum Geology & Oilfield Development in Daqing (2007). in Chinese with English abstract).

[28] Hansen, D. M, Cartwright, J. A, & Thomas, D. D Seismic Analysis of the Geometry of Igneous Sills and Sill Junction Relationships. Geological Society, London, Memoirs (2004)., 29(1), 199-208.

[29] Wan, C. L, Jin, Q, & Fan, B. J. Current Research Situation on Hydrocarbon-Generating Evolution of Volcanic Minerals upon Hydrocarbon Source Rocks. Petroleum Geology and Recovery Efficiency (2001). in Chinese with English abstract).

[30] Reuter, J. H, & Perdue, E. M. Importance of Heavy Metal-Organic Matter Interactions in Natural Waters. Geochimica Et Cosmochimica Acta (1977)., 41(2), 325-334.

[31] Mango, F. D. Transition Metal Catalysis in the Generation of Petroleum and Natural Gas. Geochimica Et Cosmochimica Acta (1992)., 56(1), 553-555.

[32] Jin, Q. Hydrocarbon Generation in Rift Basins, Eastern China: Catalysis and Hydrogenation-Interaction between Volcanic Minerals and Organic Matter. Advance in Earth Sciences (1998). in Chinese with English abstract).

[33] Zhang, J. L, & Zhang, P. Z. A Discussion of Pyrite Catalysis on The Hydrocarbon Generation Process. Advance in Earth Sciences (1996). in Chinese with English abstract).

[34] Lu, Q, & Liu, H. F. Studies of Kerogen from Baize Basin, Guangxi-Also on Relationship of Evolution of Kerogen and Clay Minerals. Acta Sedimentologica Sinica (1993). in Chinese with English abstract).

[35] Wei, Y. N, & Zhang, D. D. Mineral Petrology. Beijing: China Coal Industry Publishing House; (2007).

[36] Luo, G. F. Crystallography and mineralogy. Nanjing: Nanjing University Press; (1993).

[37] Cheng, R. H, Wang, P. J, Liu, W. Z, et al. Influence of Tectonic Activity and Volcanism on Hydrocarbon Generation of Kerogen. Geological Science and Technology Information (2003). in Chinese with English abstract).

[38] Fjeldskaar, W, Helset, H. M, Johansen, H, et al. Thermal Modeling of Magmatic Intrusions in the Gjallar Ridge, Norwegian Sea: Implications for Vitrinite Reflectance and Hydrocarbon Maturation. Basin Research (2008)., 20(1), 143-159.

[39] Zhou, Q. H, Feng, Z. H, & Men, G. T. Present Geothermal Features and Relationship with Gas in Xujiaweizi Fault Depression, Songliao Basin. Science in China (Series D: Earth Sciences) (2007). SII): 177-188 (in Chinese with English abstract).

[40] Wang, C. Q, Du, X. R, & Liu, J. X. Keluo-Wudalianchi-Erke Volcanic Cluster, Chapter 17, Characteristics of Melt Inclusions and Diagenesis Temperatures of Volcanic Rocks. Journal of East China College of Geology (Natural Science Edition) (1987). in Chinese).

[41] Wang, J. F. A Preliminary Study on the Geochemical Behaviors of Trace Elements and the Origin of Mesozoic Volcanic Rocks in North- Western Zhejiang Province. Geochimica (1992). in Chinese with English abstract).

[42] Çiftçi, N. B, Temel, R. Ö, & Iztan, Y. H. Hydrocarbon Occurrences in the Western Anatolian (Aegean) Grabens, Turkey: Is there a Working Petroleum System? AAPG Bulletin (2010). , 94(12), 1827-1857.

[43] Farrimond, P, Bevan, J. C, & Bishop, A. N. Hopanoid Hydrocarbon Maturation by an Igneous Intrusion. Organic Geochemistry (1996).

[44] Liu, J. Q, & Meng, F. C. Hydrocarbon Generation, Migration and Accumulation Related to Igneous Activity. Natural Gas Industry (2009). in Chinese with English abstract).

[45] Schimmelmann, A, Mastalerz, M, Gao, L, et al. Dike Intrusions Into Bituminous Coal, Illinois Basin: H, C, N, O Isotopic Responses to Rapid and Brief Heating. Geochimica Et Cosmochimica Acta (2009). , 73(20), 6264-6281.

[46] George, S. C. Effect of Igneous Intrusion on the Organic Geochemistry of a Siltstone and an Oil Shale Horizon in the Midland Valley of Scotland. Organic Geochemistry (1992). , 18(5), 705-723.

[47] Raymond, A. C, & Murchison, D. G. Development of Organic Maturation in the Thermal Aureoles of Sills and its Relation to Sediment Compaction. Fuel (1988). , 67(12), 1599-1608.

[48] Carslaw, H. S, & Jaeger, J. C. Conduction of Heat in Solids. New York: Oxford University Press;(1959).

[49] Dow, W. G. Kerogen Studies and Geological Interpretations. Journal of Geochemical Exploration (1977). , 7(0), 79-99.

[50] Galushkin, Y. I. Thermal Effects of Igneous Intrusions on Maturity of Organic Matter: A Possible Mechanism of Intrusion. Organic Geochemistry (1997).

[51] Chen, R. S, He, S, Wang, Q. L, et al. A Preliminary Discussion of Magma Activity on the Maturation of Organic Matter-Taking Geyucheng-Wenan Area of Hebei Province as an Example. Petroleum Exploration and Development(1989). in Chinese with English abstract).

[52] Zhu, D. Y, Jin, Z. J, Hu, W. X, et al. Effect of Igneous Activity on Hydrocarbon Source Rocks in Jiyang Sub-Basin, Eastern China. Journal of Petroleum Science and Engineering (2007).

[53] Lomonosov, M. Ma WJ (translator). Stratum. Beijing: Science Press; (1958).

[54] Tissot, B. P, & Welte, D. H. Petroleum Formation and Occurrence. Berling Heidelberg New York Tokyo: Springer; (1978).

[55] Hunt, J. M. Petroleum Geochemistry and Geology. San Francisco: W H Freeman; (1979).

[56] Liu, W. H, Huang, D. F, Xiong, C. W, et al. Advances in Hydrocarbon Generation Theory and Immature and Low Mature Oil and Gas Distribution and Research. Natural Gas Geoscience (1999). in Chinese with English abstract).

[57] Guo, Z. Q, & Wang, X. B. A Discussion of Abiogenic Natural Gas in the Songliao Basin. Science in China (Series B) (1994). in Chinese).

[58] Wang, X. B, Li, C. Y, Chen, J. F, et al. A Discussion of Abiogenic Natural Gas. Chinese Science Bulletin (1997). in Chinese)

[59] Yang, Y. F, Zhang, Q, Huang, H. P, et al. Abiogenic Natural Gases and their Accumulation Model in Xujiaweizi Area, Songliao Basin, Northeast China. Earth Science Frontiers (2000). in Chinese with English abstract).

[60] Wang, P. J, Hou, Q. J, Wang, K. Y, et al. Discovery and Significance of High CH_4 Primary Fluid Inclusions in Reservoir Volcanic Rocks of the Songliao Basin, NE China. Acta Geologica Sinica (2007)., 81(1), 113-120.

[61] Kerrick, D. M, Caldeira, K, & Metamorphic, C. O. Degassing and Early Cenozoic Paleoclimate. GSA Today (1994)., 4, 57-65.

[62] Dai, J. X. Abiogenic Gas in Oil-Gas Bearing Basins in China and Its Reservoirs. Natural Gas Industry (1995). in Chinese with English abstract).

[63] Tu, G. C. The Discussion on Some CO_2 Problems. Earth Science Frontiers (1996). in Chinese with English abstract).

[64] Yun, J. B, Pang, Q. S, Xu, B. C, et al. Research on the Formation Condition of CO_2 Gas Pool in the Southern Songliao Basin. Journal of Daqing Petroleum Institute (2000). in Chinese with English abstract).

[65] Dai, J. X. Composition Characteristics and Origin of Carbon Isotope of Liuhuangtang Natural Gas in Tengchong County, Yunnan Province. Chinese Science Bulletin (1989)., 34(12), 1027-1030.

[66] Cas RAFWright JV. Volcanic Successions: Ancient and Modern. London: Allen and Unwin; (1987).

[67] Fisher, R. V, & Schmincke, H. U. Pyroclastic Rocks. Berlin-Heidelberg-New York: Springer; (1984).

[68] Lajoie, J. Facies Models 15: Volcaniclastic Rocks. Geoscience Canada (1979). , 6(3), 129-139.

[69] Qiu, J. X. Volcanic Lithofacies and their Characteristics. Geological Science and Technology Information (1984). in Chinese with English abstract).

[70] Wang, D. Z, & Zhou, X. M. Volcanic Petrology. Beijing: Science Press; (1982).

[71] Wang, P. J, Chi, Y. L, Liu, W. Z, et al. Volcanic Facies of the Songliao Basin: Classification, Characteristics and Reservoir Significance. Journal of Jilin University (Earth Sciences Edition) (2003). in Chinese with English abstract).

[72] Li, N, Qiao, D. X, Li, Q. F, et al. Theory on Logging Interpretation of Igneous Rocks and Its Application. Petroleum Exploration and Development (2009). in Chinese with English abstract).

[73] Pan, B. Z, Li, Z. B, Fu, Y. S, et al. Application of Logging Data in Lithology Identification and Reservoir Evaluation of Igneous Rock in Songliao Basin. Geophysical Prospecting for Petroleum (2009). in Chinese with English abstract).

[74] Guo, Z. H, Wang, P. J, Yin, C. H, et al. Relationship between Lithofacies and Logging Facies of the Volcanic Reservoir Rocks in Songliao Basin. Journal of Jilin University (Earth Science Edition) (2006). in Chinese with English abstract).

[75] Wang, M, Xue, L. F, & Pan, B. Z. Lithology Identification of Igneous Rock Using FMI Texture Analysis. Well Logging Technology (2009). in Chinese with English abstract).

[76] Yao, R. S, Wang, P. J, Song, L. Z, et al. Imaging Logging Response to Volcanic Pores and Fractures of Yingcheng Formation in the Songliao Basin. Progress in Geophysics (2011). in Chinese with English abstract)

[77] Zhang, Y, Pan, B. Z, Yin, C. H, et al. Application of Imaging Logging Maps in Lithologic Identification of Volcanic. Geophysical Prospecting for Petroleum (2007). in Chinese with English abstract).

[78] Tang, H. F, Wang, P. J, Jiang, C. J, et al. Application of Waveform Classification to Identify Volcanic Facies in Songliao Basin. Oil Geophysical Prospecting (2007). in Chinese with English abstract).

[79] Xu, Z. S, Wang, Y. M, Pang, Y. M, et al. Identification and Evaluation of Xushen Volcanic Gas Reservoirs in Daqing. Petroleum Exploration and Development (2006). in Chinese with English abstract).

[80] Chen, Z. Y, Li, J. S, Zhang, G, et al. Relationship between Volcanic Rocks and Hydrocarbon within Liaohe Depression of Bohai Gulf Basin, China. Petroleum Exploration and Development (1996). in Chinese with English abstract).

[81] Zou, C. N, Zhao, W. Z, Jia, C. Z, et al. Formation and Distribution of Volcanic Hydrocarbon Reservoirs in Sedimentary Basins of China. Petroleum Exploration and Development (2008). in Chinese with English abstract).

[82] Magara, K. Volcanic Reservoir Rocks of Northwestern Honshu Island, Japan. Geological Society, London, Special Publications (2003). , 214(1), 69-81.

[83] Petford, N, & Mccaffrey, K. Hydrocarbons in Crystalline Rocks: An Introduction. Geological Society, London, Special Publications (2003). doi:GSL.SP.2003.214.01.01, 214(1), 1-5.

[84] Tang, H. F, Pang, Y. M, Bian, W. H, et al. Quantitative Analysis on Reservoirs in Volcanic Edifice of Early Cretaceous Yingcheng Formation in Songliao Basin. Acta Petrolei Sinica (2008). in Chinese with English abstract).

[85] Huang, Y. L, Wang, P. J, & Shao, R. Porosity and Permeability of Pyroclastic Rocks of the Yingcheng Formation in Songliao Basin. Journal of Jilin University (Earth Science Edition) (2010). in Chinese with English abstract).

[86] Guo, Z. Q. Volcanic Activity versus Formation and Distribution of Oil and Gas Fields. Xinjiang Petroleum Geology (2002). in Chinese with English abstract).

[87] Li, C. G. Oil and Gas Related to the Volcanic Rocks in Dongying and Huimin Depressions. Explorationist (1997). in Chinese with English abstract).

[88] Fu, G, Hu, M, & Yu, D. Volcanic Cap Rock Type and Evaluation of Sealing Gas Ability: An Example of Xujiaweizi Depression. Journal of Jilin University (Earth Science Edition) (2010). in Chinese with English abstract)

[89] Lee, G. H, Kwon, Y. I, Yoon, C. S, et al. Igneous Complexes in the Eastern Northern South Yellow Sea Basin and their Implications for Hydrocarbon Systems. Marine and Petroleum Geology (2006). , 23(6), 631-645.

Chapter 5

Volcanic Natural Resources and Volcanic Landscape Protection: An Overview

Jiaqi Liu, Jiali Liu, Xiaoyu Chen and Wenfeng Guo

Additional information is available at the end of the chapter

http://dx.doi.org/10.5772/54586

1. Introduction

A volcano is a treasure of nature; volcanic activity is an important source for creating natural wealth. Most mineral resources are related to volcanic activity (syn- and post-volcanism), such as metal deposits of gold, silver, copper, lead and zinc, and non-metal deposits of sulphur and also diatomite; some diamonds and gemstones such as sapphire, ruby, garnet, olivine, zircon, spinel, chalcedony, agate or obsidian - all the results of some specific type of volcanic activity. Every part of a volcano is valuable. Even volcanic ash, scoria and various coherent volcanic rocks are all good building materials. Natural rock fibre and composite material made of basalt have become recently the "green" building material in the 21st century. Hot springs and mineral springs associated with volcanic areas have medical value. In addition, geothermal resources are clean energy sources that could be widely used in the future. What also attracts people is that volcanic activity is a sculptor of nature - producing multiple landscape contours and diverse ecological environments. Many volcanic areas are natural heritage sites and can act as major tourist attractions, in fact, many famous tourist resorts are developed in volcanic fields.

This article attempts to reduce people's fear of volcano eruptions by viewing them from the perspective of the infinite wealth which they create for us. We will give a brief introduction and description of the gifts that volcanoes provide us, e.g., the land resources, the geothermal energy, the volcanic materials, the gemstones and mineral resources, and the volcanic landscapes. We will then introduce their characteristics, distributions and the influences on the development of human beings based on the latest research achievements. A lot of vivid examples would be presented to readers in order to allow them to appreciate the material and spiritual wealth of volcanoes.

2. Volcanic land resources

Many land resources are formed by volcanoes. In the vast oceans, there are numerous sporadic islands, big and small. Take Kosrae and Azores Archipelago for example. Kosrae (5°9′N, 163°00′E), Federated States of Micronesia, is a small (112 km^2) volcanic island in the west-central Pacific Ocean [1]. Compared with Kosrae, Azores Archipelago (36°55′~39°43′N, 25°01′~31°07′W) is a much larger volcanic island. It consists of nine volcanic islands and covers 2,247 km^2 [2].

These volcanic islands not only offer space for humans to live on, they have also become courier stations for shipping and communication. They were particularly more important in ancient times when seamanship was not well developed. Some typical lands which have been created by volcanic activities will be introduced below.

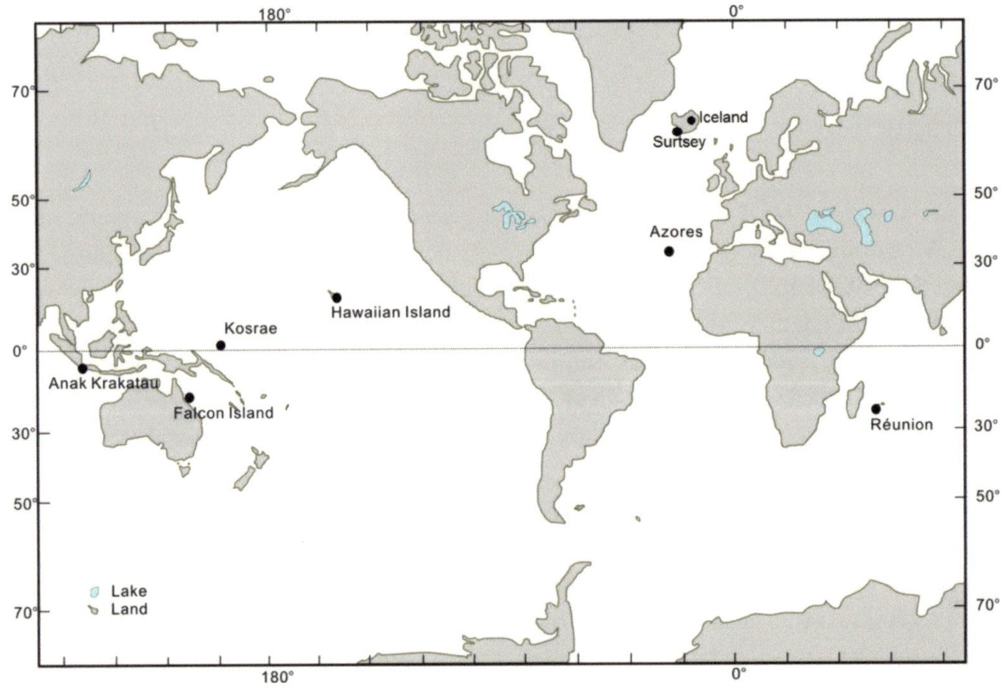

Figure 1. Locations of the volcanic islands referred to in the article

The Hawaiian Islands in the Pacific Ocean are a typical example of islands constructed by volcanoes. With continuous oceanic volcano eruptions and magma pouring into the ocean constructing islands, the young Hawaiian Islands keep growing. This situation also happens in Iceland and Reunion.

Indonesia is the biggest archipelagic state in the world and also a "volcano country", most of the islands there have been constructed by volcanoes, for example, Anak Krakatau is called Krakatoa's son. It was formed both by volcanic activity and wave-cut erosion. Eventually the growth speed of the volcano exceeded the wave-cut erosion and emerged in 1930, piled up from an ocean floor of over 100 m depth, forming islands of cinder and lava covering more than 2 km^2 in area [3]. This new island, with an elevation of 9 m in 1930, grew relatively quickly in the first decade of its existence, to 67 m in 1933, and 132 m by 1941 and subsequently to 170 m by 1966. Changes in height were accompanied by enlargement of the island's area. By 1981 the diameter of Anak Krakatau was reported to be about 2,000 m, with the highest elevation at about 200 m above sea level [4].

Falcon Island (Australia) is a small, uninhabited, rocky island, bounded by a fringing reef, at the south of the Palm Islands, formed under the joint action of submarine volcanic eruptions and waves. It is approximately 35 km offshore, 40 km from the nearest city and river estuary, and 8 km from an aboriginal settlement on Great Palm Island [5]. It is in the Pacific Ocean about 1600 km from the eastern coast of Australia. In 1915, Falcon Island disappeared suddenly due to the dominant role of seawater erosion [6]. Eleven years later, the accumulative effect of submarine volcanic eruptions made it re-emerge out of the seawater.

Surtsey (63.3°N, 20.6°W) is also a new island which was formed by volcanic eruptions in 1963-1967. It is a volcanic island off the southern coast of Iceland. It was formed in a volcanic eruption which began 130 metres below sea level and reached the surface on 15 November 1963. The eruption lasted until 5 June 1967, when the island reached its maximum size of 2.7 km^2. Since then, wind and wave erosion have caused the island to steadily diminish in size: as of 2002, its surface area was 1.4 km^2 [7]. The heavy seas around the island have been eroding it ever since the island appeared and since the end of the eruption almost half its original area has been lost. The island currently loses about 1.0 hectare of its surface area each year [7]. This island is unlikely to disappear entirely in the near future. The eroded area consisted mostly of loose tephra, easily washed away by wind and waves. Most of the remaining area is capped by hard lava flows, which are much more resistant to erosion. In addition, complex chemical reactions within the loose tephra within the island have gradually formed highly erosion resistant tuff material, in a process known as palagonitization [8]. Estimates of how long Surtsey will survive are based on the rate of erosion seen up to the present day. Assuming that the current rate does not change, the island will be mostly at or below sea level by 2100 [9].

China has few volcanic islands (Figure 2). They were mainly formed by volcanic eruptions and principally include the Penghu Islands (Figure 3), Pengjia Islet, Mianhua Islet and Huaping Islet in the north of Taiwan Island, Green Island, Orchid Island and Gueishan Island on the continental slope in the east of Taiwan, Diaoyu Islands in the East China Sea and so on.

3. Geothermal energy

In addition to the land resources introduced above, volcanoes provide us with clean energy - geothermal energy. So it has received the attention of all countries as a new energy source and the reserves are much more than the whole amount energy people are using currently. Its exploitation and utilization are developing rapidly. The definitions, distribution, present exploitation and utilization situation, and some typical examples would be introduced below.

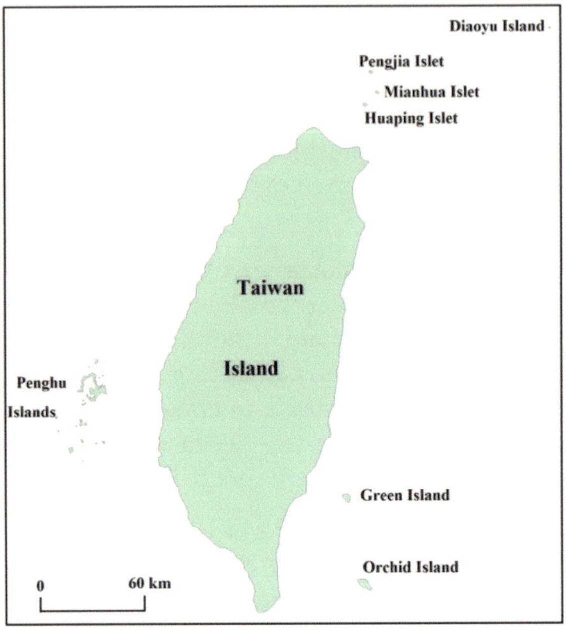

Figure 2. Location of volcanic islands in China

Figure 3. Penghu Island

Geothermal energy is the heat energy stored inside the Earth that originates from the melted magma of the Earth and the decay of radioactive substances. Most geothermal energy gathers around plate borders where most volcanic eruptions and earthquakes happen.

Volcanically active areas normally have a background of high geothermal energy. Volcanic eruption is the most violent exhibition of the internal thermal energy of the Earth on the Earth's surface [10]. Areas with more volcano activities are generally areas with high geothermal flow of geothermal energy. This is because the hot magma chamber under the volcano may heat the circulating groundwater. The heated groundwater is either stored under the ground or spurts out of the ground surface to form hot springs, boiling springs (e.g., the Sirung Volcanic Boiling Spring, Lesser Sunda Islands, Indonesia and Great Boiling Spring, Nevada, United States), geysers (e.g., the Fly Ranch Geyser, Nevada, United States, the Strokkur Geyser, Iceland), fumaroles (e.g., the Valley of Ten Thousand Smokes, Katmai National Park, Alaska, United States) and boiling mud pots (Fountain Paint Pots, Yellowstone National Park, United States, the mud pool at Orakei Korako, north of Taupo, New Zealand and the mud pool at Hverir, Iceland)[10].

The formation of a useful geothermal system needs to possess three essential conditions: an underground heat source (hot rocks), a heat transfer medium (groundwater) and a heat conducting channel (the fissures or boreholes communicating the underground heat reservoir and ground surface).

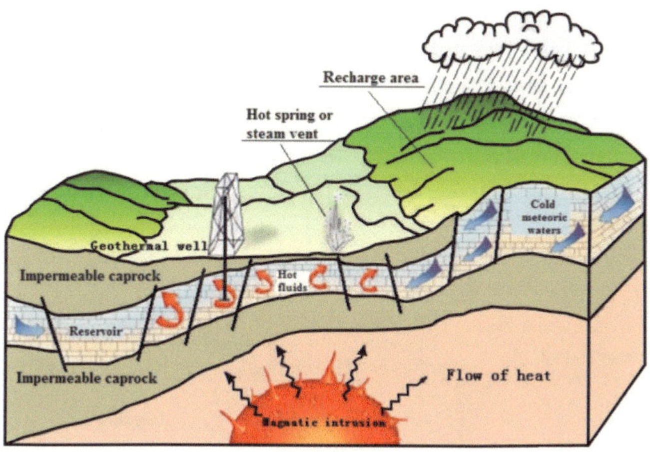

Figure 4. A natural geothermal system [11]

Geothermal energy may be classified into four categories [12]: (1) hydrothermal: the hot water or hydrothermal steam in the shallow layer with a depth of hundreds of metres ~ 2,000 metres; (2) geopressured geothermal energy: the high-temperature high-pressure fluid sealed thousands of metres under the ground in some large sedimentation basins; (3) magmatic thermal energy: the enormous thermal energy stored in magma pockets; (4) hot dry rock: it is a high-temperature rock mass stored deep underground without water or steam with the depth (thousands of metres) that can be reached with current drilling technology. Usually, the geothermal energy contained in hot dry rocks and magma pockets is much more than the geothermal energy of hydrothermal and geopressured heat reservoirs.

Hot dry rock is mainly metamorphic rock or crystalline rock. The key technology of heat collection is to build a heat exchange system in the body of hot dry rock without percolation. Generally, high-pressure water is injected into the rock stratum of 2,000~6,000 metres underground through a injection well which permeates the gaps of rock stratum and obtains geothermal energy; then the high-temperature water and steam in the gaps of rocks are picked up through a dedicated production well (at a distance of 200~600 metres) to the ground; the water and steam can be used to generate electricity after pouring into a heat exchange system; the water after refrigeration will be injected into the underground heat exchange system again via a high-pressure pump for recycling. The entire procedure is a closed system.

There are many problems in trying to tap Earth's internal heat as an alternative clean energy source. Earthquake risks and poorly understood geology are the two most important aspects. Domenico Giardini called for a better understanding of earthquake risk in pursuing deep geothermal energy using an enhanced geothermal system (EGS) [13]. In fact an EGS demonstration project in Geysers, north California, was halted by the geological anomalies. The California-based company AltaRock Energy was unable to penetrate the formation capping the hot rocks after months of drilling in 2009. Similar frustrations were encountered during EGS drilling projects at Paralana and the Cooper Basin, both in South Australia. In general, depths of 3–10 km are optimal for geothermal exploitation because they are extremely hot and accessible to modern drilling techniques. But this rule can be broken by geological surprises. In order to improve our geological understanding and enable us to find optimal drilling sites, China is launching the deep exploration technology and experimentation project, SinoProbe, to locate mineral resources and to find out more about earthquakes and volcanism. Meanwhile, the United States and China will inject US$150 million over the next five years into a joint Clean Energy Research Center [13].

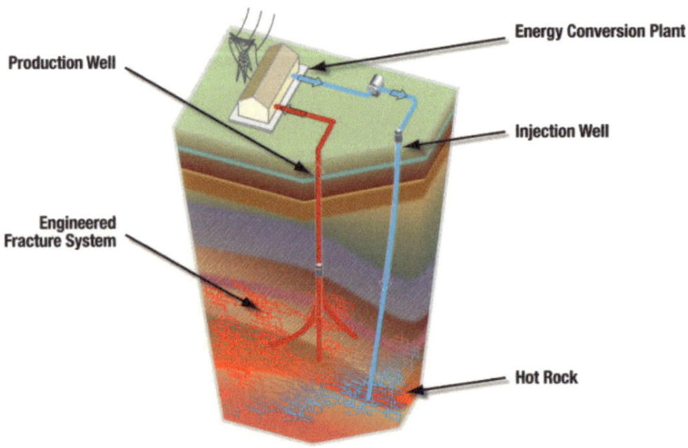

Figure 5. EGS Cutaway Diagram [14]

After the Second World War, countries around the world began to pay attention to exploiting and utilizing geothermal energy. Both the number of countries producing geothermal power and the total worldwide geothermal power capacity under development appear to be increasing significantly (Table 1 and Table 2) [15].

http://dx.doi.org/10.5772/54586

Country	1990 (MWe)	1995 (MWe)	2000 (MWe)	2005 (MWe)	2010 (MWe)
Argentina	0.7	0.6	0	0	0
Australia	0	0.2	0.2	0.2	1.1
Austria	0	0	0	1	1.4
China	19.2	28.8	29.2	28	24
Costa Rica	0	55	142.5	163	166
El Salvador	95	105	161	151	204
Ethiopia	0	0	8.5	7	7.3
France	4.2	4.2	4.2	15	16
Germany	0	0	0	0.2	6.6
Guatemala	0	33.4	33.4	33	52
Iceland	44.6	50	170	322	575
Indonesia	144.8	309.8	589.5	797	1197
Italy	545	631.7	785	790	843
Japan	214.6	413.7	546.9	535	536
Kenya	45	45	45	127	167
Mexico	700	753	755	953	958
New Zealand	283.2	286	437	435	628
Nicaragua	35	70	70	77	88
Papua New Guinea	0	0	0	39	56
Philippines	891	1227	1909	1931	1904
Portugal	3	5	16	16	29
Russia	11	11	23	79	82
Thailand	0.3	0.3	0.3	0.3	0.3
Turkey	20.6	20.4	20.4	20.4	82
USA	2774.6	2816.7	2228	2544	3093
Total	5831.7	6866.8	7974.1	9064.1	10716.7

Table 1. Installed geothermal electric generating capacity worldwide from 1995 to 2000 in five-year intervals[16]

Country	1990-2010 (increase in MW_e)	Increase (%)
Iceland	530.4	1189.24
Portugal	26	866.67
Indonesia	1052.2	726.66
Russia	71	645.45
Turkey	61.4	298.06
France	11.8	280.95
Kenya	122	271.11
Nicaragua	53	151.43
Japan	321.4	149.77
New Zealand	344.8	121.75
El Salvador	109	114.74
Philippines	1013	113.69
Italy	298	54.68
Mexico	258	36.86
China	4.8	25
USA	318.4	11.48
Thailand	0.3	0
Costa Rica	166	--
Papua New Guinea	56	--
Guatemala	52	--
Ethiopia	7.3	--
Germany	6.6	--
Austria	1.4	--
Australia	1.1	--
Argentina	-0.7	n/a
Total	4885	83.77

Table 2. Development of generating geothermal power from 1990 to 2010

Indonesia is a "volcano country". It owns 40% of the world's geothermal energy reserves [17]. According to the latest data released by the Indonesian Ministry of Energy and Mineral Resources, the potential installed capacity of geothermal power generation in the country is as much as 27,140 MW, equivalent to the power generated by 12 billion barrels of petroleum, twice the oil deposit of Indonesia and about 40% of the total reserves of geothermal resources in the world [18]. Indonesia ranks third in the world in terms of geothermal energy consumption, after the US and the Philippines. At present, the geothermal power generation capacity of Indonesia is 1,197 MW. It is also the third biggest emitter of greenhouse gases and aims to cut emissions by 16% by 2025 [15].

11.5% of the territory of Iceland is covered by modern glaciers. However, within the range of 103,000 km^2, it has at least 200 volcanoes formed in the last million years, including about 30 active volcanoes [19]. It has ample geothermal resources and more than 800 thermal fields. Called the "*Land of Ice and Fire*" it is one of the countries among a few with many thermal fields. The distribution of other thermal fields is closely related to the locations of volcanoes. In the volcanic zone running diagonally through the whole island from southwest to northeast, there are 26 high-temperature steam fields, about 250 low-temperature geothermal fields and more than 800 natural hot springs [19]. Within the range of the 0~10 km thick crust, the total amount of geothermal resources is 300 million TWh (1TWh is equivalent to 100 million kWh of electricity); within the range of the 0~3 km thick crust, the total amount is 30 million TWh; and the amount of geothermal resources technically exploitable is one million TWh [19]. Through long-term exploitation and utilization, Iceland has developed efficient geothermal utilization techniques and become the only country in the world almost non-dependent on petroleum. Heating and electric power both rely on natural geothermal energy [20].With a small population, the country is currently generating 100% of its power from renewable sources, deriving 25% of its electricity and 90% of its heating from geothermal resources [17]. Of course, the long-term exploitation and utilization of geothermal energy has also had some negative impacts on geothermal resources e.g., continual descent of the water level in some geothermal systems, temperature and chemical component changes because of the injection of cold water [21]. To address these impacts, Iceland has taken many measures, for example, according to Iceland's environmental law, any geothermal exploitation with a gross amount of above 25 MWe or a net amount of above 10 MWe must submit a detailed environmental impact assessment report [20]. This attitude of attaching importance both to resource exploitation and utilization, and to their conservation and future development is commendable.

New Zealand lies on the suture line between the southwestern margin of the Pacific Plate and the Indo-Australian Plate. This suture line extends from the sea southeast of North Island to the northwest of South Island and then goes further to the south along the western edge of South Island (see Figure 6). The Pacific Plate to its east subducts into the crust of North Island and forms the Taupo Volcanic Zone (see Figure 6), which is the home of the main active volcanoes and geothermal fields in New Zealand. Among the exploited geothermal fields, there are Wairakei, Broadlands, Rotokawa, Kawerau, Ohaaki and Mokai (see Figure 7). Wairakei Geothermal Power Station was the world's first geothermal power station which generates power from wet steam and also is the world's second largest geothermal power station, behind only Italian Larderello Geothermal Power Station [23]. This geothermal field is located in the central volcanic zone on North Island, New Zealand, and about 16 km to the northeast of Taupo Lake. It is the largest geothermal field in New Zealand. Wairakei Geothermal Power Station was developed in 1950 and started power generation in 1958. Now it has more than 100 gas wells, including 60 wells for power generation, with a total installed capacity of about 180 MW and annual power generation of 1501 GW h [23]. The "wet" steam extruded from the gas wells contains 80% water. Its temperature may reach 300℃ at most. During power generation, white water vapour spurts out of the well mouth continuously up to the sky and turns into clouds shortly after. Against the blue sky and over the green pines it looks majestic (Figure 8).

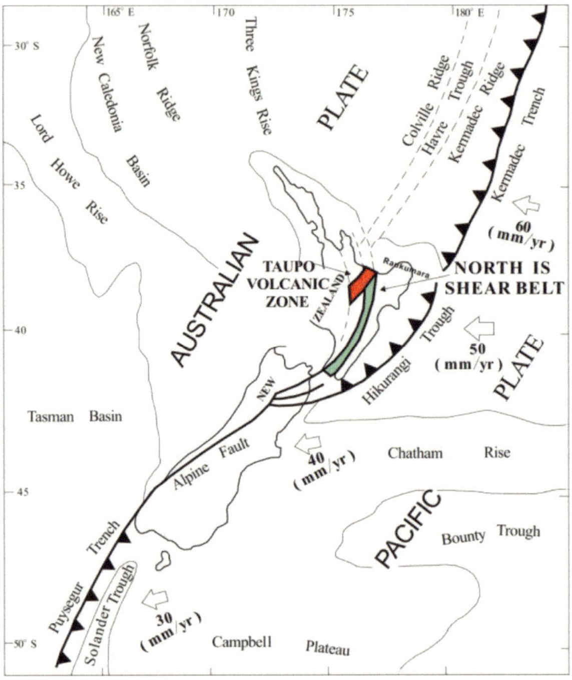

Figure 6. Location of Pacific-Australian plate boundary and Taupo Volcanic Zone[22]

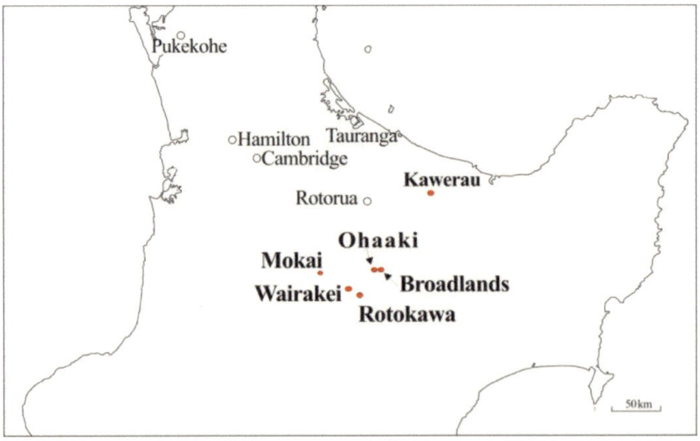

Figure 7. Location of the Wairakei Geothermal Power Station in the Taupo Volcanic Zone, New Zealand

http://dx.doi.org/10.5772/54586

Figure 8. Steam pipelines towards the Wairakei Geothermal Power Station [24]

4. Hot springs and mineral springs

4.1. Hot springs

A hot spring is a kind of spring belonging to ground water. It is called a hot spring if the temperature of the spring pouring out of the Earth's surface is higher than that of the local ground water. If the temperature of the spring is lower than that of the local ground water, it is called a cold spring. Hot springs are a display of geothermal energy.

A hot spring can be created in many ways; generally it is created by the ground water percolation cycling system effect. The average geothermal gradient of the Earth's near-suface is 3 degrees Celsius per 100 metres, the atmosphere penetrates into the underground, becomes aquifer and absorb heat from the deep underground rocks. The heated water can produce steam as well as the air included in the water expanding, that increases the pressure of the water-containing system, and then it pours out at the surface along cracks and gaps to become the hot spring.

Most hot springs are located in volcanic areas and are closely related to volcanic activities. No matter if the volcanoes are erupting or dormant (even extinct ones, under which magma pockets existing or magma activities always exist), lots of heat energy is accumulated, which heats the surrounding ground water and make it pour out as hot springs.

During the storing and moving process, due to the interaction of water and rocks, the hot spring includes many chemical components. Because of the different features of these components, the medical effects are also different.

For example [25], when people have a bath in hot spring containing carbonic acid gas, i.e., carbon dioxide gas, carbonic acid gas will adhere to the skin in bubbles and form a carbonic acid gas film which will keep exchanging with new small bubbles leading to a warm and cosy feeling. They also can stimulate the blood capillaries to expand and promote blood circulation. The carbon dioxide gas breathed into the lungs can strengthen gas metabolism and help balance acid-base organism.

Bicarbonate hot springs can soften and clean skin. People will feel smooth and cool after the bath. As the calcium in the spring can slightly dry the skin out, it is also a good medical treatment for trauma, chronic eczema and ulcers.

When people take a bath in a hot spring including hydrothion, sodium sulphide will be formed on the skin, which can stimulate the skin's blood circulation and nutrition metabolism, promote softer skin and dissolve cutin, reduce inflammation and increase immunity; it also can adjust blood pressure in two ways - improve the insufficiencies of coronary arteries. However, hydrothion is a poisonous gas; excessive hydrothion can lead to neurotoxicity, from headaches to dizziness to respiratory paralysis. Therefore, be aware when bathing in hydrothion hot springs.

Sulphate hot springs can be used for a bath and the water can be drunk; as a bath, the water stimulation from salt to skin can lead to the expansion of blood capillaries and promote the body's metabolism, it can also assist in the treatment of some skin diseases. Drinking sodium sulphate and magnesium sulphate in hot spring waters can promote the secretion and excretion of bile, clean the stagnancy of bile and prevent calculus forming, so it can act as a medical treatment for cholecystitis and gall-stones; drinking calcium sulphate hot spring water can help purine supersession and promote excretion of uric acid, so it can act as a medical treatment for gout and urethra inflammation.

A chloride hot spring is called a "nerve painkiller" as the excitement of the nerve can be reduced when bathing in chloride springs - a good medical treatment for neuralgia. The osmotic pressure of a chloride hot spring with medium concentration (content is above 5g/l) is close to a salt solution, therefore bathing in a hot spring of 36~38 degrees Celsius will help to treat trauma, haemorrhoids and skin diseases; chloride hot springs with a high concentration will help to promote the constitution, to recover the function of the ligament arthroclisis, muscle atrophy and dyskinesia.

Drinking from a hot spring with silicate (metasilicate) can help to adjust the metabolism and promote gastrointestinal motility and strength digestion; when bathing, silicate will adhere to the skin, clean the skin and skin mucosa.

In addition, hot springs with radon, arsenic, bromine, iodine and other microelements also have medically positive effects for humans.

There are hot springs on all continents and in many countries around the world. We list some of the famous hot springs with medical value around the world – see Table 3 and Figure 9.

http://dx.doi.org/10.5772/54586

Country	Hot Spring
USA	Glenwood Springs
Chile	Chihuío
Ecuador	Baños de San Vicente
China	Tengchong Hot Spring Zone Taiwai (Beitou Hot Spring, Yangmingshan Hot Spring)
Indonesia	Maribaya Hot Spring
Japan	Kusatsu Onsen, Arima Onsen, Gero Onsen
Iceland	Blue Lagoon Spring, Geysir Hot Springs
Italy	Bormio, Sondrio Province, Lombardy (geothermal spa)
New Zealand	Rotorua Volcanic Hot Spring

Table 3. Some famous hot springs with medical value around the world

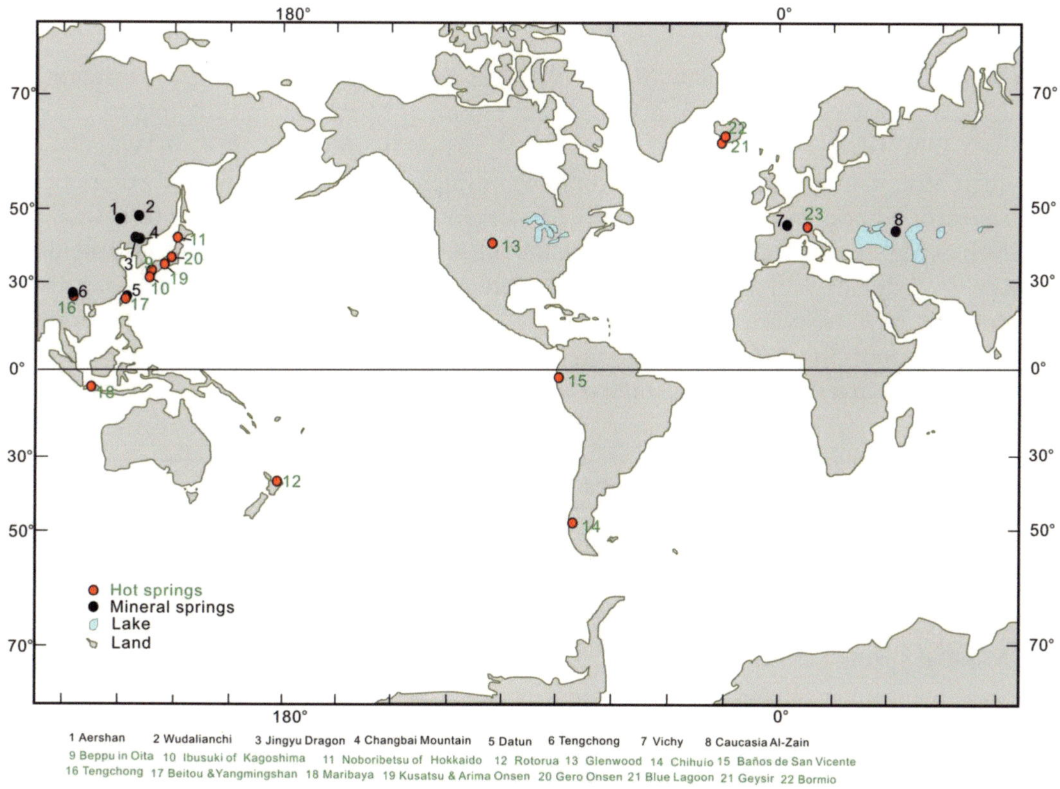

Figure 9. Location of famous hot springs and mineral springs around the world

4.2. Mineral springs

Mineral springs are springs with lots of mineral substances. They may be hot, defined by Mazor as having a temperature of >6 degrees Celsius above that of the mean annual surface temperature, or cold [26]. Different stipulations had been set up by different organizations. The element content standards of the mineral spring water which has been adopted by the Codex Alimentarius Commission in 1993 can be seen in Table 4 [27]. But according to the stipulations of the International Mineral Spring Association, in natural water, if one or more microelements and mineral substances needed for humans, such as lithium, strontium, zinc, copper, barium, cadmium, selenium, arsenic, manganese, antimony, nickel, bromide, iodide, metasilicate, free carbon dioxide, etc. reaches the standard required, it can be called "mineral springs" [28]. The French Vichy spring contains four, the Russian Caucasia Al-Zain spring contains two and the Chinese Wudalianchi mineral spring contains more than six (Figure 9). However, the Russian Caucasia Al-Zain spring is created by the melting of Mount Elbrus' glaciers, while the French Vichy and Wudalianchi springs are closely related to volcanoes.

The French Vichy Mineral Spring was formed by a volcanic eruption and can be found at 103 locations. At present, only 15 mineral spring wells have been exploited and utilized. As none of the mineral springs is an artesian spring, the extraction is not easy and wells must be dug. In the downtown of Vichy there is a mineral spring called Célestins. It is the only mineral spring for daily drinking.

In China, Wudalianchi Yaoquan (Figure 9) (literally "Medicated Spring") Mountain in Heilongjiang province, the Aershan Wuli Spring in Inner Mongolia and the Jingyu Dragon Spring in Jilin province are all high-quality volcanic mineral springs (Figure 9). In Wudalianchi Yaoquan Mountain, the mineral water contains multiple desired elements and components such as carbonic acid, iron, zinc, strontium, lithium and germanium. The water features large reserves, is high quality and has medical value. It has an obvious curative effect on stomach disease, skin disease and arthritis. The mud there may be used for pelotherapy and many sanatoriums have been built there. The former wasteland around the area has become an emerging modern city - Wudalianchi today. It is probably the only city established by relying on volcanic resources and it has become a famous geopark.

Zn	Cu	Ba	Cd	Cr_{Total}	Pb	Hg	Se	As_{Total}	Mn	Sulphide	Sb	Ni
-	-	0.7	0.003	0.05	0.01	-	-	0.01	0.5	-	0.05	0.02

Table 4. The elements content standards of the mineral spring water [27]

4.3. Mineral spring culture

The historic culture of mineral springs can be traced back to two thousand years BC. In ancient Greek stories, the goddess who can cure human diseases lives in mineral springs, which made people desperate to attain the waters [29]. In the ancient Rome era third century AD, the development and utilization of mineral springs took shape, it is said that there were more than 860 mineral spring bathing places in the city of Rome.

After the 18th century, people began to study mineral spring as a science. In 1742, a German doctor called Hoffmann confirmed some components of mineral springs based on a predecessor's research, which laid a foundation for the development of mineral spring science [29].

The 20th century saw the development of theoretical research and applied research, and a specialized agency was founded in developed countries like Japan, France, Germany and the US to develop the research and train talent, even putting crenology on required courses of advanced medical schools.

Each country of the world has its own mineral spring historical cultural expression, especially Japan which is called the country of hot springs - thousands of hot springs and mineral springs are located across the Japanese islands. Every family has bathing equipment and it seems that Japanese people like to bathe in hot springs the most. Hot spring bathing can not only reduce tiredness, cure disease and strengthen the body, it can also be a place to communicate with fellow bathers.

Volcanoes have caused natural disasters in Japan, but have also created abundant hot spring resources across the country. Among the 2,200 natural hot springs, the most favoured hot springs are located in Oita, Kagoshima and Hokkaido with different features [29].

Beppu Hot Spring in Oita (Figure 9) has been known about in Japanese since ancient times, there are more than 3,800 spring water holes and the water inflow is more than 200 thousand tons every day [29]. Known as the city of spring, it is the biggest natural hot spring area in Japan and also a world-class hot spring city. Sand Steam Hot Spring (Figure 9) in Ibusuki, Kagoshima, attracts more women (Figure 10). It is the only sand steam hot spring in Japan. To *"sand steam"* is to bury the body except the head in hot sand, termed *"sand happiness"*. It is similar to having a sauna to make you sweat - in less than five minutes your body will feel hot. The sand pressure and hot water promote blood circulation, sweating from the whole body and tiredness to reduce, which is a fantastic medical treatment for preventing rheumatism and nerve ache. There is "sodium" in the sand which makes skin fair, so it is favoured by women for cosmetic reasons. In Noboribetsu, Hokkaido (Figure 9), the hot spring has another fun aspect – the spring, rock, flowers and grass form extremely pleasant scenery. Hot spring hotels are located along the main street, the bathroom in the Noboribetsu International View Club is especially well-known - with length of 90 m and width of 20 m there are more than ten hot spring pools of different sizes and temperatures to choose from, and men and women can bathe together.

Hot springs in New Zealand are located across the country. Rotorua (Figure 9) sitting on volcanic-prone area is called the *"New Zealand Hot Spring City"* and is home to the largest hot spring waterfall in the southern hemisphere and the only mud bath pool in New Zealand - *"Wai Ora Spa"*. The mud contains abundant mineral substances, which have health benefits. In addition, the unique Maori culture all combine to make this area a thriving tourist attraction.

In China, there are fewer present-day volcanic eruptions, but the hot springs and mineral springs related to volcanoes widespread and some volcanic areas, such as Wudalianchi in Heilongjiang, Changbai Mountain in Jilin, Aershan in Inner Mongolia, Tengchong in Yunnan and Datun in Taiwan, produce hot springs and mineral springs (Figure 9). Aershan in Inner

Figure 10. Sand steam hot spring in Ibusuki, Kagoshima [30]

Mongolia in particular contains hot and cold springs, and mineral springs [31, 32]. Hot springs in Tengchong are not only located widely and with numerous spring holes, but it is also well-known that they have a high temperature and water flow [33, 34].

Figure 11. Landscape of Tengchong, China

5. Volcanic materials

Volcanic rocks forming volcanic edifices and surrounding volcanic ring plains consisting of volcanic ash, scoria, pumiceous deposits and the coherent volcanic rocks part of lava flows, lava domes or exposed subsurface facies of a core of a volcano such as dykes, sills, laccoliths are all good building materials widely used for construction such as paving and building houses. Scoria and the volcanic ash of basalt can be used directly as a filling in clinker-free cement to produce high-quality cement [35]; cast stone bricks, tiles and panels made of basalt are good fire-proof materials, which are not only heat-resistant, but also crush-resistant and corrosion-resistant, so much so that they are considered as a substitute for steel in machine tool and the chemical industry [36-38]. What attracts people's attention is basalt fibre, which takes basalt as the raw material. It has the advantage of good combination properties and cost effectiveness, and no poisonous substances, waste gas, waste water, residue or pollution in normal machining processes. It is applied widely in fire-fighting, environment protection, aviation, the arms industry, automobile and ship making, engineering plastics, the construction industry and the so-called "green industrial materials" of the 21st century. The classifications are continuous basalt fibre, thick fibre, narrow fibre, super-narrow fibre and subtle scale, among which the continuous basalt fibre has been previously studied and studies of the others are developing at present.

5.1. Development background of basalt fibre

In 1840, the trial production of rock wool with basalt as the primary raw material succeeded in Wales, UK [39]. In the early 1950s, Germany, the then Czechoslovakia, Poland and Hungary produced basalt wool with an average fibre diameter of 25 μm~30μm from basalt using the centrifugal method [40]. In the early 1960s, the USA, the former Soviet Union and Germany vigorously developed a vertical blowing production process, resulting in the rapid growth of basalt wool output. The former Soviet Union introduced a German patent for producing mineral wool using the vertical blowing method. On the basis of digestion and absorption, it succeeded in applying this technology to the production of basalt wool. The production capacity is 38~40 tons of basalt wool a day [41].

Basalt fibre was successfully developed by the Russian Moscow Glass and Plastic Research Institute in 1953~1954. The first furnace for industrial production was built and put into operation in the Ukraine Fibre Laboratory (TZI) in 1985. The fibre-drawing process of a 200-hole bushing and combination furnace was adopted with an annual output of 260 t [40]. Then, they adopted a 400-hole tank furnace fibre-drawing process to produce CBF and its products. The annual output is about 700 t. In recent years, China, Japan, South Korea and Germany have also carried out relevant research and achieved some new research results [41]. At present, the USA, Germany, China, Japan and South Korea are also carrying out relevant research and are achieving some new research results [41].

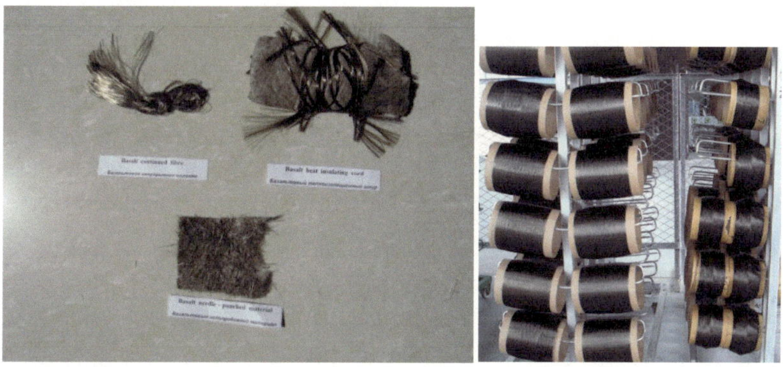

Figure 12. Basalt fibre

5.2. Properties and use of basalt fibre

Compared with glass fibre, rock wool, asbestos and chemical fibre, basalt fibre and its products have extraordinary performance and have multiple application capabilities (Table 5):

1. Good thermal property and flame retardant property: basalt fibre is amorphous state inorganic silicate substance, with good temperature-resistance and heat insulation and without thermal contraction [42]. The temperature range of its usage is -269~700 degrees Celsius, the softening point is 60 degrees Celsius, which is much higher than that of glass fibre - 60~ 450 degrees Celsius and carbon fibre - 500 degrees Celsius. Under a temperature of 500 degrees Celsius, its stability of thermal shock resistance is unchanged. Initial mass fraction loss is less than 0.02 and it is 0.03 under 900 degrees Celsius. Under a temperature of 600 degrees Celsius, its breaking strength can still keep as 80% as the original intensity, and it will not shrink under 860 degrees Celsius while mineral wool with good temperature-resistance can only keep 50%~60%. Carbon fibre will produce CO and CO_2 under 300 degrees Celsius [43], while glass fibre is crushed completely. Therefore, basalt fibre can be used for the manufacture of flame retarding materials such as fire-proof suits, blankets and curtains, and high-temperature filters such as scrim, material and high-temperature resistant felt. In addition, with regard to low-temperature resistance, the intensity of basalt fibre will not change under the medium of low-temperature (-196 degrees Celsius) liquid nitrogen. So it is an effective low-temperature heat insulation material [41].

2. Good chemical stability and corrosion resistance: basalt fibre maintains favourable chemical stability and powerful acid and alkali resistance. With natural consistency and high stability in alkali medium such as cement, it can be used as the enhancement material for concrete building structures to replace rebar. Flake coating made of basalt can offer protection to building and objects under water, including warships, to strengthen the capacity and life of corrosion prevention.

3. Good stretch ability and modulus of elasticity: the density of CBF is 2.65~3.00g/cm^3 and its hardness is very high, Mohs hardness scale 5~9, so it has excellent wear resistance and tensile reinforcement [43]. The tensile strength of CBF is 3800~4800 MPa, higher than that of large-tow carbon fibre, aramid fibre, PBI fibre, steel fibre, boron fibre and aluminium oxide fibre, and equivalent to S glass fibre. The strength of basalt fibre may be maintained for 1200 h in 70 degrees Celsius water, while ordinary glass fibre will lose its strength at 200 h. At 100~250 degrees Celsius, the tensile strength of basalt fibre may be increased by 30%, while the tensile strength of ordinary glass fibre will decrease by 23% [41]. Therefore, basalt fibre material has a huge advantage in bridge building as well as in sport material usage such as fishing rods, hockey sticks, antennae, skis, umbrella handles, poles, bows and arrows, and crossbows [43].

   Meanwhile, basalt fibre keeps high modulus of elasticity: 9100 kg/mm2 ~11000 kg/mm^2, higher than that of alkali-free glass fibre, asbestos, aramid fibre, PP fibre and silicon fibre, and close to that of expensive S glass fibre. It may replace S glass fibre in the making of heat insulating products and composite materials, for example, hard armour and GFRP products [40, 41].

4. Low coefficient of heat conduction, good heat-proof quality: the conductivity factor of basalt fibre is 0.031 W/m•K~0.038 W/m•K, lower than that of aramid fibre, aluminium silicate fibre, alkali-free glass fibre, rock wool, silicon fibre, carbon fibre and stainless steel [40]. It may be used for heat preservation of heat treatment equipment, heat insulation of automobiles and ships, and heat preservation of pipelines.

5. High acoustical absorption coefficient and good stealth performance: the acoustic absorption factor of CBF is 0.9~0.99, higher than that of alkali-free glass fibre and silicon fibre. Basalt fibre has good wave permeability, certain wave absorbability, excellent acoustical absorption and insulation. In addition, it may be used as a stealth material [42].

6. High dielectric coefficient and good insulating property: the specific volume resistance of basalt fibre is 1x1012Ω•m, much higher than that of alkali-free glass fibre and silicon fibre. By relying on its good dielectric performance, low hydroscopicity and good temperature resistance, basalt fibre may be made into high-quality PCB and blades [40]. After basalt fibre is treated with a special impregnating compound, its dielectric loss angle tangent is 50% lower than that of ordinary glass fibre, and it may be used to produce new-type heat-resistant dielectric materials [41].

7. Low hygroscopicity and good seepage-proof and anti-crack property: the hygroscopicity of basalt fibre is below 0.1%, lower than that of aramid fibre, rock wool and asbestos [40]. Compared to glass fibre, the hydroscopicity of basalt fibre is 6~8 times lower [44]. It has strong seepage control and crack resistance, and may be widely used in expressways, runways, port terminals, hydroelectric engineering buildings and other infrastructure fields.

The features of basalt decide the wide usage of basalt fibre, not only for aviation, the arms industry, fire-fighting, traffic, energy, environment protection and construction industry, but also espionage, communication and special materials under thermal shock. With abundant basalt resources, its future in industrial production and marketing promotion is bright.

6. Gemstones and other mineral resources

Many gemstones are related to or result from volcanic processes and therefore they are hosted in volcanic deposits and rocks. Gemstones in volcanic rocks include sapphire and ruby, and sometimes adamas can also be found; garnet, pyroxene and olivine with good crystals and bright colours can be found in basalt, which can be used directly as gemstones. Obsidian itself erupted from a volcano can be high quality pure volcanic glass with gem-quality. Crystal, agate and aragonite produced by volcano action are all valuable gemstones. Besides gemstones, many metal and non-metal minerals are related to volcano action, such as gold, silver, copper, lead, zinc and sulphur and diatomite.

Properties	Performance	Basalt fibre	E-glass fibre	S-glass fibre	Carbon fibre-HS	Aramid fibre 1313
Physical properties	Density/g·m^{-3}	2.65-3.00	2.55-2.62	2.46-2.49	1.78	1.44
	Tensile strength/MPa	3000-4840	3100-3800	4590-4830	3100-5000	2758-3034
	Elasticity modulus/GPa	79.3-93.1	76-78	88-91	230-240	124-131
	Elongation after fracture/%	3.15	4.70	5.60	1.20	230
	Conductivity factor /w·(m·kg)$^{-1}$	0.031-0.038	0.034-0.04	0.036-0.04	5-185	Low
	Volume resistivity /Ω·m	1×10^{12}	1×10^{11}	1×10^{11}	2×10^{-5}	>1×10^{11}
	Acoustic absorption factor	0.9-0.99	0.8-0.93	0.8-0.93	Small	Small
Chemical properties	Softening point	960	850	1056	—	—
	Maximum operating temp/℃	700	380	300	2000	250
	Chemical resistance	Acid and alkali resistance	Moderate resistance to alkalis	Moderate resistance to alkalis	No effect	Moderate resistance to acids
	Price/yuan·kg^{-1}	38	17	20	300	200

Table 5. Comparison of the performance of the basalt fibre with other fibres [43]

6.1. Diamonds

Diamonds are commonly hosted in kimberlite pipes that are commonly looked upon as a specific type of maar diatreme volcano (Figure 13). Kimberlite is a type of potassic volcanic rock best known for sometimes containing diamonds. It is named after the town of Kimberley in South Africa, where the discovery of an 83.5 carat (16.7 g) diamond in 1871 spawned a diamond rush, eventually creating the Big Hole [45].

Diamonds form at a depth greater than 93 miles (150 kilometres) beneath the Earth's surface. After their formation, diamonds are carried to the surface of the Earth by volcanic activity. As this molten mixture of magma (molten rock), minerals, rock fragments and diamonds approaches the Earth's surface it begins to form an underground structure (pipe) that is shaped like a champagne flute. These pipes can lie directly underneath shallow lakes formed in the active volcanic calderas or craters [46].

http://dx.doi.org/10.5772/54586

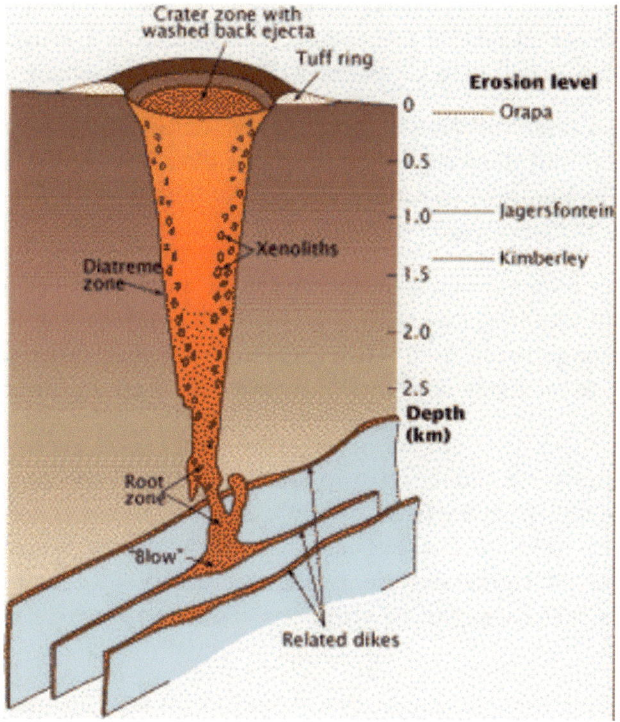

Figure 13. Kimberlite pipe [47]

Kimberlite occurs in the Earth's crust in vertical structures known as kimberlite pipes. According to the descriptions of Volker Lorenz [48], "the formations of maars and diatremes suggest a specific process. Magma rises along a fissure and contacts ground - or surface-derived water. The resulting phreatomagmatic eruptions give rise to base surge and air-fall deposits consisting of juvenile and wall-rock material. Spalling of the wall-rocks enlarges the fissure into an embryonic vent. At a critical diameter of the vent large-scale spalling at depth and slumping near the surface gives rise to a ring-fault of large diameter and subsidence of the enclosed wall-rocks and overlying pyroclastic debris. This subsidence leads to a maar crater at the surface. Various features of kimberlite diatremes seem to be consistent with this model. They extend into fissures along which hot kimberlite magma rise. The diatremes, however, indicate emplacement by a cool gas phase, probably steam. Indicators for subsidence along ring-faults may be diatremes with large diameter, slickenside on walls, saucer-shaped structures, subsided "floating reefs", concentration of xenoliths from specific horizons within certain areas, and zoning of diatreme rocks. It is suggested that formation of kimberlite diatremes may have been influenced by non-juvenile water. For example, the Premier kimberlite pipe and Jagersfontein kimberlite pipe in South Africa, and the Mwadui kimberlite pipe in Tanzania are all famous diamond mines around the world.

Lamproite pipes produce diamonds to a lesser extent than kimberlite pipes. Lamproite pipes are created in a similar manner to kimberlite pipes, except that boiling water and volatile compounds contained in the magma act corrosively on the overlying rock, resulting in a broader cone of eviscerated rock at the surface. This results in a martini-glass shaped diamondiferous deposit as opposed to kimberlite's champagne flute shape [46].

The Argyle diamond mine in Western Australia is one of the first commercial open-cast diamond mines that is dug along an olivine lamproite pipe. The Argyle pipe is a diatreme, or breccia-filled volcanic pipe that is formed by gas or volatile explosive magma which has breached the surface to form a "tuff" cone [46].

The complex volcanic magmas that solidify into kimberlite and lamproite are not the source of diamonds, only the elevators that bring them with other minerals and mantle rocks to the Earth's surface. Although rising from much greater depths than other magmas, these pipes and volcanic cones are relatively small and rare, but they erupt in extraordinary supersonic explosions [49].

Kimberlite and lamproite are similar mixtures of rock material. Their important constituents include fragments of rock from the Earth's mantle, large crystals and the crystallized magma that glues the mixture together. The magmas are very rich in magnesium and volatile compounds such as water and carbon dioxide. As the volatiles dissolved in the magma change to gas near the Earth's surface, explosive eruptions create the characteristic carrot- or bowl-shaped pipes [49].

Diamonds also can be formed by subduction (Figure14). When the ocean floor slides under the mantle, the basaltic rock becomes eclogite, and organic carbon in sediments may become diamonds [49].

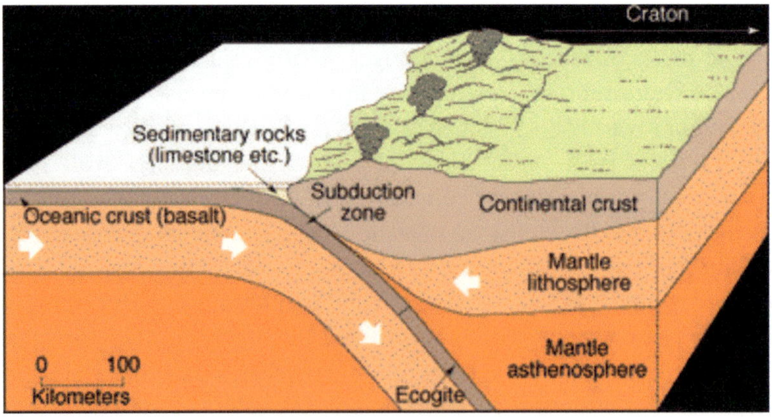

Figure 14. Diamonds can be formed by subduction [50]

6.2. Sapphires

Sapphires are commonly hosted in dykes, volcanic necks and lava flows which are composed by alkali basalts. Many sapphires have been found in Changle of Shandong, Liuhe of Jiangsu and Muling of Heilongjiang in China. Sapphires are also formed in basalts in Kanchanaburi and Banca in Thailand, Pailin in Cambodia and Anaky in Australia [51].

6.3. Garnets

Garnets are a group of silicate minerals that have been used since the Bronze Age as gemstones and abrasives. Garnets possess similar physical properties and crystal forms but different chemical compositions. The different species are pyrope, almandine, spessartine, grossular (varieties of which are hessonite or cinnamon-stone and tsavorite), uvarovite and andradite [52].In these species, only the pyrop has a relationship with the volcanic activities. They are mainly produced in kimberlite, basalt and mantle xenoliths. A famous pyrop, Bohemian garnet in the Czech Republic, is known to be associated with maar diatreme volcanoes of the Czech Republic.

In the abyssal lherzolite inclusions of the basalts, there are olivine, pyroxene, garnet, spinel and other minerals which have fine crystals and bright colours. They have been widely used as gems. This type of inclusion can be easily found in the countries which have volcanoes [51].

6.4. Opal and others

Australia is the opal capital of the world. Opals are often host in fractures, voids and primary stomata of the volcanic rocks (basalts, andesites, rhyolites and tuffs). Besides Australia, opals have been found in the Czech Republic, Mexico, Honduras and other countries. The compositions and formations of the agate and chalcedony are similar to opal, but their distributions are much wider than opal. A lot of agates and chalcedonies have been produced in India, China, Brazil, the United States, Russia, Madagascar, Ireland, Namibia and Egypt [51].

Opal, agate and chalcedony are semi-precious stones that come under the category of quartz. These stones are characterized by fine granular texture and bright colours. They are processed and marketed in different shapes and sizes [53].

Gems with a variety of colours and textures can be formed by the aragonites that fill in the stomata of the basalts and andesites. The basalt of the Penghu Islands in China is rich in this precious gem. This type of gem has been produced in Spain, Italy, Austria, Chile, the United States and other countries [51].

6.5. Obsidian

Obsidian is a naturally occurring volcanic glass formed as an extrusive igneous rock. It is produced when felsic lava extruded from a volcano cools rapidly with minimum crystal growth. Obsidian is commonly found within the margins of rhyolitic lava flows known as obsidian flows, where the chemical composition (high silica content) induces a high viscosity and polymerization degree of the lava. Obsidian can be found in locations which have experienced rhyolitic eruptions. It can be found in Argentina, Armenia, Azerbaijan, Canada, Chile, Greece, El Salvador, Guatemala, Iceland, Italy, Japan, Kenya, Mexico, New Zealand, Peru, Scotland and the United States [54]. Anatolian sources of obsidian are known to have been the material used in the Levant in modern-day Iraqi Kurdistan from a time beginning sometime about 12,500 BC [55]. In Ubaid in the 5th millennium BC, blades were manufactured from obsidian mined in what is now Turkey [56]. Now obsidian has been used for blades in surgery. Obsidian is also used for ornamental purposes and as a gemstone. It possesses the property of presenting a different appearance according to the manner in which it is cut: when cut in one direction it is jet black, in another it is glistening grey. Plinths for audio turntables have been made of obsidian since the 1970s [54].

6.6. Gold deposit

Gold deposits attract the constant attention of every country around the world. People have found that the formation of gold deposits has a close relationship with volcanic activities in terms of time and space. Hu et al. studied the classification of gold deposits relevant to volcano, sub-volcano, intrusion and hydrotermalism. He introduced classification methods of gold deposits according to the volcanic internal ore-controlling structures [57]: 1) gold deposits from bedrock, such as the super large gold bearing porphyry copper deposits in Bingham in the United States and Chuquicamata in Chile; 2) gold deposits from the contact zone between bedrock and volcanic caprock, such as the gold ore in Tuanjiegou, Heilongjiang, China; 3) gold deposits from the volcanic edifice of the caprock of volcanic rock series, such as Conrad in Kazakhstan; 4) gold deposits from hot spring accumulation near volcanic craters, such as Round Mountain in Nevada, California, the United States.

Jiang et al. [58] undertook an in-depth study on volcanism and gold deposits, and claimed that volcanism is one of the optimal geologic conditions for the formation of gold deposits. They classified gold deposits into three categories [58]. The first category is volcanic sediment deposits in tensioned structures which were formed by submarine volcanism. The second category is plutonic volcanic gold deposits formed due to plate activities and collisions during the orogenic and post-orogenic period. They are classified into three sub-categories: 1) gold deposits in intrusive rocks or contact zones including the vein deposits in intrusive rocks and porphyric gold deposits, e.g., Barrick (gold reserves: ~420t), Lihir Island (gold reserves: ~500t) and other huge sub-volcanic porphyry gold deposits in Papua New Guinea, which were formed along with the intrusive magmatism in the Cenozoic island-arc volcanic rock zone subducted by the Western Pacific Plate; 2) gold deposits from continental volcanic rocks; they are mainly developed in the Circum-Pacific Island Arc and related to Middle Cenozoic volcanic activities; 3) metasomatism and filling gold deposits related to plutonic volcanic magmatism, e.g., the Carlin type gold deposits in Nevada, the United States [58] and the Munmtau gold deposit belt in Uzbekistan [59]. The third category is the placer gold deposits formed due to weathering and sedimentation under hypergene conditions.

6.7. Diatomite

Diatomite is a naturally occurring, soft, siliceous sedimentary rock that is easily crumbled into a fine white to off-white powder. It has a particle size ranging from less than 1 micrometre to more than 1 millimetre, but typically 10 to 200 micrometres. Diatomite is formed by the accumulation of the amorphous silica (opal, $SiO_2\ nH_2O$) remains of dead diatoms (microscopic single-celled algae) in lacustrine or marine sediments [60]. So diatomite can be found in exposed maar crater lakes. Where there are maar crater lakes, there are diatomite (e.g., Germany, China, Hungary, Slovakia and so on). It is used as a filtration aid, mild abrasive, mechanical insecticide, absorbent for liquids, matting agent for coatings, reinforcing filler in plastics and rubber, anti-block in plastic films, porous support for chemical catalysts, cat litter, an activator in blood clotting studies and a stabilizing component of dynamite. As it is heat-resistant, it can also be used as a thermal insulator [60].

Most of these deposits are related to calc-alkali volcanic rock, mainly andesite. Some of these deposits are replacement and vein deposits. Stratiform lead and zinc deposits are found at Nong Phai and Song Thoin Kanchanaburi in middle Ordovician limestone. The zinc deposits at Pha Daeng, Mae Sot is the largest zinc deposit in Thailand. The ore are zinc carbonate and zinc silicate in the supergene enrichment in the Jurassic Kamawkala limestone near the Thai-Myanmar border.

7. Volcanic landscapes

Volcanoes are nature's sculptors, making numerous beautiful scenic spots and natural landscapes, which are not only tourist attractions but also ideal places for scientific research. Many famous landscapes and tourist attraction are also volcanic areas, and all of them are colourful and charming. Many of the world's geoparks and natural heritage sites are related to volcanoes.

Heritage is our legacy from the past, what we live with today, and what we pass on to future generations. The United Nations Educational, Scientific and Cultural Organization (UNESCO) seeks to encourage the identification, protection and preservation of cultural and natural heritage around the world considered to be of outstanding value to humanity. This is embodied in an international treaty called the Convention concerning the Protection of the World Cultural and Natural Heritage, adopted by UNESCO in 1972 [61]. The World Heritage List includes 936 sites forming part of the cultural and natural heritage which the World Heritage Committee considers as having outstanding universal value [61].These include 725 cultural, 183 natural and 28 mixed properties in 153 states. Some of them are closely related to volcanic activities and the unique geological landscape and ecological systems in these places illustrate the charm of volcanic activities for humans.

A geopark is defined by UNESCO in its International Network of Geoparks' program as "a territory encompassing one or more sites of scientific importance, not only for geological reasons but also by virtue of its archaeological, ecological or cultural value" [62]. A global geopark is a unified area with geological heritage of international significance and where that heritage is being used to promote the sustainable development of the local communities that live there [63]. The key heritage sites within a geopark should be protected under local, regional or national legislation as appropriate.

We will now introduce some typical heritage sites and geoparks which are associated with volcanic activities (Table 6, Table 7, Figure 15).

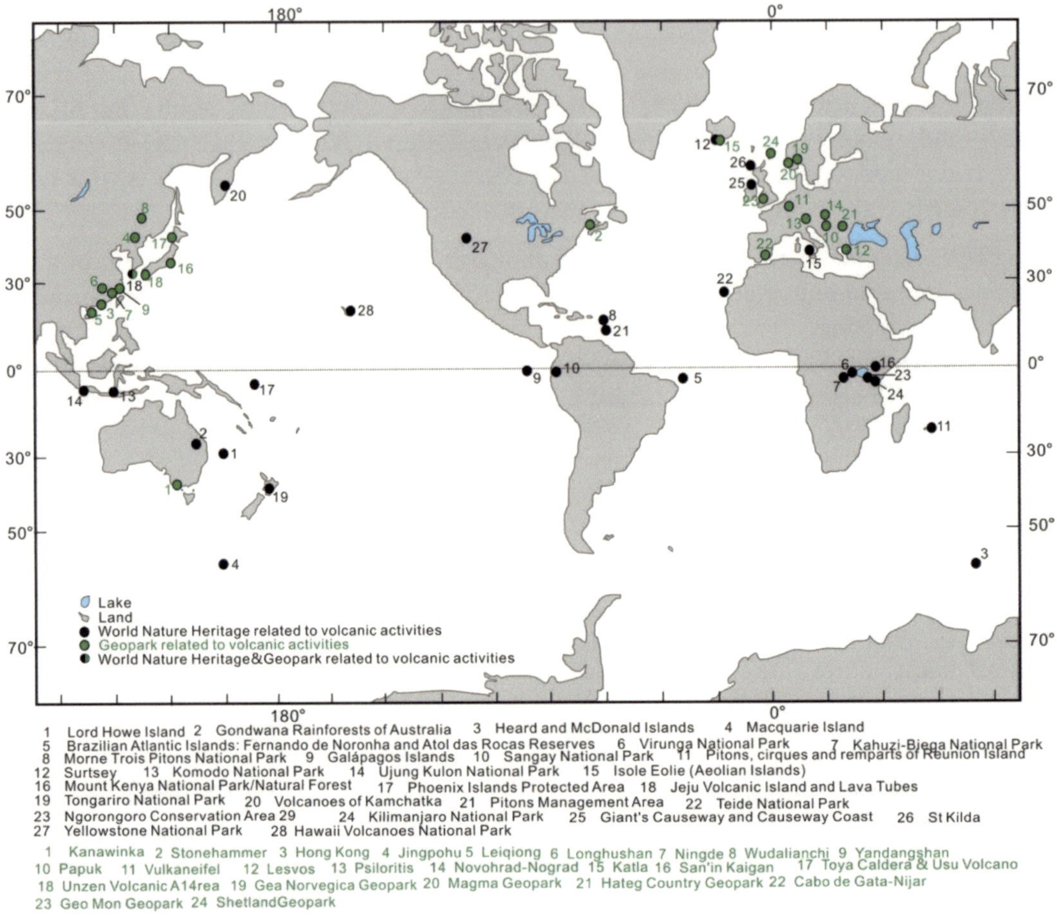

Figure 15. Location of the world nature heritage sites and geoparks which are related to volcanic activities

Country		World Nature Heritage Sites
Australia	Lord Howe Island Group	Located in the South Pacific, 700 km north-east of Sydney, the property is included administratively in New South Wales. Lord Howe Island is the eroded remnant of a large shield volcano which erupted from the sea floor intermittently for about 500,000 years in the late Miocene (6.5-7 million years ago). The entire island group has remarkable volcanic exposures not known elsewhere.
	Gondwana Rainforests of Australia	The outstanding geological features displayed around shield volcanic craters and the high numbers of rare and threatened rainforest species are of international significance for science and conservation.
	Heard and McDonald Islands	Located in the Southern Ocean, approximately 1,700 km from the Antarctic continent and 4,100 km south-west of Perth. As the only

http://dx.doi.org/10.5772/54586

Country		World Nature Heritage Sites
		volcanically active sub-Antarctic islands they 'open a window into the Earth', thus providing the opportunity to observe ongoing geomorphic processes and glacial dynamics.
	Macquarie Island	Macquarie Island is an oceanic island in the Southern Ocean, lying 1,500 km south-east of Tasmania and approximately halfway between Australia and the Antarctic continent. It is the only island in the world composed entirely of oceanic crust and rocks from the Earth's mantle deep below the surface.
Brazil	Brazilian Atlantic Islands: Fernando de Noronha and Atol das Rocas Reserves	The Fernando de Noronha Archipelago and Rocas Atoll represent the peaks of a large submarine mountain system of volcanic origin, which rises from the ocean floor some 4,000 m in depth. The Fernando de Noronha volcano is estimated to be between 1.8 million and 12.3 million years old.
Democratic Republic of the Congo	Virunga National Park	Virunga National Park is notable for its chain of active volcanoes and the greatest diversity of habitats of any park in Africa. Features include hot springs in the Rwindi plains and the Virunga Massif volcanoes, such as Nyamulagira (3,068 m) and Nyiragongo (3,470 m), are still active, which alone account for two-fifths of the historical volcanic eruptions on the African continent. They are especially notable because of their highly fluid alkaline lavas. The activity of Nyiragongo is globally significant for its demonstration of lava lake volcanism, with a quasi-permanent lava lake at the bottom of its crater, periodic draining of which has been catastrophic to the local communities.
	Kahuzi-Biega National Park	A vast area of primary tropical forest dominated by two spectacular extinct volcanoes, Kahuzi and Biega, the park has a diverse and abundant fauna.
Dominica	Morne Trois Pitons National Park	Luxuriant natural tropical forest blends with scenic volcanic features of great scientific interest in this national park centred on the 1,342-m-high volcano known as Morne Trois Pitons. With its precipitous slopes and deeply incised valleys, 50 fumaroles, hot springs, three freshwater lakes, a 'boiling lake' and five volcanoes, located on the park's nearly 7,000 ha.
Ecuador	Galápagos Islands	The islands were formed by volcanic processes and most represent the summit of a volcano, some of which rise over 3,000 m from the Pacific floor. Ongoing seismic and volcanic activities reflect the processes that formed the islands. The larger islands typically comprise one or more gently sloping shield volcano, culminating in collapsed craters or calderas. Other noteworthy landscape features include crater lakes, fumaroles, lava tubes, sulphur fields and a great variety of lava and other ejects such as pumice, ash and tuff.
	Sangay National Park	The site is situated in the Cordillera Oriental region of the Andes in central Ecuador. The park is dominated by three volcanoes, Tungurahua (5,016 m)

Country	World Nature Heritage Sites	
		and El Altar (5,139 m) to the north-west and Sangay (5,230 m) in the central section of the park.
France	Pitons, cirques and remparts of Reunion Island	The Pitons, cirques and remparts of Reunion Island site covers more than 100,000 ha or 40 % of La Réunion, an island comprising two adjoining volcanic massifs located in the south-west of the Indian Ocean. Dominated by two towering volcanic peaks, massive walls and three cliff-rimmed cirques, the property includes a great variety of rugged terrain and impressive escarpments, forested gorges and basins creating a visually striking landscape.
Iceland	Surtsey	Surtsey is a new island formed by volcanic eruptions in 1963-1967. It has been legally protected from its birth and provides the world with a pristine natural laboratory. There is a recent tendency to promote Surtsey Island as a geopark.
Indonesia	Komodo National Park	The generally steep and rugged topography reflects the position of the national park within the active volcanic 'shatter belt' between Australia and the Sunda shelf. These volcanic islands are inhabited by a population of around 5,700 giant lizards, whose appearance and aggressive behaviour have led to them being called 'Komodo dragons'. They exist nowhere else in the world.
	Ujung Kulon National Park	This national park, located in the extreme south-western tip of Java on the Sunda shelf, includes the Ujung Kulon peninsula and several offshore islands and encompasses the natural reserve of Krakatoa. In addition to its natural beauty and geological interest – particularly for the study of inland volcanoes – it contains the largest remaining area of lowland rainforests in the Java plain.
Italy	Isole Eolie (Aeolian Islands)	The Aeolian Islands provide an outstanding record of volcanic island-building and destruction, and ongoing volcanic phenomena. Studied since at least the 18th century, the islands have provided the science of vulcanology with examples of two types of eruption (Vulcanian and Strombolian).
Kenya	Mount Kenya National Park/Natural Forest	At 5,199 m, Mount Kenya is the second-highest peak in Africa. It is an ancient extinct volcano, during whose period of activity (3.1-2.6 million years ago) it is thought to have risen to 6,500 m.
Kiribati	Phoenix Islands Protected Area	The Phoenix Island Protected Area (PIPA) is a 408,250 sq.km expanse of marine and terrestrial habitats in the Southern Pacific Ocean. PIPA conserves one of the world's largest intact oceanic coral archipelago ecosystems, together with 14 known underwater sea mounts (presumed to be extinct volcanoes) and other deep-sea habitats.

Country	World Nature Heritage Sites	
Korea, Republic of	Jeju Volcanic Island and Lava Tubes	Jeju Volcanic Island and Lava Tubes together comprise three sites that make up 18,846 ha. It includes Geomunoreum, regarded as the finest lava tube system of caves anywhere, with its multicoloured carbonate roofs and floors, and dark-coloured lava walls; the fortress-like Seongsan Ilchulbong tuff cone, rising out of the ocean, a dramatic landscape; and Mount Halla, the highest in Korea, with its waterfalls, multi-shaped rock formations, and lake-filled crater.
New Zealand	Tongariro National Park	Tongariro National Park is situated on the central North Island volcanic plateau. The park's volcanoes, which are outstanding scenic features of the island, contain a complete range of volcanic features. The mountains at the heart of the park have cultural and religious significance for the Maori people and symbolize the spiritual links between this community and its environment.
Russian Federation	Volcanoes of Kamchatka	This is one of the most outstanding volcanic regions in the world, with a high density of active volcanoes, a variety of types, and a wide range of related features.
Saint Lucia	Pitons Management Area	Dominating the mountainous landscape of St Lucia are the Pitons, two steep-sided volcanic spires rising side by side from the sea. The Pitons are part of a volcanic complex, known to geologists as the Soufriere Volcanic Centre which is the remnant of one (or more) huge collapsed stratovolcano. The Pitons occur with a variety of other volcanic features including cumulo-domes, explosion craters, pyroclastic deposits (pumice and ash) and lava flows
Spain	Teide National Park	Teide National Park, dominated by the 3,781 m Teide-Pico Viejo stratovolcano, represents a rich and diverse assemblage of volcanic features and landscapes concentrated in a spectacular setting. Mount Teide is a striking volcanic landscape dominated by the jagged Las Cañadas escarpment and a central volcano that makes Tenerife the third tallest volcanic structure in the world. Teide National Park is an exceptional example of a relatively old, slow moving, geologically complex and mature volcanic system.
Tanzania, United Republic of	Ngorongoro Conservation Area	The Ngorongoro Conservation Area spans vast expanses of highland plains, savannah, savannah woodlands and forests. It includes the spectacular Ngorongoro Crater, the world's largest caldera. Ngorongoro Crater is one of the largest inactive unbroken calderas in the world which is unflooded. It has a mean diameter of 16-19 km, a crater floor of 26,400 ha, and a rim soaring to 400-610 m above the crater floor. The formation of the crater and other highlands are associated with the massive rifting which occurred to the west of the Gregory Rift Valley.

Country		World Nature Heritage Sites
	Kilimanjaro National Park	Mount Kilimanjaro is one of the largest volcanoes in the world. It has three main volcanic peaks, Kibo, Mawenzi and Shira. With its snow-capped peak and glaciers, it is the highest mountain in Africa.
United Kingdom of Great Britain and Northern Ireland	Giant's Causeway and Causeway Coast	The Giant's Causeway lies at the foot of the basalt cliffs along the sea coast on the edge of the Antrim plateau in Northern Ireland. It is made up of some 40,000 massive black basalt columns sticking out of the sea. The Causeway Coast has an unparalleled display of geological formations representing volcanic activity during the early Tertiary period some 50-60 million years ago. The most characteristic and unique feature of the site is the exposure of a large number of regular polygonal columns of basalt in perfect horizontal sections forming a pavement. Tertiary lavas of the Antrim Plateau, covering some 3,800 km^2, represent the largest remaining lava plateau in Europe. The coastline is composed of a series of bays and headlands consisting of resistant lavas. The average height of the cliffs is 100 m, and has a stepped appearance due to the succession of five or six lava flows through geological time.
	St Kilda	This volcanic archipelago, with its spectacular landscapes, is situated off the coast of the Hebrides and comprises the islands of Hirta, Dun, Soay and Boreray. It has some of the highest cliffs in Europe, which have large colonies of rare and endangered species of birds, especially puffins and gannets.
United States of America	Yellowstone National Park	Yellowstone National Park, established in 1872, covers 9,000 km^2 of a vast natural forest of the southern Rocky Mountains in the North American west. The park has a globally unparalleled assemblage of surficial geothermal activity, thousands of hot springs, mudpots and fumaroles, and more than half of the world's active geysers. The world's largest recognized caldera (45km by 75km – 27 miles by 45 miles) is contained within the park. It boasts an impressive array of geothermal phenomena, with more than 3,000 geysers, lava formations, fumaroles, hot springs and waterfalls, lakes and canyons. The park is part of the most seismically active region of the Rocky Mountains, a volcanic 'hot spot'. The latest eruptive cycle formed a caldera 45 km wide and 75 km long, when the active magma chambers erupted and collapsed. The crystallizing magma is the source of heat for hydrothermal features such as geysers, hot springs, mud pots and fumaroles.
	Hawaii Volcanoes National Park	Lies in the south-east part of the island of Hawaii (Big Island), the easternmost island of the State of Hawaii, and includes the summit and south-east slope of Mauna Loa and the summit and south-western, southern, and south-eastern slopes of the Kilauea Volcano. Mauna Loa and Kilauea are two of the world's most active and accessible volcanoes where ongoing geological processes are easily observed. This property

Country	World Nature Heritage Sites
	serves as an excellent example of island building through volcanic processes. It represents the most recent activity in the continuing process of the geologic origin and change of the Hawaiian Archipelago.

Table 6. World nature heritage sites which are related to volcanic activities [61]

Country	Geopark	
Australia	Kanawinka Geopark	The Kanawinka Geopark is located in south-eastern Australia. The area covers 26,910 square kilometres, seven local government areas with 58 protected geosites within Australia's most extensive volcanic province. The surface geology of Kanawinka is a striking contrast of sweeping plains and spectacular cones which are largely the product of volcanic activity. The Blue Lake is one of the most significant volcanic sites within the Kanawinka Global Geopark. Mount Elephant is considered as "one of the highest and one of the major scoria cones in the largest homogenous volcanic plains on earth". The Geopark displays geological history and the nature of volcanoes in the region.
Canada	Stonehammer Geopark	Stonehammer Geopark is located in Southern New Brunswick on the East Coast of Canada. The Geopark encompasses 2500 square kilometres and extends from Lepreau Falls to Norton and from the Fundy Trail to the Kingston Peninsula. The landscape of the Stonehammer Geopark has been created by the collision of continents, the closing and opening of oceans, volcanoes, earthquakes, ice ages and climate change. It shows a billion years of Earth's history.
China	Hong Kong Geopark	Situated in close proximity to Hong Kong's world-famous financial and business centre, Hong Kong Global Geopark of China is lauded as a unique "geopark in the city". Hong Kong Geopark's major geo-attractions include well-outcropped acidic volcanic hexagonal rock columns, of which the average diameter is 1.2 metres and which are distributed over a land-and-sea area of 100 square kilometres. The Geopark also has comprehensive sedimentary rock formations formed about 400 to 55 million years ago. In addition, It is integrated with diverse ecological resources and historical relics, all combining to form the unique natural landscape of Hong Kong Geopark.
	Jingpohu Geopark	The Jingpohu Geopark is located in the upper-middle reaches of the Mudanjiang River, Northeast China's Heilongjiang Province. Jingpo lake covers an area of 79 square kilometres with a total water volume of 1.62 billion cubic meters. It is the largest lava barrier lake in China, and the second largest in the world. About 1 million years ago, Lava as a result of volcanic eruption blocked the river bed of the Mudanjiang River forming a lava dam. As a result, the incipient Jingpo Lake began to take shape. In the

Country		Geopark
		late Glacial Period, larger volcanic eruptions helped build a more voluminous lava dam. Hence the Jingpo Lake as we see it today. The granite rocks after experiencing various earth movements are in the shape of cliffs which are called a "geological corridor" by visitors. The Diaoshuilou Waterfall is the only large collapsed-lava waterfall of China which is also rarely seen in the world.
	Leiqiong Geopark	Leiqiong Global Geopark is located in the southern margin of Chinese Mainland, straddling Qiongzhou Strait. More than 100 volcanoes are densely distributed across the Geopark. These include examples of nearly all volcanic types. Judging from the number, variety and completeness of the volcanoes, the park is considered topmost among the Quaternary volcanic belts of China. Two districts of Leiqiong Geopark feature several 'maar' craters – broad, low-relief craters caused by groundwater coming into contact with hot lava. The Geopark is extremely rich in volcanic landscapes and lava structures, such as different kinds of lava flows and tunnels. The territory of the park spans Haikou City and Zhanjiang City, which are famous cultural centres. Also, the Geopark is positioned in an ecological transition area characterized by a rich diversity of flora and fauna.
	Longhushan Geopark	Located in China's Jiangxi Province, Longhushan Geopark holds geological, geomorphological, human and natural ecological landscapes. The major landform in the park is the Danxia landform and others but minor are volcanic landforms.
	Ningde Geopark	Ningde Geopark is located in the southeastern part of the Eurasian Plate, belonging to the continental marginal active zone adjacent to the Pacific Ocean. Ningde Geopark is an integration of various landforms such as miarolite landform, volcanic landform, erosion riverbed landform, and erosion coastal landform. The occurrence of many landforms in a geopark suggests that the area has a complicated geological history. At the same time, the park landform landscape also has the very high ornamental value.
	Wudalianchi Geopark [66] (Figure 16)	Wudalianchi Geopark is located in the north central Heilongjiang Province, China. Wudalianchi World Geopark is such a precious legacy left by the volcanic actions in the Quaternary period. Fourteen young and old volcanoes, the world's top volcanic resources, stand in the 1,060 km^2 geopark, with their eruption ages ranging from 2.07 million years prehistory to 280 years before now. They are the world's best-conserved volcanic landforms with the most varieties and most typical forms. The 14 volcano cones that sprung from the ground add beauty to the mountains and rivers, composing a remarkable picture; the micro-topographic landscapes in various forms such as lava platform, lava sea, lava cascade,

http://dx.doi.org/10.5772/54586

Country	Geopark	
		lava postern, lava stalactite, lava Vortex, trunk lava, flower-like lava, fumarolic cone and disc, lapilli and volcanic bomb are named "Natural Museum of Volcano" and "An Open Textbook on Volcano". The five connected lakes, like a string of beads, are formed by the latest volcanic magma filling the ancient Baihe valley, thus named Wudalianchi (five connected lakes). The mineral water here is renowned as "The World's Three Cold Springs" together with Vichy mineral water of France and Ciscaucasia mineral water of Russia.
	Yandangshan National Geopark	Yandangshan National Geopark is located in Zhejiang Province in China, which covers an area of 294.6 square kilometres. The Mount Yandang Shan is a natural park based on its geological landform of Cretaceous volcanic rhyolite. The geoheritage in the Park is referred to as a typical example of the formation and evolvement mode of reviviscent caldera on the edge of Asian Plates in late Mesozoic. From the Geoheritage, we can see the complete geo-evolving process of the eruption, subsiding and reviviscent apophysis of volcanoes. The Geoheritage provides a permanent sample for studying the volcanoes in Mesozoic.
Croatia	Papuk Geopark	Papuk Geopark is located in Slavonia, in eastern Croatia. Mount Papuk has a "stormy" geological history going as far back as the Precambrian with metamorphic rocks that date back as far as the Precambrian, more than 540 million years ago. More recent geological features result from mountain building with Mesozoic sediments 260 and 65 million years old, including carbonate rocks, mostly dolomites, as well as the Cretaceous volcanic sedimentary complex of Rupnica. The Rupnica geosite is famous due to a characteristic exposure of the columnar jointing developed in the albite rhyolite volcanic rock.
Germany	Vulkaneifel Geopark	Located in the middle of Central Europe, at the northwestern part of the 'Rheinish Slate Mountains', the rolling Eifel highlands are a hilly landscape with deep, glacially carved valleys cut into old Devonian sediments (360-415 million years old). Volcanoes dot the landscape, with 350 known eruption centres, and give the area its name – Vulkaneifel. In some craters, bogs and lakes have formed while others remain dry. Known as 'maar' craters, these bodies reveal a nearly uninterrupted stack of sediments dating back to 150,000 years ago that provides data for the reconstruction of past climate, vegetation and ecology. Similarly, fossils found in 43 million years old sediments of Eckfeld Maar are of worldwide importance, since they contained an archetypal horse and the oldest known honey bee. The Vulkaneifel has attracted geo-scientists for 200 years and many international research projects have been conducted here.
Greece	Lesvos Petrified Forest Global Geopark	Located in the western part of the Greek island of Lesvos, north-east Aegean Sea, the Lesvos Petrified Forest Geopark features rare and

Country		Geopark
		impressive fossilised tree-trunks. Formed some 15 to 20 million years ago, due to intense volcanic activity, the trees were covered by lava, ashes and other materials that were spewed into the atmosphere. The Lesvos Petrified Forest Global Geopark brings visitors to an ancient forest preserved by a massive volcanic eruption 20 million years ago.
	Psiloritis Geopark	Psiloritis Monts rose up through the sea a few million years ago when the African continent encroached on Europe. Psiloritis Geopark is characterized by its superb geodiversity. This is reflected by a great variety of volcanic, sedimentary and metamorphic rocks aging from Permian to Pleistocene (300 to 1 million year ago), outstanding folds and faults, fascinating caves and deep gorges with rich biodiversity.
Hungary-Slovakia	Novohrad-Nograd	Novohrad-Nograd Geopark lies at the border of Hungary and Slovakia. Being transnational, the Geopark's name comes from the Slovak and Hungarian names of the very county, where the Novohrad - Nograd Geopark is located. The geology of the Geopark includes diverse past volcanic events and a geological history dating back the last 30 million years from the birth of the Pannonian basin. Within a small area, the Geopark contains a wide spectrum of volcanic sites of spectacular sights, and several landscape protection areas and other territories. The area is also recognized as an important centre for the Palóc ethnic group's folk art and living traditions. Recently people pay more attention to the Maar diatreme volcanism and Miocene andesite volcanism in the region.
Iceland	Katla Geopark	The Katla Geopark is named after the volcano Katla that has for centuries had great impact on Icelandic nature and people living in the area. Katla is one of the largest central volcanoes in Iceland, covered by the glacier Mýrdalsjökull. Katla Geopark is located in the southern part of Iceland. The Geopark consists of three municipalities: Skaftárhreppur, furthest east; Mýrdalshreppur, in the middle; and Rangárþing eystra, in the west. The volcanic activity of Eyjafjallajökull, Katla, Grímsvötn, Lakagígar and Eldgjá and its widespread effect on the landscape in the area provide the geological basis for the Geopark. Apart from the ice-capped volcanoes and lava streams, sandur plains with their black beaches and rootless vents (pseudocraters) are prominent features in the landscape.
Japan	San'in Kaigan Geopark	The San'in Kaigan Geopark is located in the west of Japan and stretches from the eastern Kyogamisaki Cape, Kyoto to the western Hakuto Kaigan Coast, Tottori. The San'in Kaigan Geopark is home to a diversity of geological sites related to the formation of the Sea of Japan, including granite outcrops formed when Japan was part of the Asian continent (70 million years ago), as well as sedimentary and volcanic rocks accumulated when Japan rifted away from Asia (25 to 15 million years ago) to form the Sea of Japan, a geological process still on-going today. It also contains

http://dx.doi.org/10.5772/54586

Country		Geopark
		geographical features, such as ria type coasts, sand dunes, sandbars, volcanoes and valleys. Thanks to such diversity, the Geopark is home to rare plants like Pseudolysimachion ornatum and Ranunculus nipponicus, as well as Ciconia boyciana (Oriental White Storks) - a symbol of biodiversity.
	Toya Caldera and Usu Volcano Global Geopark	The Toya Caldera and Usu Volcano Global Geopark is located in Hokkaido, northern Japan and displays a unique showcase of active volcanism on the Pacific Rim. This is a unique region which has a wealth of characteristic geological relics in a relatively compact area, from the 110,000-year-old Toya Caldera to the 20,000 to 10,000-year-old Nakajima domes and a strato-volcano Usu, as well as the recent history of eruptions. The recent eruptive stage of Usu volcano started in 1663, and repeated nine times, creating lava domes, such as "Showa-Shinzan" that was born in a wheat field during 1943-1945. There are also precious fauna and flora living in the conserved thick forest and abundant water resources.
	Unzen Volcanic Area Global Geopark	Unzen volcano gives us a lot of gifts: outstanding landscapes, hot springs, spring water, fertile agricultural soil, and so on. On the other hand, Unzen volcano erupts repeatedly and causes serious disasters. In 1792, about 15,000 people were killed by the tsunami derived from the strong earthquakes due to sector collapse of an old lava dome, one of the worst volcanic disasters in Japan. In 1991-1995, a part of lava dome at the summit of the mountain collapsed. The pyroclastic surges containing giant hot ash clouds took 44 lives. Many buildings, houses, and school were also completely burned or buried by the flows and many residents lost their property. The burned school building and destroyed houses of the last eruption have been preserved just as they were. Now, these facilities are the main geosites of the Geopark and have been utilized widely for local disaster prevention educational programs in Japan. The theme of Unzen Volcanic Area Global Geopark is "the coexistence of an active volcano and human beings".
Korea	Jeju Geopark	See Table 6.
Norway	Gea Norvegica Geopark	Gea Norvegica Geopark is located in southeastern Norway, in the counties of Vestfold and Telemark. The story told in this geopark is a 1.5 billion year-long journey, from old mountain chains, the tropical sea, strange volcanoes, rifting of a continent and a glaciated surface -and how we all depend on this geological diversity and natural resources.
	Magma Geopark	Magma Geopark is situated in southwest Norway. The story began as early as 1.5 billion years ago when red-hot magma and sky-high mountains characterized the region. Through millions of years, glaciers helped to form today's characteristic landscape. Although the magma has cooled down and solidified and the mountains have been worn away, the area

Country	Geopark	
		offers a glimpse into the roots of an ancient mountain chain. Here is a rock type called anorthosite that is more common on the moon than on Earth.
Romania	Hateg Country Geopark	The Hateg Country Geopark is located in the central part of Romania, in a very fertile region surrounded by mountains. The region is world famous for its dwarf dinosaurs from the end of Cretaceous, 65 million years ago. Also well documented at the Geopark are the volcanic rocks-tuffs, lavas and craters that mark eruptions that took place during the age of the dinosaurs.
Spain	Cabo de Gata-Nijar Global Geopark	The Geopark's geodiversity derives from ancient Miocene volcanic substrata emplaced between 16 and 8 million years ago. In fact, the Geopark represents the most extensive and complex calco-alcaline fossil volcanism in the Iberian Peninsula. Visitors can walk through an open air geological museum with lava flows, volcanic domes, calderas, columnar joints, and fossilized sand beaches with tropical fossil reefs. This semi-arid climate and poor soil supports a surprising richness of plant species which ranks among the most diverse in Europe.
United Kingdom	Geo Mon Geopark (Wales)	The beautiful Isle of Anglesey lies off the west coast of Wales, UK. The island is renowned for its diverse tectonic geology. Brilliantly coloured Precambrian (about 800 Ma) 'pillow' lavas (erupted on the deep ocean floor with a characteristic shape) and deep ocean sediments are exposed at the western end whilst on the north coast is the world type locality for melange containing blocks of limestone with 800 million year old fossils. The remarkable folds and faults of South Stack date from the Cambrian period (520 Ma). Carboniferous limestones deposited about 350 Ma, crowded with fossil coral and shells show how ancient ice ages, sea level changes and plate movements affected the world long ago.
	Shetland Geopark	This tiny windswept archipelago has played host to tropical seas, volcanoes, deserts, ice ages and ancient rivers. The islands can boast the best section through the flank of a volcano in the UK, the best exposure of one of Europe's major tectonic faults, and are one of the best places in the world to see a compact vertical section through ancient oceanic crust.

Table 7. Global Geoparks which are related to volcanic activities [64, 65]

There are many other volcanic landscapes around the world besides the heritage sites and geoparks which have not been listed in Table 6 and Table 7. Take the Llancanello Volcanic Field in Argentina as an example, together with the nearby Payun Matru Field, there are at least 800 scoria cones and voluminous lava fields that cover an extensive area behind the Andean volcanic arc, six volcanoes show evidence of explosive eruptions involving magma-water interaction. Tuff rings and tuff cones can also be seen in this field. The diversity of volcanic landforms is so well-preserved that some people are promoting it as a geopark.

Figure 16. Landscape of Wudalianchi, China

Figure 17. Landscape of Changbai Mountain, China

Changbai Mountain volcano area (Figure 17) in northeast China is also a famous resort around the world. In Changbai Mountain, the forest is boundless, waterfalls are plentiful, the mountain peaks poke into the clouds and steamy hot springs flow along the canyons. Tianchi is embraced by a group of peaks. It is a paradise for scientific research because of rich biological resources, forest resources, mineral resources and typical volcanic geomorphologic landscape. We believe that all volcanic heritage sites should be protected under local, regional or national legislation as appropriate.

8. Conclusion

Volcanoes have provided us with material and spiritual wealth. The land resources enlarge our habitats; geothermal activity is a clean and regenerative energy source; hot springs and mineral springs are beneficial to our health; volcanic materials had become new and popular materials of the 21st century; gemstones and mineral resources are symbols of wealth; volcano landscapes provide us with an opportunity to experience the rich and extraordinary natural world of the volcanic zones. It is said that volcanoes have had a far-reaching impact upon our lives and participated in the progress of our society. So we should protect volcanic resources, exploit them reasonably and appreciate all the gifts which volcanoes give us.

Author details

Jiaqi Liu*, Jiali Liu, Xiaoyu Chen and Wenfeng Guo

*Address all correspondence to: liujq@mail.iggcas.ac.cn

Institute of Geology and Geophysics, Chinese Academy of Sciences, Beijing, China

References

[1] Allen, J. A, Ewel, K. C, & Jack, J. Patterns of natural and anthropogenic disturbance of the mangroveson the Pacific Island of Kosrae. Wetlands Ecology and Management (2001)., 9, 279-289.

[2] Louvat, P, & Allègre, C. J. Riverine erosion rates on Sao Miguel volcanic island, Azores archipelago. Chemical Geology (1998)., 148, 177-200.

[3] Fridriksson, S, & Magnússon, B. Development of the Ecosystem on Surtsey with References to Anak Krakatau. GeoJournal (1992)., 28(2), 287-291.

[4] Partomihardjo, T, Mirmanto, E, & Whittaker, R. J. Anak Krakatau's Vegetation and Flora circa 1991, with Observations on a Decade of Development and Change. GeoJournal (1992)., 28(2), 233-248.

[5] Klumpp, D. W, & Burdon-jones, C. Investigations of the potential of bivalve molluscs as indicators of heavy metal levels in tropical marine waters. Australian Journal of Marine and Freshwater Research (1982)., 33(2), 285-300.

[6] BaiduBaike: Volcanic island. http://baike.baidu.com/view/15872.htmaccessed 16 June (2012).

[7] The Surtsey Research SocietySurtsey-The Surtsey Research Society: Geology: Erosion of the Island. http://www.surtsey.is/pp_ens/geo_2.htmaccessed 16 June (2012).

[8] The Surtsey Research SocietySurtsey-The Surtsey Research Society: Geology: Formation of Palagonite Tuffs. http://www.surtsey.is/pp_ens/geo_4.htmaccessed 16 July (2012).

[9] Garvin, J. B, Williams, J, Frawley, R. S, & Krabill, J. J. W. B. Volumetric evolution of Surtsey, Iceland, from topographic maps and scanning airborne laser altimetry: In: Surtsey Research 11. Reykjavil: Gutenberg; (2000). p127-134. Available from http://www.surtsey.is/pp_ens/report/report_XI.htm (accessed 16 July 2012)., 1968-1998.

[10] Liu, R. X, & Li, N. Review on Volcano Resources. Bulletin of Mineralogy Petrology and Geochemistry (2000). , 19(3), 172-174.

[11] INFORSE- International Network for Sustainable EnergyGeothermal Energy: Geothermal System. http://www.inforse.dk/europe/dieret/Geothermal/geotermal.htmlaccessed 12 May (2012).

[12] Wang, J. Y, Liu, S. B, & Zhu, H. Z. Development Strategy of China's Geothermal Energy in 21st Century. Electric Power (2000). , 33(9), 85-94.

[13] Huang, S. P, & Liu, J. Q. Geothermal energy stuck between a rock and a hot place. Nature (2010).

[14] Department of Energy US. An evaluation of enhanced geothermal systems technology. http://www1.eere.energy.gov/geothermal/enhanced_systems.htmlaccessed 9 May (2012).

[15] Alison, H, Leslie, B, Dan, J, & Karl, G. Geothermal Energy: International Market Update. Geothermal Energy Association. http://www.geo-energy.org/pdf/reports/gea_international_market_report_final_may_2010.pdf (accessed 12 June (2012).

[16] International Geothermal AssociationGeothermal in the world: Electricity Generation: Installed Generating Capacity. http://www.geothermal-energy.org/226,installed_generating_capacity.htmlaccessed 12 June (2012).

[17] Desy, N. Alfian. Indonesia set to become user of geothermal. The Jakarta Post. http://www.thejakartapost.com/news/2010/04/27/indonesia-set-become-no-user-geothermal.htmlaccessed 12 June (2012). (1)

[18] Si, W. Development and utilization of geothermal resources in Indonesia. Geothermal Energy (2010). , 5, 31-32.

[19] Wu, X. M, Bo, Q, Yuan, H. M, Wan, X. Z, & Zhao, X. J. Present situation of the development and utilization of geothermal resources in Iceland. Hydrogeology and Engineering Geology (2007).

[20] Chu, J. The country which using geothermal energy instead of the oil. Life & Disaster (2009).

[21] Ragnarsson, A, Du, S. P, & Zhao, P. Geothermal energy in Iceland. Geothermal Energy (1999).

[22] Cole, J. W. Structural control and origin of volcanism in the Taupo volcanic zone, New Zealand. Bulletin of Volcanology (1990). , 52, 445-459.

[23] Zhang, D. Z, & Yan, Y. H. The coexistence of the high temperature geothermal power generation and volcanic geothermal landscape in New Zealand. In: The proceedings of the Meeting of Development and Protection of Geothermal Resources in China: the National Protection Research Seminar on the Exploitation and Protection of the Geothermal Resources: October (2007). Tengchong, Yunnan Province, China., 42-44.

[24] WikipediaSteam pipelines towards Wairakei geothermal power station.jpg. http://en.wikipedia.org/wiki/File:Steam_pipelines_towards_Wairakei_geothermal_power_station.jpg (accessed 16 June (2012).

[25] (Yang X. China Daily: The type and effect of hot spring. Http://www.chinadaily.com.cn//hqylss/2007-01/15/content_783931.htm (accessed 20 May 2012). , 2007-01.

[26] Cartwright, I, Weaver, T, Tweed, S, Ahearne, D, Cooper, M, Czapnik, K, & Tranter, J. Stable isotope geochemistry of cold CO_2-bearing mineral spring waters, Daylesford, Victoria, Australia: sources of gas and water and links with waning volcanism. Chemical Geology (2002). , 185, 71-91.

[27] BaiduBaike: Mineral spring water. http://baike.baidu.com/view/35726.htmaccessed 16 June (2012).

[28] BaiduBaike: The mineral spring water of Wudalianchi. http://baike.baidu.com/view/877619.htmaccessed 16 June (2012).

[29] An, K. S. Health Preserving and Culture of mineral spring. In: International Forum on hot spring culture of China conference proceedings, May, (2008). Xi'an, China., 27-28.

[30] Japan-iExplore Japan: Kyushu: Kagoshima: Ibusuki: Ibusuki Hot Spring. http://www.japan-i.jp/explorejapan/kyushu/kagoshima/ibusuki/d8jk7l000002rrsa.htmlaccessed 16 June (2012).

[31] Chen, Q. Q, Du, G. Y, & Yang, Y. Q. Discussion on designing of tourism products in the Aershan volcano and warm spring national geopark. Resources & Industries (2006). , 8(4), 85-88.

[32] Han, X. J, Jin, X, & Sun, C. H. Geothermal structure of Aershan hot springs, Inner Mongolia. Acta Geoscientica Sinica (2001). , 22(3), 259-264.

[33] Li, S. X, Bin, D. Z, & Chen, J. P. Development and protection suggestion of the volcanic warm spring resources in Tengchong. Geothermal Energy (2007).

[34] Dong, W, & Zhang, M. T. editor. Geothermal Energy in Tengchong. Beijing: Science Press;(1989).

[35] Li, X. H. Volcanic ash used as cement admixture and its effect on the cement performance. Wealth Today (2011).

[36] Su, X. H. A Study on rapid annealing technology and thermo-instrument in the control system in industrial test for energy saving tunnel furnace in cast basalt production. China Building Materials Science (1992).

[37] Xiao, G. P, Li, J. X, & Wang, Z. D. Application of Casting Basalt Technology in the Sluiceway of BF. Shandong Metallurgy (2003).

[38] Bai, S. L, Zhao, L, & Su, X. H. Basalt cast stone annealing process parameters and yield. China Building Materials Science & Technology (2000).

[39] Qi, F. J, Li, J. W, Li, C. X, Wei, H. X, & Gao, Y. Z. Summary on the research of continuous basalt fiber. Hi-Tech Fiber & Application (2006). , 31(2), 42-46.

[40] Wen, F. Basalt fiber: the new environmentally friendly fiber in 21st century. China Building Materials News: ((2009). April 16), , A3.

[41] Lei, J, Dang, X. A, & Li, J. J. Characteristic, application and development of basalt fiber. New Chemical Materials (2007). , 35(3), 9-11.

[42] Guo, H, Ma, Y, & Chen, Z. N. Development and application prospect of continuous basalt fiber. China Fiber Inspection (2010).

[43] Wang, R. S. Discuss of Basalt Filament's Property and Application. Progress in Textile Science & Technology (2010).

[44] Xie, E. G. Characteristics of basalt fiber and its application prospects in China. Fiber Glass (2005).

[45] WikipediaKimberlite. http://en.wikipedia.org/wiki/Kimberliteaccessed 24 June (2012).

[46] AllAboutGemstonesDiamonds: Diamond Geology & Kimberlites. http://www.allaboutgemstones.com/diamond_geology.htmlaccessed 24 June (2012).

[47] Diamond Source of VirginiaInc. Education: Diamonds Education: What is a diamond? http://www.diamondsourceva.com/Education/diamonds-what-is-a-diamond.aspaccessed 24 June (2012).

[48] Lorenz, V. Formation of phreatomagmatic maar-diatreme volcanoes and its relevance to kimberlite diatremes. Physics and Chemistry of the Earth (1975).

[49] American Museum of Natural HistoryOrigins: The nature of diamond: Kimberlite and lamproite. http://www.amnh.org/exhibitions/ diamonds/kimberlite.html (accessed 24 June (2012).

[50] American Museum of Natural HistoryOrigins: The nature of diamond: What does the carbon come from? http://www.amnh.org/exhibitions/diamonds/kimberlite.html (accessed 24 June (2012).

[51] Luan, B. A. Gem. Beijing: Metallurgical Industry Press; (1993).

[52] WikipediaGarnet. http://en.wikipedia.org/wiki/Garnetaccessed 24 June (2012).

[53] India martCompany Directory: Gems & Jewelry: Precious Gemstones. http://www.indiamart.com/gem-world/precious-gemstones.html (accessed 24 June (2012).

[54] WikipediaObsidian. http://en.wikipedia.org/wiki/Obsidianaccessed 24 June (2012).

[55] Pollard, A. M, & Heron, C. History of obsidian. In: Archaeological Chemistry. Royal Society of Chemistry. (2008).

[56] (Wilford J. N. In Syria, a Prologue for Cities. The New York Times. http://www.nytimes. com/2010/04/06/science/06archeo.html?pagewanted=all (accessed 24 June 2012).

[57] Hu, S. X, & Wang, W. B. Inquiring into the classification of gold deposits related to volcanism-subvolcanism-intrusion-hydrothermal processes. Gold Geology (1997). , 3(3), 25-29.

[58] Jiang, F. Z, & Wang, Y. W. Volcanism and gold deposit. Chinese Geology (2003). , 30(1), 84-92.

[59] Dru, L. J, & Zhang, Q. M. Geological and tectonic evolution of Muruntau gold deposit in Uzbekistan. Geological Science and Technology Abroad (1997).

[60] WikipediaDiatomaceous earth. http://en.wikipedia.org/wiki/Diatomiteaccessed 24 June (2012).

[61] UNESCOUNESCO World Heritage Centre: World Heritage List. http://whc.unesco.org/en/listaccessed 20 May (2012).

[62] Great Arc InitiativeNiagara Escarpment Resource Network. http://www. escarpmentnetwork.org/ great-arc-initiative.php (accessed 16 June (2012).

[63] UNESCONatural Sciences: Environment: EARTH SCIENCES: Geoparks: What is a Global Geopark? http://www.unesco.org/new/en/natural-sciences/environment/earth-sciences/geoparks/some-questions-about-geoparks/what-is-a-global-geopark/accessed 16 June (2012).

[64] UNESCONatural Sciences: Environment: EARTH SCIENCES: Geoparks: Members. http://www.unesco.org/new/en/natural-sciences/environment/earth-sciences/geoparks/members/accessed 16 June (2012).

[65] GeoparksNature Park Eisenwurzen. http://www.geoparks.it/node/123accessed 16 June (2012).

[66] WudalianchiAbout us: instruction. http://english.wdlc.com.cn/yyweb/show.asp?scid=7accessed 18 June (2012).

Section 4

Non-Magmatic Volcanism

Chapter 6

An Overview of Mud Volcanoes Associated to Gas Hydrate System

Umberta Tinivella and Michela Giustiniani

Additional information is available at the end of the chapter

http://dx.doi.org/10.5772/51270

1. Introduction

Humankind has been aware of volcanic activity in this planet since ancient times. This can be inferred from the remains of human settlements giving evidence of destruction by volcanic activity, and by the many myths around the world describing events that can be interpreted in relation to volcanic eruptions. Furthermore, such occurrences are evidenced by the special words that some human groups created to designate the special cases of "fire mountains", thus distinguishing these from other (non-volcanic) mountains found in the same region. More recently, as suggested by [1] Cañón-Tapia and Szakács, films and television shows devoted to exploring volcanoes have become very common, making it easier for the general public to gain access to "firsthand" experiences concerning this type of natural phenomenon. Consequently, it is only fair to say that at present almost everyone has an "intuitive" knowledge of what a volcano is [1]. In last years, many authors have devoted efforts to provide some additional theoretical aspects about volcanism, and one of the most recent publications is the book title "What is a volcano?" [1], addressing also a phylosophical question about this topic.

The definition of volcano is often missed or just defined as "opening-in-the-ground" in the textbooks. Starting from the classical Aristotelian requirements of a definition, it is shown that only a definition that is part of a hierarchically organized system of definitions can be accepted. Thus, conceptual constructs should reflect the same type of makeup as nature's processes, which are hierarchically organized. Such a line of reasoning implies that a volcano should be defined by making an explicit mention of the hierarchy of systems to which it belongs. Therefore, volcano can be defined as either a subsystem (i.e., the eruptive subsystem) of the broader igneous system or as a particular type of igneous system (i.e., one reaching the surface of Earth). A volcano, viewed as a volcanic system, is composed of a magma-generation subsystem, a magma transport subsystem, magma storage subsystem(s), and an eruptive subsystem. The accurate definition and identification of each subsystem should allow distinction between individual volcanoes in both space and time. Minimal conventional requirements need to be agreed upon by volcanologists to identify and recognize a particular volcano from other volcanoes, including those partially occupying the same space but separated in time, or those partially overlapping in both space and time. Conceptual volcanology can be envisaged as addressing the issue of accurate definition of basic terms and concepts, besides nomenclature and systematics, aiming at reaching the conceptualization level of more basic sciences. In contrast to volcanologists, sedimentologists are not only interested in "classical" volcanoes, but also in a second type, sedimentary volcanoes. This type of volcano is helpful for sedimentologists in understanding the processes that occur in the commonly unconsolidated subsoil, even after deep burial. Sedimentary volcanoes can be grouped in three classes: mud volcanoes, sand volcanoes, and associated structures such as water-escape and gas-escape structures. Sedimentary volcanoes have several characteristics in common with "classical" volcanoes, including their shapes and the processes that contribute to their genesis, as well described by [2] and here reported. These two type of volcanoes have largely similar morphologies, they overlap each other in size (large sedimentary volcanoes reach size that compete with those of many classical volcanoes), and they have a genesis that is comparable in many respect. For instance, comparison of the material flowing out from classical and sedimentary volcanoes shows that the sedimentary outflows resemble basaltic lava in the case of mud volcanoes, and acid lava in the case of sand volcanoes; this results in sedimentary volcanoes that morphologically resemble shield volcanoes and stratovolcanoes respectively.

It could be established that sedimentary volcanoes have, like the "classical" one, some kind of magma chamber in the form of a gas-and/or liquid-bearing layer with increasing pressure (either continuously, for instance, from the weight of the ever thickening sedimentary overburden, or incidentally, for instance, from an earthquake-induced shock wave). If the pressure exceeds a threshold, the pressured gas and/or liquid (commonly pore water with dissolved air or hydrocarbons) breaks upward through the overlying sediment, often following an already existing zone of weakness, to finally flow out at the sedimentary surface, either subaerially or subaqueously. In fact, the material that rises up through the connection between the source in the subsoil and the sedimentary surface behaves in several respects as magma on its journey from the magma chamber to the vent or crater opening: Magma becomes more fluid and develops gas bubbles while rising owing to decreasing pressure, and in sedimentary volcanoes with a deep-seated source the formation of bubbles also take place. These similarities should be promoted a fruitful collaboration between the researchers that can be mutually beneficial.

Among the sedimentary volcanoes, the mud volcanoes are the most interesting form many points of view, as described in this chapter. Mud volcanoes and mud volcanism are some of nature's most anonymous, mysterious and undiscussed geological features, which have been studied for millennia. This is strong interested and is explained considering a number of facts. First of all, thousands of mud volcanoes exist worldwide, defining and affecting the habitat and the daily lives of the millions of people living amongst them. Secondly, mud volcanism and mud volcano distribution is strongly connected to the formation and the distribution of the world's petroleum assets, thus serving as an indicator for valuable natural resources [3]. Thirdly, mud volcanoes offer an insight into otherwise hidden deep structural and diagenetic processes, such as the formation of gas hydrates, mineral dissolution and transformation, degradation of organic material and high pressure/temperature-reactions [4]. Lastly, mud volcanism generally involves voluminous generation and emission of both methane and carbon dioxide whereby most mud volcanoes serve as an efficient, natural source of greenhouse gases and, consequently, play an important role in global climate dynamics [3,5-10].

Among fluid venting structures, mud volcanoes are the most important phenomena related to natural seepage from the earth's surface [11]. Their geometry and size are variable, from one to two meters to several hundred meters in height, and they are formed as a result of the emission of argillaceous material and fluids (water, brine, gas, oil; [3,12-13]). They occur globally in terrestrial and submarine geological settings: most terrestrial mud volcanoes are located in convergent plate margin with thick sedimentary sequences within the Alpine-Himalayan, Carribean and Pacific orogenic belts [14-21]. Mud volcanoes and mud diapirs are responsible for the genesis of many chaotic deposits, such as mélanges, chaotic breccias and various deformed sediments [22-24].

The normal activity of mud volcanoes consists of gradual and progressive outflows of semiliquid material called mud breccia or diapiric mélange. Explosive and paroxysmal activities are interpreted as responsible for ejecting mud, ash, and decimetric to metric clasts. Mud volcano breccias are composed of a mud matrix, which supports a variable quantity of chaotically distributed angular to rounded rock clasts, ranging in diameter from a few millimeters to several meters [i.e., 14,18]. Clasts are of various lithologies and provenances, derived from the rocks through which the mud passed on its way to the surface or to the sea-floor. Slumps, slides and sedimentary flows can also affect the entire structure of mud volcanoes, even if gradients are very low.

The occurrence of mud volcanoes is controlled by several factors, such as tectonic activity, sedimentary loading due to rapid sedimentation, the existence of thick, fine-grained plastic sediments and continuous hydrocarbon accumulation [13,14,25-28].

A comprehensive study of submarine mud volcanoes is increasing in the last decades due to the wide use of geophysical methods, in particular side scan sonar, and the increased accuracy of the positioning of bottom samplers. Reference [12] presented an up-to-date list of known and inferred submarine mud volcanoes, describing their distribution, the mechanisms by which they form, and associated gas hydrates accumulations (Figure 1). Reference [12] clearly summarized the importance of research on submarine mud volcanoes, which is:

1. they are a source of methane flux from lithosphere to hydrosphere and atmosphere (greenhouse effect and climatic change);
2. they may provide evidence of high petroleum potential in the deep subsurface;
3. useful data about the sedimentary section in mud volcanic areas can be determined by examination of rock fragments incorporated in mud volcanic sediments (breccia);
4. submarine mud volcanic activity may impact drilling operations, ring installations and pipeline routings;
5. gas hydrates associated with deep-water mud volcanoes are a potential energy resource.

In this Chapter, we review the mud volcanism related to gas hydrate system. It is well known that natural gas hydrate occurs worldwide in oceanic sediment of continental and insular slopes and rises of active and passive margins, in deep-water sediment of inland lakes and seas, and in polar sediment on both continents and continental shelves (Figure 2). In marine sediments, where water depths exceed about 300 m and bottom water temperatures approach 0° C, gas hydrate is found at the seafloor to sediment depths of about 1100 m. In polar continental regions, gas hydrate can be present in sediment at depths between about 150 and 2000 m. Thus, natural gas hydrate is restricted to the shallow geosphere where its presence affects the physical and chemical properties of near-surface sediment [29].

Finally, we show an interesting case study in Antarctic Peninsula, where an important gas hydrates reservoir and mud volcanism are associated [30-32].

2. The mud volcanoes

Mud volcanism is not one specific process and mud volcanoes are not uniform feature settings; in fact, driving forces, activity, materials and morphologies may vary significantly [3,12-13,35-37]. About 2000 mud volcanoes have been identified in the worldwide; however, as exploration of the deep seas continues, this number is expected to increase substantially. It is estimated that the total number of submarine mud volcanoes is between 7,000 and 1,000,000 [10,12].

The geographical distribution of mud volcanoes is strongly controlled by geological environments in which they occur, as pointed out by reference [13]. In fact, mud volcanoes are localized within the compressional zones, such as accretionary complexes, thrust and overthrust belts, forelands of Alpine orogenic structures, as well as zones of dipping noncompensating sedimentary basins, which coincide with the active plate boundaries. The few mud volcano areas out of these belts are attached to zones with high rates of recent sedimentation, such as modern fans (including underwater deltas of large rivers) or intensive development of salt diapirism.

http://dx.doi.org/10.5772/51270

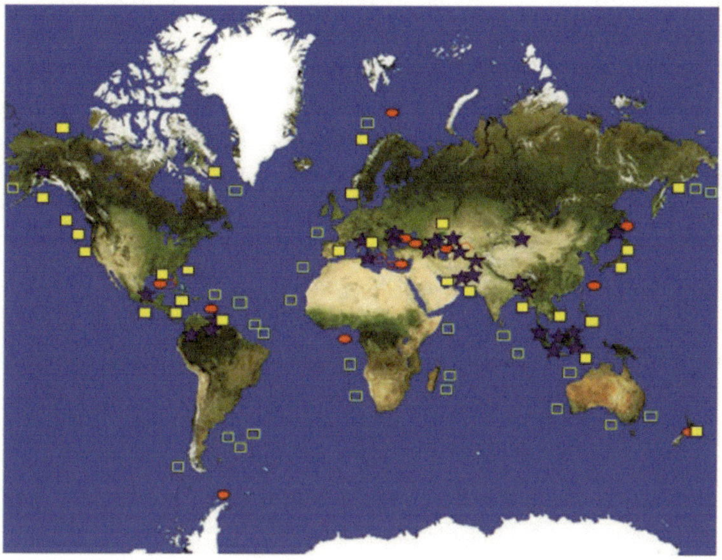

Figure 1. Map showing the worldwide locations of onshore (blue stars, after [33], with additions), known (red open oval, without gas hydrates; solid red oval hydrate bearing), and inferred (solid yellow rectangle) submarine mud volcanoes. The "possible sediment diapirs" mapped by [34] are also shown (open yellow rectangle). Modified after [12].

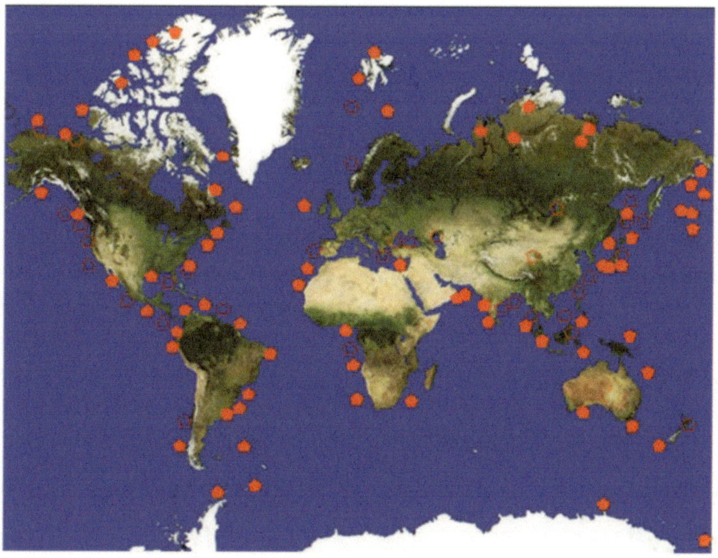

Figure 2. Map showing the worldwide locations of gas hydrate obtained by direct (open red pentagons) and indirect (solid red pentagons) measurements. Modified after [29].

Moreover, in all the mud volcano areas, there are suitable "source" layers of muddy sediments in the deeper part of the sedimentary basins or in the vicinity of the decolements in the accretionary complexes. Usually, this source is composed of fine-grained, soft material of low density overlain by at least 1–1.5 km of denser sediments. All collision complexes, where mud volcanoes are abundant, are characterized by thick sediment sequences caused by thickening of accreted sediments. The same is valid for forearc and outer orogenic basins where thickening is caused by thrusting and overthrusting. Enormous thickness is established in the noncompensating sedimentary basins.

Finally, given that the mud volcanoes are situated in areas where hydrocarbons have been, or are, actively being generated, there is a strong connection between the world mud volcano distribution and industrial oil and gas concentrations. In fact, the surrounding facies, below and laterally adjacent to mud volcanoes, may be particularly favorable as both reservoir and source environments for hydrocarbons, very often resulting in multilevel fields. Although this relationship is not valid for many mud volcano areas, for modern accretionary complexes in particular, it is established that present-day or recently active oil and especially gas generation are characteristic features for all of them.

Summarizing these brief comments the occurrence of mud volcanoes is strongly controlled by [12]:

- recent tectonic activity, particularly compressional activity;
- sedimentary or tectonic loading due to rapid sedimentation, accreting or overthrusting;
- continuous active hydrocarbon generation;
- existence of thick, fine-grained, soft, plastic sediments deep in the sedimentary succession.

Here, we summarize the main locations of the mud volcanoes as described in [12-13] and reported in Figure 1. The known or supposed mud volcanoes are irregularly clustered in separated areas forming belts, which almost totally coincide with active areas of the plate boundaries and zones of young orogenic structures. More than half mud volcanoes can be related to the Alpine-Himalaya active belt, where the largest and best cone-shaped mud volcanoes occur, as resumed in [13]. The most active terrestrial mud volcano area with the greatest number of mud volcanoes in the world is the Baku region of the Caspian Coast, Eastern Azerbaijan. Along the Alpine-Himalaya active belt, mud volcanism has been recognized in:

1. Mediterranean Ridge [28] and adjacent land areas (Sicily, Albania and Southern, Central and Northern Italy [38]),
2. the forelands of Eastern Carpathians in Romania, Kerch and Taman Peninsulas [39,40],
3. Great Caucasus [40] and the Black Sea [27],
4. the area of Southern Caspian Sea (Azerbaijan and Turkmenistan [26,41], South Caspian Basin [42], and Gorgon Plain in Iran),

5. the Makran coast of Pakistan [43],
6. Southern Himalayas (India and China), and Burma.

Furthermore, the Alpine–Himalayas mud volcano belt continues to the south in the most NE part of Indian Ocean on and around numerous forearc islands situated along the Indonesia and Banda Arcs [22], Indonesia – Australia accretion and collision complexes [44], as well as within the Banda accretionary complex offshore [22]. The greatest number of mud volcanoes seems to be known on Timor Island at the southeast end of this belt.

The western flank of the Pacific ocean – from the Sakhalin Island/Sea of Ochotsk-area in the north via Japan, Taiwan, the Marianas, Melanesia, Samoa and Australia to New Zealand in the south – holds some 150 onshore individuals [3,7]. The total number of offshore mud volcanoes along this belt is not yet fully determined but can be expected to be even higher, while the eastern flank of the Pacific Ocean is markedly less dense in mud volcanoes. Examples are known from and around the Aleutian Trench, Alaska, British Columbia, California, Costa Rica, Ecuador and inland Peru [3,7].

In the Atlantic Ocean, several hundreds of both onshore and offshore mud volcanoes have been recognized. The vast majority is concentrated along the Caribbean thrust belts and within the Barbados accretionary complex [7], although smaller clusters/individual features have been confirmed in connection to the Amazon and the Niger deltas [7,35], along the Gulf of Cadiz [45], within the southern Canary basin [46] and offshore Portugal and Morocco in the Alboran Basin [47].

Smaller numbers of mud volcanoes have also been described from the Mississippi delta [14], Lake Michigan [3], Greenland [3], the North Sea [48], the Netherlands [49], and areas of salt diapirism, such as in the Gulf of Mexico [50], Buzachi Peninsula (North-Eastern Caspian Sea), where they are related to salt diapirism, and Alboran basin in the Western Mediterranean [47].

The main components, that contribute to mud volcanoes formation and activities, are: mud breccia, water and gas. The relative quantities and the exact qualitative properties of these components vary, depending on local geology and processes at work.

Mud breccia is basically clasts in a clay mineral-rich matrix and is what makes up most mud volcanic features. Whereas the mud typically stems from one specific carrier bed and thus is characterized by a distinct geochemical signature and clast fragments. The first reflects subsurface mud volcanic conditions and processes (clay mineral dehydration/ transformation processes), while the latter is derived from units through which the mud pass on its way to the surface and are consequently of variable lithologies, sizes (up to 5 m) and shapes. Young and forceful mud volcanoes generally extrude mud breccias with a very high clast-matrix ratio (virtually clast-supported deposits), whereas the mud breccia of older mud volcanoes may be virtually clast-free with a mud content of up to 99% [3,13,35]. The latter are often related to the final phase of an eruptive cycle, when loose wall rock along the conduit has already been removed by the ascending mud [3].

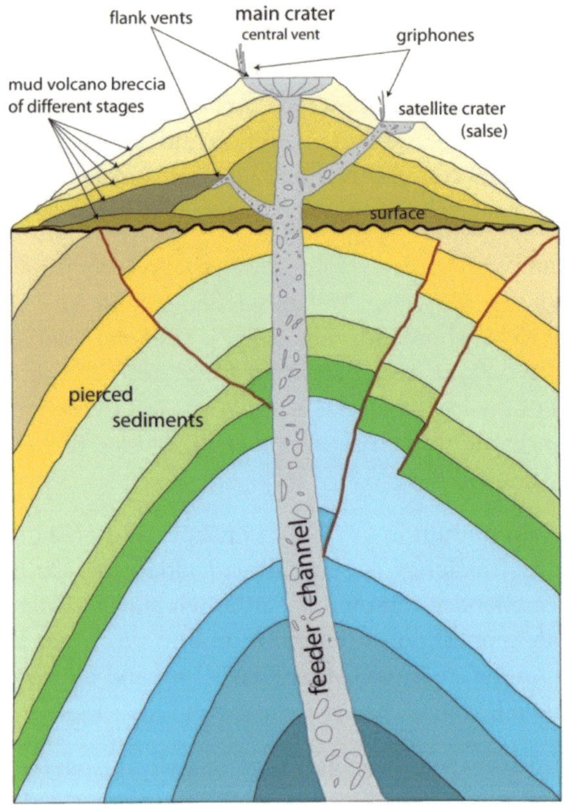

Figure 3. Basic structure and main elements of a conical mud volcano. Gryphones are small secondary vents shorter than 3 m, which may form around the craters and in many places on the mud volcano body. These commonly emit gas, mud and water and are characterized by complete absence of solid rock fragments. Modified after [13].

The water in mud volcanic extrusions typically stems from both shallow and deep sources and is normally derived through a variety of processes. Consequently, exact geochemical properties may vary virtually indefinitely [4,51]. However, clay mineral-dehydration water often makes up a significant proportion [3]. Mud breccia and mud volcanic water commonly mix whereby mud volcanic flows of different viscosities may form. During fierceful mud volcanic eruptions, up to 5 million cubic metres of such flow-material can be expelled [35].

Methane is almost always the dominated gas (70 - 99%) produced and emitted through mud volcanism. Moreover, since most mud volcanoes are very deeply rooted, thermogenic, 14C-depleted (fossil) methane is more common than biogenic [3,10,52]. Remainders typically include (in falling order) carbon dioxide, nitrogen, hydrogen sulfide, argon and helium [10,13,37].

As clearly explained in [13], a mud volcano comprises two main morphological elements: an internal feeder system and an external edifice (Figure 3). The characteristics of these elements are highly dependent on prevailing mud volcanic processes and in some cases, vice versa. Usually, mud volcano breccia is extruded from one major funnel called central or feeder channel (as summarized in Figure 3). Near the surface, several accompanying smaller flank or lateral pipes may split off the feeder channel. The outcrop of the feeder channel (usually situated on the summit of mud volcano) is called the main vent or central crater being of varied shapes: from planoconvex or flat and bulging plateau circled by a bank to deeply sunk rim depression (a caldera-type crater). Calderas form when a volcano collapses because a large mass of mud volcano breccia has drained through a lower vent, or because of the expulsion of a massive amount of material in an explosive eruption. Such eruptions have been known to destroy the entire structure of a volcano. Craters associated with the lateral vents are called satellite, parasite or secondary craters (Figure 3). Sometimes they collapse and are filled up by water that collects to form small lakes. Such a pool bubbling clay and gas is called salses. Numerous small secondary vents called gryphons may form around the craters and in many places on the mud volcano body. These commonly emit gas, mud and water and are characterized by complete absence of solid rock fragments. Although the numerous visual observations show that the gas seeps are very common on submarine mud volcanoes [53-55], the migrated gas is often captured in the near bottom sediments as gas hydrate [12,56-58] or trapped in shallow reservoirs to erupt when overpressured to form pockmarks on the seafloor [59]. The extruded mud volcano breccia spills in relatively thin sheets from the craters over the landscape in the form of broad fan-shaped or tongue-like flows up to several hundred meters wide and some kilometers long, as explained in reference [41]. This builds up the body of the mud volcano, typically covering some thousands of square meters with each phase of activity, totaling up to few tens of square kilometers. The fluid behavior of the mud volcano breccia is attributed to its high water content, which on land rapidly evaporates to drain the mud over a period of several days. Slumps and slides often form on the entire structure of the mud volcanoes even in very low gradients.

The internal feeder systems of mud volcanoes are not well known. Although studies imply rather large variabilities, typically, they consist of one main, central and deeply (km-scale) rooted feeder channel through which most mud volcanic material is transported. Feeder channels can be everything from cylindrical to irregular shaped to mere slits [3]. Near the surface, feeder channels tend to thin off and split into smaller flanking/lateral pipes [21]. The diameters of volcanic conduits may have a profound impact on mud volcanic activity. Generally, the wider is the conduit, the more voluminous is the expulsions [3].

The external morphology and expression of a mud volcano may vary almost indefinitely. The outcrops (vents/craters) of feeder channels may take on a variety of shapes; from planoconvex or flat and bulging to concave collapse structures of caldera-type [13]. Some mud volcanoes are in fact rather anonymous and quiescent features appearing merely as solitary, mm-scale openings in the ground surface, gently seeping small amounts of high-viscosity mud breccias and/or gas [14,35]. However, some mud volcanoes are really hazardous and expel voluminous amounts of low-viscosity mud-flows through frequent, short but fierce and explosive eruptions. This type of mud volcanoes typically evolve into kilometre-scale, chaotic and complex landscapes that comprise anything from clusters of cone-shaped morphologies rising hundreds of meters above ground to mounds, ravines, pools of bubbling mud and/or water (salses), mud cracks and clastic lobes [14,37,60]. During and following this type of active, hazardous mud volcanism, combustion of emitted gases may produce columns of flames rising up to several hundreds of meters, potentially burning for months or even years [37].

Mud volcanism and mud volcanoes have repeatedly been suggested to be a natural way of degassing the Earth's interior [13,61-63]. Although mud volcanism typically does involve thermogenic formation and expulsion of gas (a natural process which to a certain extent independently would be able to force deeply buried material to the surface), such processes can hardly serve to explain the truly vast extent and scope of worldwide mud volcanism. As resumed by [35], based on the large differences observed in shape, size and eruption styles of mud volcanoes, it is clear that there is no unique model that can explain them all. Ultimately, mud volcanoes form either as clay diapirs that reach and pierce the ground-surface or as fluidized argillaceous sediments, together with water and various amounts of hydrocarbon gases, which are extruded along structural weaknesses (conduits) within subsurface sediments/rocks [12].

Either way, a fundamental requisite for mud volcanism is the existence of a potential source domain; solitary or interconnected argillaceous carrier beds for migrating fluids and gases. Yet, for the actual volcanic processes to commence and continue – for gases to form and/or for the source material to move, rise and eventually extrude from the subsurface – additional forces are needed.

Since a vast majority of the mud volcanoes that are known today exist along active plate boundaries and, more specifically, along the anticlinal crests of accretionary prisms (the major depositional centres), compression through convergent tectonics and associated high sediment accumulation rates are generally considered the major mechanisms of mud volcanic initiation and sustenance. Argillaceous sediments and rocks are typically very weak and therefore, under the influence of compressive forces, prone to various clay mineral alteration and dehydration processes [4] and to brittle deformation through e.g. faulting. Moreover, under these very conditions, thermal and/or biogenic formation of hydrocarbon gases typically increases. Together, this implies formation of potential volcanic conduits, liquefaction, fluidization, gasification, density inversion, pore pressure increase and focused migration of mud volcanic material – i.e. mud volcanism – either through diapirism or along newly created faults/conduits [3,13-14,35].

Finally, the same forces and processes may explain mud volcano formation along passive continental margins. Although tectonic forces are lacking in such settings, compression, fluidization, gasification, overpressuring and mud volcanism may take place due to loading through rapid deposition of large amounts of (argillaceous) sediments [12,35]. A common characteristic for regions of mud volcanism located outside convergent plate boundaries is that mud volcanoes measure greatly in the vertical section (at least 2 km) and that they are a compound of undercompacted sedimentary sequences [7]. Consequently, although local settings may vary, the main mechanism of formation for mud volcanoes and mud volcanism is compression – either through tectonic forces or through high sediment accumulation rates – eventually leading to overpressuring through in situ gas generation, fluidization and liquefaction.

The fact that most mud volcanoes present regular, distinct seasonal changes in activity on a range from weeks to tens of years suggests an influence of more than one external agent, initiating and sustaining some kind of continuous, cyclic, natural pressure-recharging process within the mud volcanoes themselves. Astronomical cycles – e.g. orbital forcing – undoubtedly serve as one explanation. Through altering atmospherical and hydrospherical PT-conditions over a great variety of time-scales, such cycles may also affect and alter PT-conditions in the sediments and thereby mud volcanic processes via e.g. fluid access and bacterial activity (gas formation; [6]). As an example, after studying mud volcanism in the south Caspian Basin, reference [37] concluded that as much as 60% of all eruptions took place during either new or full moon. Moreover, reference [35] suggested a relationship between an 11- year cycle of the sun's activity and the initiation of mud volcano eruptions.

Even though astronomical cycles may explain most of the steady variations in mud volcanic eruption frequencies, they do not explain the rather frequent, more irregular eruptions. These are rather a result of ample, sudden seismic activity. If earthquake hypocenters are located within/in connection to potential carrier beds, shaking of the sediments may induce liquefaction and faulting as well as a significant increase in gas formation and dissociation. Consequently, rather sudden, eruptive mud volcanism may be generated in a normally quiescent or even dormant mud volcanic area [3,13,37,64].

Mud volcanoes can be related also to volcanic basins, which are sedimentary basins with a significant amount of primary volcanic rocks (e.g. sills and dykes). Pierced basins are sedimentary basins with many piercement structures such as mud volcanoes, dewatering pipes and hydrothermal vent complexes. Sills are tabular igneous intrusions that are dominantly layer parallel. They are commonly subhorizontal. Sills may locally have transgressive segments (i.e. segments that cross-cut the stratigraphy). Hydrothermal vent complexes are pipe-like structures formed by fracturing, transport and eruption of hydrothermal fluids. These complexes are dominated by sedimentary rocks with a negligible content of igneous material. Sediment volcanism is surface eruption of mud, sand or sediment breccias through a vent complex [65].

Hydrothermal and phreatomagmatic vent complexes are recognized from several sedimentary basins associated with large igneous provinces, including the Vøring and Møre basins off mid-Norway [66-68], the Faeroe–Shetland Basin [e-g- 69], the Karoo Basin in southern Africa [e.g. 70-75], in the Karoo-equivalent basins of Antarctica [76-78], and the Tunguska Basin in Siberia, Russia [e.g. 79]. Generally, the hydrothermal vent complexes represent conduit zones up to 8 km long rooted in contact aureoles around sill intrusions, where the upper part of the vent complexes comprise eyes, craters or mounds, up to 10 km in diameter [68]. An important consequence of intrusive activity in sedimentary basins is that the magma causes rapid heating of the intruded sediments and their pore fluids, causing expansion and boiling of the pore fluid [75], and metamorphic dehydration reactions. These processes may lead to phreatic volcanic activity by the formation of cylindrical conduits that pierce sedimentary strata all the way to the surface. The hydrothermal vent complexes thus represent pathways for gases produced in contact aureoles to the atmosphere, with the potential to induce global climate changes [80]. Consequently, constraints on processes leading to the formation of hydrothermal vent complexes in sedimentary basins, their abundance and structure may lead to a better understanding of the causes of the abrupt climate changes that are associated with many large igneous provinces [e.g. 81, 82]. References [65, 83]have analyzed the presence of voluminous basaltic intrusive complexes, extrusive lava sequences and hydrothermal vent complexes in the Karoo basin. In this area, the hydrothermal vents pierce the horizontally stratified sediments of the basin. They study have documented that the hydrothermal vent complexes were formed by one or a few phreatic events, leading to the collapse of the surrounding sedimentary strata. They proposed a model in which hydrothermal vent complexes originate in contact metamorphic aureoles around sill intrusions. Heating and expansion of host rock pore fluids resulted in rapid pore pressure build-up and phreatic eruptions. The hydrothermal vent complexes represent conduits for gases and fluids produced in contact metamorphic aureoles, slightly predating the onset of the main phase of flood volcanism.

Reference [84] investigated and understood the mechanisms responsible for the formation of piercement structures in sedimentary basins and the role of strike-slip faulting as a triggering mechanism for fluidization. For this purpose four different approaches were combined: fieldwork, analogue experiments, and mathematical modeling for brittle and ductile rheologies. The results of this study may be applied to several geological settings, including the newly formed Lusi mud volcano in Indonesia [84], which became active the 29th of May 2006. Their integrated study demonstrates that the critical fluid pressure required to induce sediment deformation and fluidization is dramatically reduced when strike-slip faulting is active. The proposed shear-induced fluidization mechanism explains why piercement structures such as mud volcanoes are often located along fault zones. Their results support a scenario where the strike-slip movement of the Watukosek fault triggered the Lusi eruption and synchronous seep activity witnessed at other mud volcanoes along the same fault. The possibility that a drilling, carried out in the same area, contributed to trigger the eruption cannot be excluded. However, so far, no univocal data support the drilling hypothesis, and a blow-out scenario can neither explain the dramatic changes that affected the plumbing system of numerous seep systems on Java after the 27-05-2006 earthquake. Reference [85] have combined satellite images with fieldwork and extensive sampling of water and gas at seeping gryphones, pools and salsa lakes at the Dashgil mud volcano in Azerbaijan in order to investigate the fluid–rock interactions within the mud volcano conduit. The gas geochemistry suggested that the gases migrate to the surface from continuously leaking deep seated reservoirs underneath the mud volcano, with minimal oxidation during migration. However, variations in gas wetness can be ascribed to molecular fractionation during the gas rise. In contrast, the water shows seasonal variations in isotopic composition and surface evaporation is proposed as a mechanism to explain high water salinities in salsa lakes. By contrast, gryphones have geochemical signals suggesting a deep-seated water source. This study has demonstrated that the plumbing system of dormant mud volcanoes is continuously recharged from deeper sedimentary reservoirs and that a branched system of conduits exists in the shallow subsurface. While the gas composition is consistently similar throughout the crater, the large assortment of water present reflects the type of seep (i.e. gryphones versus pools and salsa lakes) and their location within the volcano.

In last decade, many researchers conducted in Karoo have pointed out an interesting link between hydrothermal venting (potential sand/mud volcanoes on the surface) and maar-diatreme volcanism (eg. magma - water interaction driven explosive volcanism). The hydrothermal vent complexes identified in this area have previously been termed diatremes and volcanic necks, and have, since the pioneer work of references [70,71], been interpreted as the result of phreatic or phreatomagmatic activity [e.g. 72,74,86-90]. Some hydrothermal vent complexes are spatially associated with sill intrusions, but a direct relationship between conduit zones and contact aureoles cannot be demonstrated because of lack of exposures and boreholes. A general genetic relationship between sills and vent complexes is, however, supported by interpretations of seismic data from the Vøring and Møre basins of offshore mid-Norway, where it is shown that hydrothermal vent complexes are rooted in aureole segments of sill intrusions [68,75]. The general lack of igneous material in the hydrothermal vent complexes strongly suggests that they are rooted in a zone without major magma disintegration. Reference [65] have recently proposed a model of vent complex formation by heating and boiling of pore fluids in contact aureoles around shallow sills [75]. In this model, boiling of pore fluids may occur at depths as great as c. 1 km, and overpressure and possibly venting occur if the local permeability is low. Thick sills are common in the Stormberg Group sediments, at least in the Molteno Formation, which can be assumed to have caused shallow (1 km) boiling and expansion of pore fluids in contact aureoles. A high-permeability host rock requires a very rapid pressure build-up compared with permeability to initiate hydrofracturing. Following hydrofracturing, the gas phase may expand and lead to a velocity increase during vertical flow through the conduit zone. Thus, the vent formation mechanism bears resemblance to shallow breccia-forming processes in hydrothermal and volcanic systems [e-g- 76,91,92].

In systems dominated by fragmentation of magma (e.g. kimberlite pipes and diatremes), the resulting conduit zone will comprise mixtures of sediments and igneous material, and associated surface deposits dominated by pyroclastic material [e.g. 91-94]. Kimberlite pipes are generally formed from fragmentation of deep dyke complexes [e.g. 92,93], and this mechanism may also explain the formation of the phreatomagmatic complexes in the Karoo Basin [e.g. 95,96]. As very well explained by [97], kimberlitic diatremes are the most important economically, but despite decades of research, numerous open pit and underground mines, and hundreds of kilometers of diamond drilling, they remain poorly understood in volcanological terms, with multiple and strongly conflicting models in place. Reference [97] attempted an evenhanded review of maar diatreme volcanology that extends from mafic to kimberlitic varieties, and from historical maar eruptions to deeply eroded or mined diatreme structures.concentrated their study to convinced that increased understanding of other maar-diatremes will drive advances in kimberlite volcanology, and is best accomplished by integrating information from all parts of all types of maar-diatreme volcanoes, and from both subsurface and surface observations.

Finally, it is important to mention that several study [i.e. 98,99] have highlighted the importance of the emplacement environment of volcanism in causing global environmental climate changes. They suggested that an understanding of the triggering mechanism and consequences of previous climatic changes driven by carbon gas emissions is highly relevant for predicting the consequences of current anthropogenic carbon emissions, as these events are likely of similar magnitude and duration.

3. The gas hydrates

Natural gas hydrates are a curious kind of chemical compound called a clathrate. Clathrates consist of two dissimilar molecules mechanically intermingled but not truly chemically bonded. Instead one molecule forms a framework that traps the other molecule. Natural gas hydrates can be considered modified ice structures enclosing methane and other hydrocarbons, but they can melt at temperatures well above normal ice [i.e., 100]. At about 3 MPa pressure, methane hydrate begins to be stable at temperatures above 0 °C and at about 10 MPa it is stable at 15 °C [101]. This behavior has two important practical implications. First, it is a nuisance to the gas company. They have to dehydrate natural gas thoroughly to prevent the formation of methane hydrates in high pressure gas lines. Second, methane hydrates will be stable on the sea floor at depths below a few hundred meters and will be solid within sea floor sediments [102]. Masses of methane hydrate "yellow ice" have been photographed on the sea floor. Chunks occasionally break loose and float to the surface, where they are unstable and effervesce as they decompose.

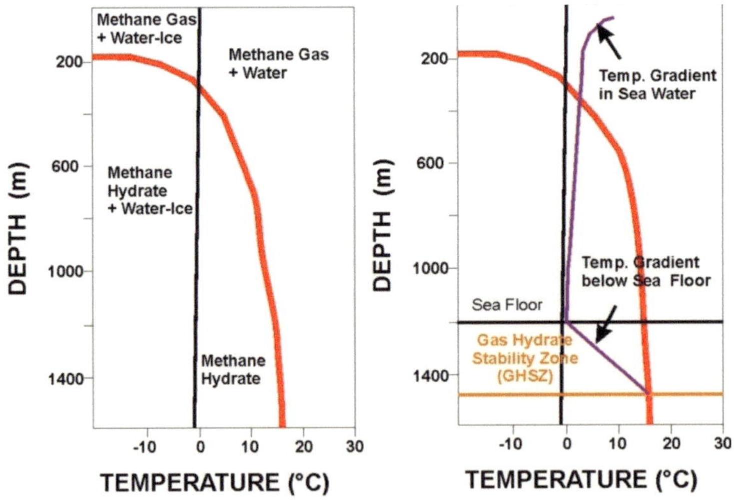

Figure 4. Left: Methane hydrate phases. Right: Typical occurrence of the gas hydrate stability zone on continental margins. A water depth of 1,200 m is assumed. The viola line represents the geothermal curve, while the red line is the gas hydrate stability curve. The orange line is the base of gas hydrate stability zone (GHSZ).

Figure 4 shows the combination of temperatures and pressures (the phase boundary) that marks the transition from a system of co-existing free methane gas and water/ice solid methane hydrate, which forms at low temperature and relative high pressure. When conditions move to the left across the boundary, hydrate formation will occur. Moving to the right across the boundary results in the dissociation of the gas hydrate, releasing free water and methane.

The phase diagram reported in Figure 4 shows a typical situation on continental shelves. Assuming a seafloor depth of 1,200 m, temperature steadily decreases with water depth, and a minimum value near 0° C is reached at the ocean bottom. Below the sea floor, temperatures steadily increase, so the top of the gas hydrate stability zone (GHSZ) occurs at roughly 400 m, while the base of the GHSZ is at 1,500 m. From the phase diagram, it appears that hydrates should accumulate anywhere in the ocean-bottom sediments where water depth exceeds about 400 m. Very deep (abyssal) sediments are generally not thought to house hydrates in large quantities. In fact deep oceans lack both the high biologic productivity (necessary to produce the organic matter that is converted to methane) and rapid sedimentation rates (necessary to bury the organic matter) that support hydrate formation on the continental shelves. Note that the conditions for gas hydrate formations are present also in sea water, but gas concentration is always not sufficient for their formations.

The gas hydrate phase is affected by gas mixture and pore-fluids composition (salinity). It is known that the presence of only a small percentage of higher hydrocarbons (such as ethane and propane) shifts the phase boundary to higher temperature (at constant pressure). The effect is that the base of GHSZ is shifted to greater depths [100,103]. Analogous to the effect of salt on the freezing point of water, if pore-fluids composition is brine, the phase boundary is shifted to lower temperatures at a given pressure and thus base of GHSZ will be shallower, as demonstrated by different authors [i.e., 204. Figure 5] shows a comparison among several gas hydrate stability models proposed in literature. The black line indicates the methane hydrate stability in brine water with a salinity of 3.5 % [105]. The red lines are evaluated by using the equations reported in [100] which consider fresh water. The solid red line indicates the methane hydrate stability curve and the dashed red line the hydrate stability curve considering a mixture of methane (90%), ethane (5%) and propane (5%). The blue lines are the hydrate stability in fresh water considering pure methane (solid line) and a mixture of methane (90%) and ethane (10%) by using the empirical expression proposed by [106]. Reference [107] obtained empirical equation for methane hydrate system in function of salinity. The green lines represent the gas hydrate stability considering the following salinity values: 0.0% (i.e. fresh water; solid line), 2 % (dotted line), 3.5 % (dashed line), and 5 % (dashed-dotted line). Finally, the magenta line indicates the stability for system of pure methane and water, following the approach of Reference [108]. Note that the methane hydrate stability curves in fresh water obtained from the proposed models are almost coincident, while the case considering salt water shows differences between the considered approaches.

The stability of methane hydrates on the sea floor has several implications [i.e., 109,110]. First, they may constitute a huge energy resource [111]. Second, natural and man-made disturbances may cause their destabilization causing the release of huge amounts of fluids (gas and water) and affecting slope stability. Finally, methane is an effective greenhouse gas (26 times more powerful than carbone dioxide), and large methane releases may explain sudden episodes of climatic warming in the geologic past. Some authors suggested that gas hydrate dissociation influenced significantly climate changes in the late Quaternary period [112-115]. The Clathrate Gun Hypothesis [116] suggests that past increases in water temperatures near the seafloor may have induced such a large-scale dissociation, with the methane spike and isotopic anomalies reflected in polar ice cores and in benthic foraminifera [115]. Reference [117] suggested that methane would oxidize fairly quickly in the atmosphere, but could cause enough warming that other mechanisms (for example, release of carbon dioxide from carbonate rocks and decaying biomass) could keep the temperatures elevated.

Gas hydrates in marine environments have been mostly detected from analysis of seismic reflection profiles, where they produce remarkable bottom-simulating reflectors (BSRs; [100,118]). Generally, the BSR is a very high-amplitude reflector that is associated with a phase reversal that approximately parallels the seafloor [119]. This phase reversal, which results from a strong acoustic impedance contrast between the layers, may indicate that sediments above the BSR are extensively filled with gas hydrates and sediments below it are filled with free gas in the pore space [i.e., 120-122]. Because the BSR follows a thermobaric surface rather than a structural or stratigraphic interface, it is normally observed to crosscut other reflectors [123].

Several studies [i.e. 30] revealed a seismic reflector below the BSR that can be associated with the base of the free gas zone, called base of the free gas reflector (BGR). The scientific community have been devoted much effort in studying marine sediments containing gas hydrates to characterize the hydrate reservoir and to quantify the gas trapped within sediments from seismic data analysis [i.e., 31,122,124,125]. To reach this goal advanced techniques have been developed. In fact, the BSR, detected from seismic data, is an easily recognizable indicator of the presence of hydrate, but it does not provide information directly on the concentration of hydrate and free gas or their distribution. One approach to estimate hydrate and free gas concentration is from seismic velocity (primarily P-wave velocity, Vp), obtained through advanced seismic analysis and/or modeling of data from a multichannel seismic streamer, using techniques such as common-image gathers analysis [i.e., 32], one dimensional waveform inversion [i.e., 126], and amplitude versus offset analysis [i.e., 127]. The obtained velocity can be translated in terms of concentration by using theoretical models [i.e., 127,128]. Figure 6 reports an example of compressional and the shear (Vs) velocity versus gas hydrate and free gas saturation in pore space by using two models: the Biot theory [129] and the approximation for seismic frequency [128]. Note that in presence of high hydrate concentration, the velocity increases significantly, while, if we suppose uniform distribution of free gas in the pore space, it is sufficient a small content of free gas to reduce drastically the velocity.

Recently, the international community has considered CO_2 sequestration as a possible means of offsetting the emission of greenhouse gases into the atmosphere [130]. Some studies have considered confining CO_2 hydrate directly to shallow sediments on the deep sea floor, but this approach would not be permissible under the above international conventions. In the case of hydrates, several studies have investigated the use of injected CO_2 to liberate methane gas from hydrate in sediments, and in the process lock up CO_2 in CO_2 hydrate [i.e., 131]. The CO_2 storage program is a further reason to assess the feasibility of mapping and monitoring the reservoir by means of an efficient seismic analysis [132,133] and to obtain information about hydrate and free gas concentrations in a time-effective way.

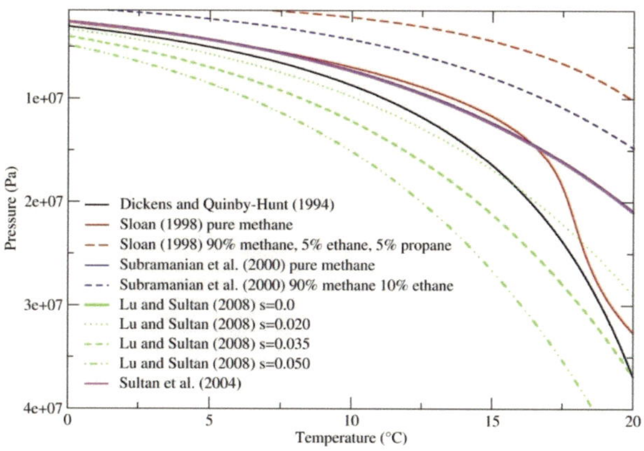

Figure 5. Comparison among different gas hydrate stability models. See text for details.

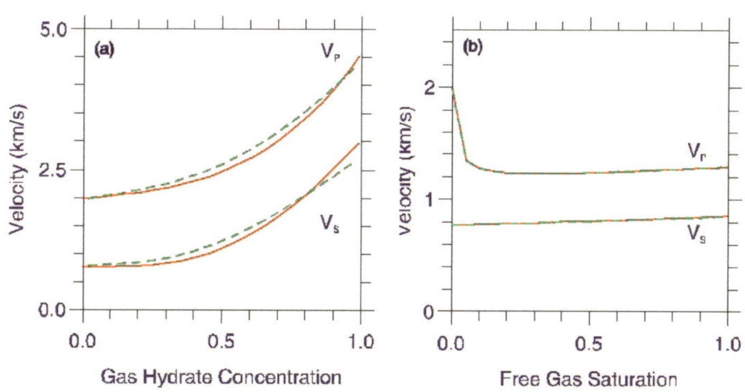

Figure 6. Comparison between two models: the Biot theory (solid red lines; [129]) and the approximation for seismic frequency (green dashed lines; [128]). (a) Compressional (Vp) and shear (Vs) velocity versus gas hydrate concentration in pore space. (b) Vp and Vs versus free gas saturation in pore space, supposing uniform distribution.

Finally, it is worth to mention that gas hydrates cannot survive for long geological times either because buried sediments are submitted to increasing geothermal conditions or are tectonically uplifted. In fossil sediments, the study of the interaction between seep-carbonates, hydrate destabilization and sediment instability is particularly difficult, owing to the lack of a direct recognition of fluid seepages and to the absence of precise quantifications of paleoenvironmental factors (pressure, temperature, paleodepth) conditioning hydrate stability conditions [134]. Following present-day analogues, the only means to infer a possible role of gas hydrates in fossil seep-carbonates are geochemical (oxygen isotope signature) and textural (presence of distinctive sedimentary features such as breccias, pervasive non-systematic fractures, soft sediment deformation) described in clathrites. Additional evidences can derive from the close association between seep-carbonates and sedimentary instability, and the large dimensions of seep-carbonate masses bearing brecciated structures. Recently, [134] studied cold seep-carbonates and associated lithologies in the northern Apennines and highlighted the seepage activity and the possible relationships with gas hydrate destabilization. In this geological context, many seep-carbonates are characterized by negative ∂13C and positive ∂18O values, by various types of brecciated structures and fluid-flow conduits, and are associated with intense sediment instability such as slumps, intraformational breccias and olistostromes.

Many authors have focused their attention on the possible modes of gas hydrate formations. Here, we report the study proposed by [135], which clearly summarized the possible mode of gas hydrate formation and produced a cartoon of gas hydrate system (Figure 7). They envisioned three possible modes of hydrate formation.

First, dipping permeable layers may focus gas flow and drive large amounts of free gas into the regional gas hydrate stability zone (Figure 7, number 1, inset). This is illustrated with a dipping stratigraphic layer in Figure 7; however, the permeability conduit could also be a fault or fracture. Beneath the GHSZ, the permeable layer draws gas from the surrounding material over an extensive source region, because of its high permeability and resultant low capillary entry pressure [136]. In this environment, gas rapidly enters the GHSZ and salinity rises as hydrate forms. The increased salinity inhibits further hydrate formation, which allows free gas to coexist with hydrate within the GHSZ. This process is repeated and the gas chimney rapidly propagates to the seafloor. Within the chimney, hydrate concentration increases upward toward the seafloor, where the system is furthest from equilibrium (Figure 7, number 1, inset). At the base of the gas chimney, the BSR will be diminished because gas is continuously present across the base of the GHSZ. These types of gas chimneys may be present, for example, at South Hydrate Ridge [109,137,138], at Blake Ridge [139-142] and along the Norwegian margin [143,144].

A second form of focused gas flow is illustrated in Figure 7 (number 2, inset). In this case, gas concentrates beneath the topographic crest of the seafloor structure. On the flanks of the structure, gas is trapped beneath the low-permeability base of the hydrate stability zone. Buoyancy drives the gas laterally toward the shallowest zone beneath the regional hydrate stability zone. The gas pressure is at a maximum at this location and ultimately the gas will drive its way through the GHSZ creating a gas chimney. As illustrated, the gas vent does not penetrate to the seafloor.

At the flanks of the topographic structures, a low permeability hydrate seal rapidly develops at the base of the GHSZ as illustrated in case 3 (Figure 7, number 3, inset). Hydrates are formed when water flows up through the GHSZ. Even in this low-flux example,

http://dx.doi.org/10.5772/51270

the gas supply is large enough to create a separate gas phase that migrates upward by buoyancy. The changes in salinity during hydrate formation are too small for a three-phase zone to develop, and hence all free gas is crystallized as hydrate at the base of the GHSZ (Figure 7, number 3, inset). As hydrate forms, the permeability drops and a capillary seal to gas is formed. In these circumstances, either the gas pressure will build until it fractures the overlying column, or if there is another pathway present, the gas will flow upward but underneath the low-permeability cap of the base of the GHSZ (Figure 7). Finally, far from where any methane gas flow is focused but where there is upward flow of water, low concentrations of hydrate may be deposited within the RHSZ but not at its base (Figure 7, number 4, inset).

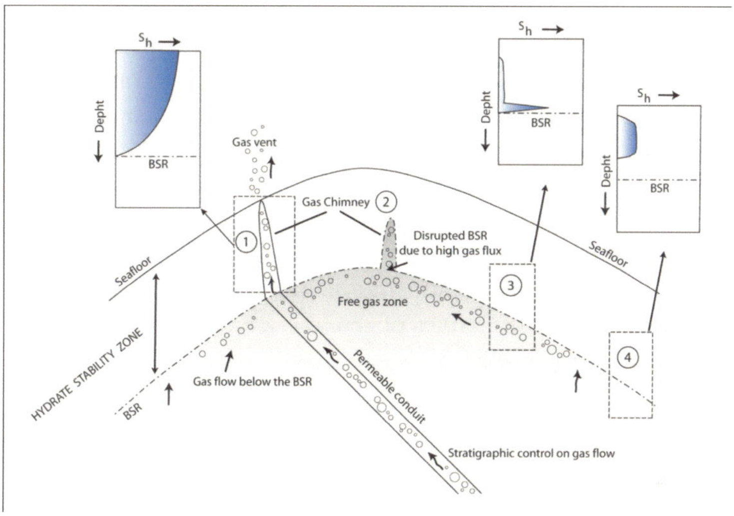

Figure 7. Cartoon of the gas hydrate reservoir system. Four characteristic forms of hydrate deposit are shown. (1) Gas chimney sourced by permeable conduits. Gas is focused along permeable conduits beneath the hydrate stability zone. Focused flow penetrates the GHSZ and self-generates a three-phase pathway to vent gas to the seafloor. (2) Gas chimney sourced by gas trapped beneath the GHSZ. Gas is focused along the base of the GHSZ and trapped beneath the crest of the structure. Gas builds up until it begins to form a chimney through the GHSZ. (3) Capillary sealing and lateral migration. On the flanks of the structure, hydrate formation rapidly forms capillary seals to gas and the gas is driven laterally to the highest structural point. (4) Aqueous flow and hydrate formation. Far from the crest, water with dissolved methane migrates upward and deposits hydrate within the GHSZ. Modified after [109].

The simulations of [109] provide insight into how gas chimneys form and sustain themselves within the GHSZ. The penetration of gas into the GHSZ is controlled by a competition between the basal supply of gas and the lateral diffusion of salt. The gas flow is driven primarily by buoyancy: as a result, the natural tendency for gas is to flow vertically even when permeability is reduced by hydrate formation. In addition, salt diffusivity is extremely low; thus high salinity zones within chimneys can be maintained for long times, particularly if there is continued supply of gas to form more hydrate and maintain salinity. Finally, lateral salt diffusion concentrates hydrate at the margins of the chimney; this further lowers the salt diffusivity and further limits salt loss.

Reference [135] showed that at South Hydrate Ridge, gas is supplied at a rate 10 times greater than is depleted by hydrate formation due to salt diffusion. Salt loss by diffusion, and hence the amount of methane needed to form hydrate to replace the salt, is independent of the vent half-width. In general, if the flux of methane supplied is greater than the loss due to diffusion, a chimney will be created and maintained at three-phase equilibrium (Figure 7, number 1, inset). However if the gas flux supplied is less than the loss due to diffusion, the chimney will only penetrate a short distance within GHSZ and free gas will not reach the seafloor (Figure 7, number 2, inset).

4. The gas hydrate in submarine mud volcanoes

The jointly occurrence of submarine mud volcanoes and gas hydrate has been reported by many authors in world-wide [i.e. 26,42,145-148]. For example, on the upper continental slope of the Gulf of Mexico, active gas migrations along faults or at mud volcanoes have been identified and their sources attributed to accumulated gas hydrates [53,149]. Reference [12] estimated that methane accumulated in gas hydrate associated with mud volcanoes is about $10^{10} \sim 10^{12}$ m^3 at normal temperature and pressure. Reference [150] estimated that up to 40% of total United Kingdom methane emission was from the continental shelf around UK. It is therefore important to investigate marine gassy sediments and submarine mud volcanoes to better understand the dynamics of shallow-water methane transport, fluid migration and the relationship of these phenomena to gas hydrate.

Firstly, we recall an important review of submarine mud volcanoes reported in Reference [12], reporting the main points. Evidence for submarine mud volcanoes exists in many regions showing the following features:

1. subcircular structures up to several kilometers in diameter elevated above the surrounding seafloor and visible on bathymetric maps and/or sonar images;
2. seafloor-piercing shale diapirs visible on seismic profiles;
3. fluid expulsion above elevated seafloor structures revealed by acoustic profiles and through visual observations from a submersible, remote operated vehicle (ROV), or by underwater video-camera;
4. transient mud islands in shallow waters;
5. gas bubbles at the surface of water that may be related to mud volcanoes.

Submarine mud volcanoes occur world-wide on continental shelves, slopes and in the abyssal parts of inland seas. Some studies (i.e. from the Barbados accretionary complex [54,151]) have linked the morphology of submarine mud volcanoes to different development stages and processes of mud liquefaction. Conical-shaped mud volcanoes ('mud-mounds' or gryphons), which do not have any central summit 'mud lakes' (or salses), are formed by the expulsion of plastic mud breccia in concentric radial flows. In contrast, shearing with the feeder conduit liquefies the mud leading to the formation of flat-top mud volcanoes (mudpies) with central 'mud lakes' and elongated, radial mud-flow tongues. In both types, the mud is found to have a plastic behavior in which its yield strength decreases with increasing porosity. Thixotropy is associated with high porosity (e.g. more than 70%), which is often related to the dissociation of gas hydrate [151]. Often, mud volcanoes are associated with methane fluxes, either as free gas or, depending on ambient temperature and pressure conditions, as gas hydrate [54,152]. On this basis, Reference [14] argue that the global flux of methane to the atmosphere from the world's terrestrial and submarine mud volcanoes is highly significant. The relative difficulty in studying submarine mud volcanoes, compared with their terrestrial counterparts, leaves substantial gaps in our knowledge about their modes of formation, the duration and frequency of eruptions and the fluxes of mud and volatile phases from the subsurface. For this reason, several efforts are spend to simulate numerically the formation of submarine mud volcanoes [i.e., 153].

At the present time, evidence for submarine mud volcanoes has been found in all oceans (Figure 1). For instance, in shallow water areas (shelves) where mud islands are recorded, submarine mud volcanoes are likely to be present. However, there are regions where the existence of submarine mud volcanoes is unexpected, such as in the Baltic Sea, where sediments are only 10 m thick, but miniature mud diapirs/volcanoes (1.5 m in diameter, 30 cm high above the surrounding seafloor) are reported [154,155]. Figure 1 includes sediment diapirs reported in [33]; however, it is unclear whether all these sediment diapirs are mud volcanoes. Submarine mud volcanoes are more extensive than their sub-aerial analogs.

All the regions where submarine mud volcanoes or evidence for them have been observed are confined to shelves, continental and insular slopes, and abyssal parts of inland seas (e.g. Black and Caspian). The examination of geologic and tectonic features of these areas is crucial for the understanding of the mechanisms of mud volcanic activity. In the abyssal parts of inland seas, mud volcanoes have been found in the Caspian Sea and in the Black Sea where the sedimentary cover is typically thick (10–20 km) for these regions. Sediments are mostly terrigenous and were deposited during Tertiary and recent times under high subsidence and accumulation rates. Shale diapirs and faults deform many sedimentary sequences.

On the continental slopes of passive margins, submarine mud volcanoes have been found in the Norwegian Sea, offshore Nigeria and in the Gulf of Mexico. In the first two cases, mud volcanic activity is confined to submarine fans (the Bear Island Trough mouth fan and the Niger Delta are notable examples), composed of Tertiary terrigenous sediments deposited at a high accumulation rate. In the Niger Delta there are many diapirs and faults [156]. The continental slope of the Gulf of Mexico is an extremely complex deep-water region characterized by a combination of rapid sediment influx, faulting, and diapiric (salt and shale) tectonism [157].

At active margins, submarine mud volcanoes have been reliably identified in the Mediterranean Sea and offshore Barbados, both characterized by different geologic and tectonic settings. In the Mediterranean Sea, mud volcanoes have been found only within the limits of the accretionary prism, characterized by a complex fabric composed of many thrusts [28]. Offshore Barbados mud volcanoes are located within the limits of the accretionary prism as well as in front of the prism where the thickness of sedimentary cover is only 2.3 km [158]. The majority of reported evidence of submarine mud volcanoes presents at active margins.

The association of gas hydrates with submarine mud volcanoes was first noted by Reference [145] and has since observed in the Caspian [42], Black [27,146-148], Mediterranean [57,148], Norwegian seas [48,58], offshore Barbados [159], offshore Nigeria [160] and in the Gulf of Mexico [161,162].

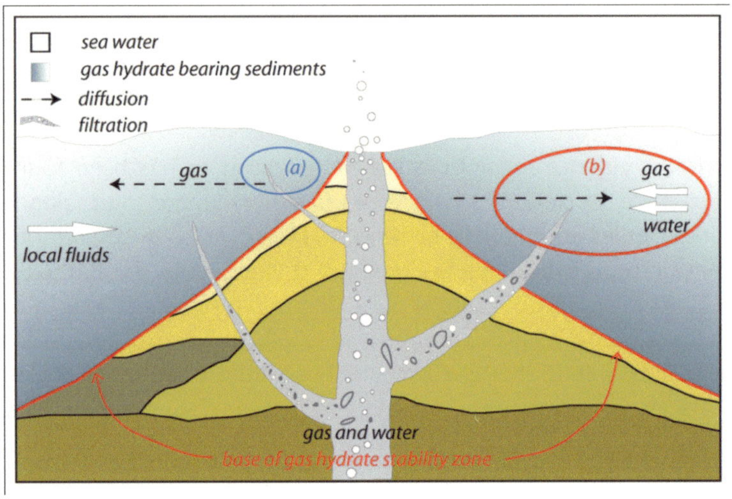

Figure 8. Cartoon showing the proposed model of the formation of gas hydrates within a mud volcano: (a) hydrothermal process dominates around the central part of the mud volcano; (b) metasomatic process dominates at the peripherical part of the mud volcano. Modified after [12].

There are many common features of gas hydrates associated with mud volcanoes. Gas hydrate content in sediments varies from 1–2% to 35% by volume and changes through a mud volcano area as well as through depth. Methane is the major gas component of gas hydrates and can be thermogenic, biogenic or mixed in origin. Reference [12] proposed a model for the formation of gas hydrates within a mud volcano: (a) hydrothermal process dominates around the central part of the mud volcano; (b) metasomatic process dominates at the peripherical part of the mud volcano (Figure 8). This generalized model is based mainly on data from the Haakon Mosby mud volcano in the Norwegian Sea, which is the most famous mud volcano characterized by a concentric-zonal distribution of gas hydrates [58,163]. Gas hydrate accumulation is controlled by the ascending flow of warm fluids. The water from the mud volcanic fluid as well as from the surrounding recent sediments is involved in the formation of gas hydrates. So, gas hydrates can occur within the edifice of a mud volcano (crater and hummocky periphery) as well as outside in the host marine sediments. However, the processes that are responsible for the formation of gas hydrates differ from point to point. Note that in Figure 8 there are no gas hydrates in the central part of the model mud volcano (usually in the central part of a crater where mud and fluid flow out) because of the high temperature. Around the central part of the mud volcano, gas hydrates form from the fluids that have risen from the deep subsurface. This fluid is warmer than the surrounding sediments (by up to 15–20° C at the sub-bottom depth of 1 m) and contains gas in solution and perhaps as a free phase. Gas hydrates crystallize from this warm fluid when it becomes cold and the solubility of gas decreases [164]. Both water and gas participating in the formation of gas hydrates have come from the deep, external fluid that filters through mud volcanic sediments. This process is analogous to the conventional low-temperature hydrothermal process of mineral formation [165].

At the peripheral part of the mud volcano, gas hydrates form from the gas that emanates from the central part of the mud volcano, and is transported in solution by diffusion. On the other hand, the water participating in the formation of gas hydrates is contained in the host sediments (local water). In addition, some local biochemical gas may be captured in gas hydrates. Thus, in this case the local water is partly replaced by gas hydrates due to the supply of gas from an external source (mud volcanic fluid). This process of gas hydrate formation is analogous to the conventional metasomatic process of mineral formation [165]. At any point between the central and peripheral parts of the mud volcano, mixing of hydrothermal and metasomatic processes is possible. The source of water (mud volcanic or local) determines which of these two processes is dominant.

The close proximity of mud volcanoes to zones where BSRs crop out on the seafloor deserves particular attention. Seismic records strongly suggest that much of the gas in mud volcanoes originates from levels deeper than that of the gas hydrates and faulting could be responsible for this unique situation. It has been argued that, in the case that a concentric zonal distribution of hydrate is present, the gas hydrates have probably been formed by gas emanating from the central part of the mud volcano, and transported into solution by diffusion [12]. For example, a strong BSR and the presence of mud volcanoes have recently been detected by seismic data along the southwest African margin, which is a passive margin [166]. This region, located in the distal part of the Orange River delta, is also characterized by overpressure which results in active fluid expulsion, as shown by the existence of mud volcanoes, pockmarks, and possibly cold-water corals thriving on methane gas seeps [167].

5. An example: The Antarctic Peninsula

The global climate change is particularly amplified in transition zones, such as the peri-Antarctic regions. For this reason, the gas hydrate reservoir present offshore Antarctic Peninsula was studied in the last 20 years acquiring a quite extensive geophysical dataset. The presence of a diffused and discontinuous BSR was discovered during the Italian Antarctic cruises of 1989–1990 [i.e., 30] and 1996–1997 [122], onboard the R/V OGS Explora. Seismic data showed the existence of a potential gas hydrate reservoir [32,168] along the South Shetland margin. Along this margin, the extent of the BSR was mapped based on about 1,000 km of seismic lines [e.g.,168 and references therein]. Ocean bottom seismometers (OBSs) deployed during the 1996–1997 cruise provided energy arrivals from the BSR and the refraction and the converted waves from the base of the free-gas zone, the so-called base of the free gas reflector or BGR [122]. During the austral summer 2003-2004, additional data were acquired in the same area: multibeam bathymetry, seismic profiles, chirp, and sediment gravity cores [31]. Figure 9 summarizes the position of the three acquisition legs.

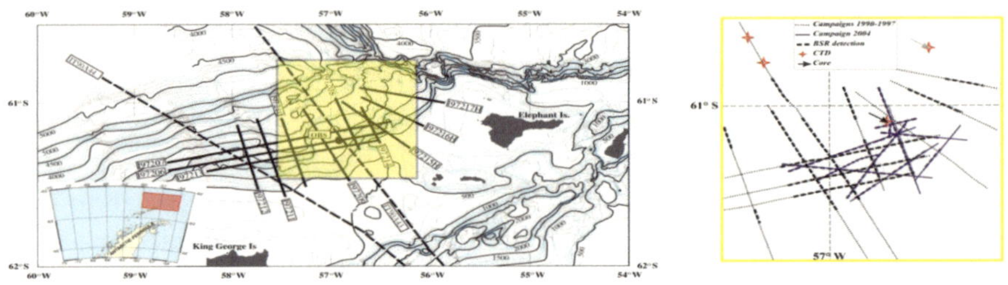

Figure 9. Left: Multibeam bathymetry map after [169], showing the locations of the seismic lines acquired in 1990 (dashed lines) and 1997 (continuous lines). The dot indicates the location of the OBS data acquired in 1997. The yellow box presents the 2004 study area. Right: 2004 study area. Continuous blue lines: locations of the airgun seismic lines. Dotted black lines: location of seismic lines of 1990 and 1997 legs. Thick dashed segments indicate the presence of particularly well developed BSRs. Also annotated are the CTD measurements (red stars), and the two coring sites (grey and black arrows cores GC01 and GC02 respectively). Modified after [31].

As already explained, seismic velocity obtained from advanced seismic analysis can be translated in terms of concentrations of gas hydrate and free gas. This procedure has been applied in the Antarctic Peninsula, where seismic velocities obtained from advanced analysis of multichannel seismic data were analyzed to determine gas hydrate and free-gas distributions and to estimate the methane volumetric fraction trapped in the sediments [170]. The elastic properties of the layers across the BSR were modeled applying the approximation of the Biot equations for seismic frequency in order to quantify the concentrations of gas hydrate and free gas in the pore space [128]. This theory considers two solid phases - grains and hydrates - and two fluid phases - water and free gas - including an explicit dependence on differential pressure and depth, and the effects of cementation by hydration on the shear modulus of the sediment matrix. So, the seismic velocities of the 2D seismic lines were translated in terms of concentrations of gas hydrate and free gas in the pore space, obtaining 2D models. The jointly interpolation of the 2D models allowed obtaining a 3D model of gas hydrate concentration from the seafloor to the BSR. The total volume of hydrate, estimated in the area (600 km^2) where the interpolation is reliable, is 16×10^9 m^3. The gas hydrate concentration is affected by errors that could be equal to about ±25%, as deduced from sensitivity tests [31,171] and from error analysis related to the interpolation procedure. The estimated amount of gas hydrate can vary in a range of 12×10^9 - 20×10^9 m^3. Moreover, considering that 1 m^3 of gas hydrate corresponds to 140 m^3 of free gas in standard conditions, the total free gas trapped in this reservoir ranges between 1.68×10^{12} and 2.8×10^{12} m^3. This estimation does not take into account the free gas contained within pore space below the hydrate layer, so this values could be underestimated.

http://dx.doi.org/10.5772/51270

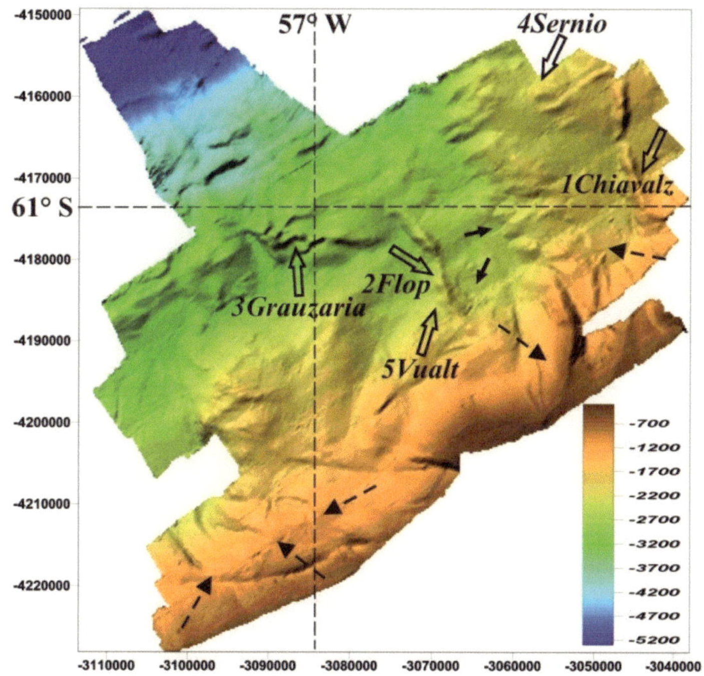

Figure 10. Multibeam bathymetry map of the study area, showing evidence of mud volcanoes (open arrows), collapse troughs (closed arrows) and slides (dashed arrows). The numbers indicate the four mud volcano ridges described in the text and in Table 1. Modified after [31].

Reference [31] reported the main results obtained by the analysis of bathymetric data, CHIRP data and gravity core analysis. In particular, the bathymetric map provided the evidences of mud volcanoes, collapse troughs and slides (Figure 10). It is well known [i.e., 172] that these features are generally associated to the presence of gas hydrate, as already explained. Reference [31] have recognized five main mud volcano ridges, named Chiavalz, Flop, Grauzaria, Sernio and Vualt (see location map in Figure 10). Table 1 reports the main characteristics of the mud volcanoes (Figures 11-15). The Vualt mud volcano is the highest detected in our study area; its top is at 2,216 m below sea level, with an elevation of about 255 m above the seafloor and an extension of 9.4 km^2 (Figure 10, label 5; Figure 11). On the flank, a gravity core (GC02) was recovered. The Flop mud volcano has its top at 2,363 m below sea level and a relief of about 115 m. Its extension is 7.5 km^2 (Figure 10, label 2; Figure 12). The Grauzaria ridge, oriented W–E and located at around 61 S–57 W (Figure 10; label 3; Figure 13), can be considered an alignment with several culminations and a total extension of 45.9 km^2. The highest culmination is at 2,594 m below sea level and exhibits a relief of about 185 m above the seafloor. The Sernio mud volcano ridge is located in the proximity of core GC01 and is oriented SW–NE (Figure 10, label 4; Figure 14). Its top is at 1,990 m below sea level with an elevation of about 185 m. It presents several culminations for a total extension of 23.9 km². Finally, the Chiavalz mud volcano ridge is located in the northeast of our survey area (Figure 10; label 1; Figure 15) and is oriented S–NE. It has its top at 1,615 m below sea level, a maximum elevation of about 210 m, and an extension of 14.5 km².

Mud Volcano	Lat. (WGS84)	Lon. (WGS84)	Meter below seafloor of the top	Elevation (m)	Extension (km²)
1 Chiavalz	60 52 29.31 S	56 18 46.88 W	1615	210	14.5
2 Flop	61 01 40.52 S	56 45 11.88 W	2363	115	7.5
3 Grauzaria	61 01 31.44 S	56 56 36.64 W	2594	185	45.9
4 Sernio	60 51 54.73 S	56 28 19.10 W	1990	185	23.9
5 Vualt	61 04 30.63 S	56 43 02.71 W	2216	255	9.4

Table 1. Details of the mud volcanoes offshore Antarctic Peninsula: latitude and longitude of the midpoint, water depth at the top, elevation with respect to the bathymetry and extension. Grauzaria is a group of several mud volcanoes.

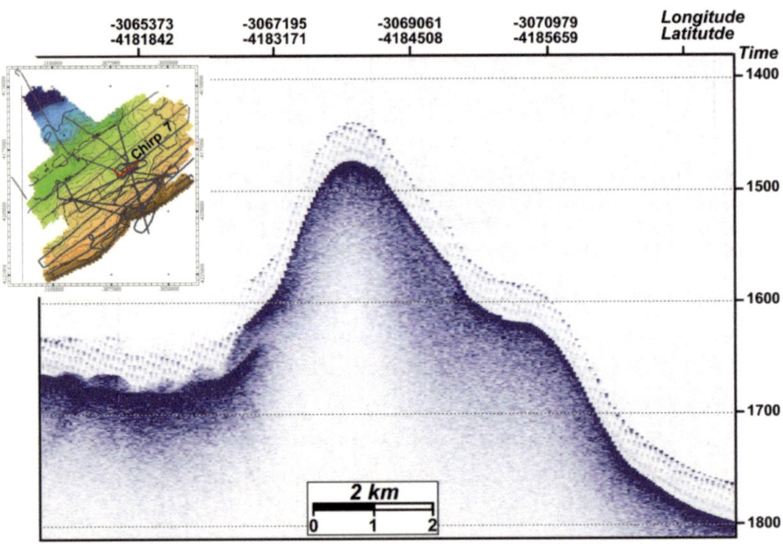

Figure 11. Comparison between chirp (a) and airgun (b) data across the Vualt mud volcano, in which the BSR is evident. The arrow indicates the location of core GC02. CHIRP image of the Vualt. Insert: Bathymetric map after BSR project. The red line indicates the CHIRP location. After [31].

Fluid analyses performed on the two gravity cores [31] revealed the presence of several hydrocarbon gases, i.e. methane, ethane, propane, butane, pentane and hexane, and traces of aromatic hydrocarbons of > C12 carbon chain length, suggesting a thermogenic origin of the gas. The major difference in gas contents between the two cores is that methane and propane are totally absent in core GC02. On the contrary, pentane is present at all analyzed depths in both cores, with quite similar contents. Below the upper 1 m of sediment in core GC02, the interstitial gases are essentially composed of pentane. The average total gas content amounts to 150.54 and 49.30 μg/kg for the two cores, respectively. The gas content measured in core GC01 is therefore about three times higher than that measured in core GC02. Downcore profiles for specific gases showed that core GC01 has a quite uniform gas type and content along the whole core; on the contrary, core GC02 has variable gas content. Even if both cores are located in the proximity of mud volcanoes, Reference [31] suggested that the sediment permeability of core GC01 is lower than that of core GC02, in which the fluids can easily escape and produce a collapse trough (see closed arrows in Figure 10). Moreover, the sediment stiffness in core GC01 is higher than that of core GC02, as suggested also by the different core length (1.07 and 2.98 m respectively); this is in agreement with the hypothesis of different permeability values between the two cores.

In conclusion, interpretation of the data acquired on the South Shetland margin confirmed the crucial role of tectonics controlling the extent of the hydrate reservoir, and active venting of fluids and mud through faults bordering and crossing the gas hydrate field. Mud volcanoes and fluid expulsion events are likely located in close association with faults, through which they are connected to the reservoir located beneath the BSR. Their activity is probably episodic [31]. Moreover, the different sediment stiffness at the two coring sites can be related to the temporal frequency of expulsion events, where the hardness of the mud volcano flanks is directly proportional to the interval between expulsion events, as suggested by [173]. Finally, the hydrocarbons trapped in our sediment cores possibly indicate the existence of deeper reserves.

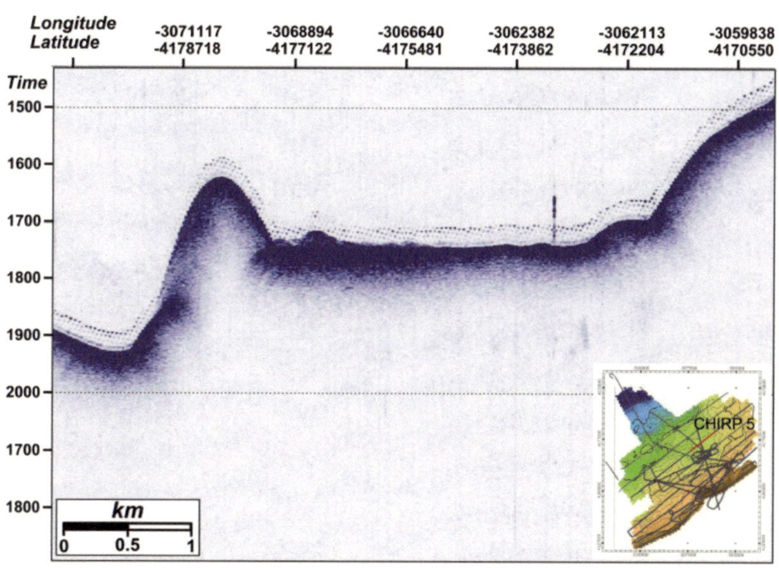

Figure 12. CHIRP image of the Flop. Insert: Bathymetric map after BSR project. The red line indicates the CHIRP location.

Figure 13. CHIRP images of the Grauzaria group. Insert: Bathymetric map after BSR project. The red line indicates the CHIRP location.

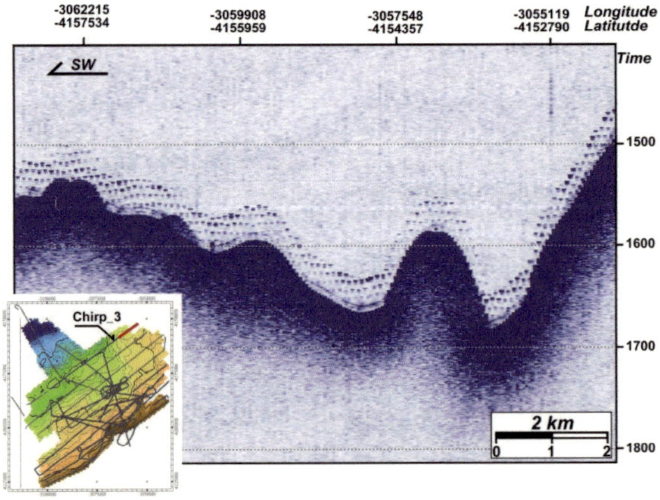

Figure 14. CHIRP images of the Sernio mud volcano. Insert: Bathymetric map after BSR project. The red line indicates the CHIRP location.

6. Conclusions

Knowledge of natural occurring gas hydrate is increasing rapidly in the last years; however commercialization of gas hydrate remains unproven. Great uncertainty of the global gas hydrate resource and imitated estimates of hydrate system retard economic analysis of hydrate recovery [174]. In this context, the gas hydrate associated to mud volcanoes is a very interesting topic because this system contains high gas hydrate concentration in a very small area. In addition, gas hydrate accumulations related to fluid discharges sites (including mud volcanoes) occur at very shallow depths or on the seafloor and show the maximum hydrate content in their upper parts. These features may consider as natural reactors, in which part of the migrating gas from the surrounding areas is stabilized in gas hydrates. Gas resources in such accumulations are therefore renewable and could become important gas hydrate formations to be exploited [11].

Moreover, the wide and extensive literature about hydrate, mud volcanism and their interaction suggests that this topic is timely because gas hydrates may play an important role in the global carbon cycle and global climate dynamics through emissions of methane and in affecting stability of geological features, including mud volcanoes. The role of gas hydrates in above-mentioned processes cannot be assessed accurately without a better understanding of the hydrate reservoir and their interactions with geological features and meaningful estimates of the amount of methane it contains. In conclusion, lack of knowledge hampers the evaluation of the resource potential of gas hydrates and the hazards related to gas hydrates, requiring efforts to improved knowledge about gas hydrate and their interaction with mud volcanoes.

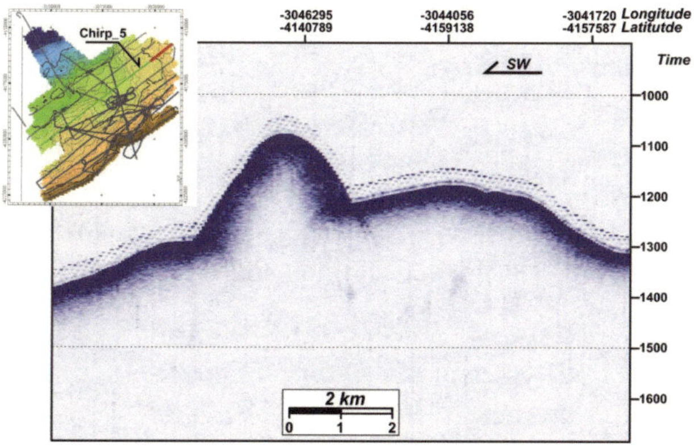

Figure 15. CHIRP images of the Chiavalz mud volcano. Insert: Bathymetric map after BSR project. The red line indicates the CHIRP location.

Acknowledgements

We wish to thank Manuela Sedmach for graphic support. This work is partially supported by Programma Nazionale di Ricerche in Antartide, project CLISM.

Author details

Umberta Tinivella* and Michela Giustiniani

*Address all correspondence to: utinivella@inogs.it

OGS – National Institute of Oceanography and Experimental Geophysics, Borgo Grotta Gigante 42C, 34010, Trieste, Italy

References

[1] Canon-Tapia, E., & Szakacs, A. (2010). What is a volcano? Boulder, Colo.: Geological Society of America. 140.

[2] van Loon, A. J. (2010). Sedimentary volcanoes: Overview and implications for the definition of a volcano on Earth. *Canon-Tapia E, Szakacs A, Editors. What is a volcano?. Boulder, Colo.: Geological Society of America*, 31-41.

[3] Kopf, A. J. (2011). Significance of mud volcanism. *Reviews of Geophysics*, 40(2).

[4] Hensen, C., Nuzzo, M., Hornibrook, E., Pinheiro, L.M., Bock, B., Magalhães, V. H., & Brückmann, W. (2007). Sources of mud volcano fluids in the Gulf of Cadiz- indications for hydrothermal imprint. *Geochimica et Cosmochimica Acta*, 71, 1232-1248.

[5] Etiope, G., & Klusman, R. W. (2002). Geologic emissions of methane to the atmosphere. *Chemosphere*, 49, 777-789.

[6] Judd, A. G., Hovland, M., Dimitrov, L. I., García, Gil. S., & Jukes, V. (2002). The geological methane budget at Continental Margins and its influence on climate change. *Geofluids*, 2, 109-126.

[7] Dimitrov, L. I. (2003). Mud volcanoes-a significant source of atmospheric methane. *Geo-Mar Lett*, 23, 155-161, doi: s00367-003-0140-3.

[8] Milkov, A. V., Sassen, R., Apanasovich, T. V., & Dadashev, F. G. (2003). Global gas flux from mud volcanoes: a significant source of fossil methane in the atmosphere and the ocean. *Geophysical Research Letters*, 30, doi:10.1029/2002GL016358.

[9] Etiope, G., & Milkov, A. V. (2004). A new estimate of global methane flux from onshore and shallow submarine mud volcanoes to the atmosphere. *Environmental Geology*, 46, 997-1002.

[10] Judd, A. (2005). Gas emissions from mud volcanoes. Significance to Global Climate Change. *Martinelli G., Panahi B., (ed.) Mud Volcanoes, Geodynamics and Seismicity*, 51, chapter 4, 147-157.

[11] Mazurenko, L. L., & Soloviev, V. A. (2003). Worldwide distribution of deep-water fluid venting and potential occurrences of gas hydrate accumulations. *Geo-Marine Letters*, 23, 162-176.

[12] Milkov, A. V. (2000). Worldwide distribution of submarine mud volcanoes and associated gas hydrates. *Marine Geology*, 167, 29-42.

[13] Dimitrov, L. I. (2002). Mud volcanoes-the most important pathway for degassing deeply buried sediments. *Earth-Science Reviews*, 59, 49-76.

[14] Hovland, M., Hill, A., & Stokes, D. (1997). The structure and geomorphology of the Dashgil mud volcano, Azerbaijan. *Geomorphology*, 21, 1-15.

[15] Kopf, A., Klaeschen, D., & Mascle, J. (2001). Extreme efficiency of mud volcanism in dewatering accretionary prisms. *Earth and Planetary Science Letters*, 189, 295-313.

[16] Delisle, G., von, Rad. U., Andruleit, H., von, Daniels. C. H., Tabrez, A. R., & Inam, A. (2002). Active mud volcanoes on- and offshore eastern Makran, Pakistan. International Journal of Earth Sciences. 91, 93-110.

[17] Etiope, G., Caracausi, A., Favara, R., Italiano, F., & Baciu, C. (2002). Methane emissions from the mud volcanoes of Sicily (Italy). *Geophysical Research Letters*, 29(8), 56-1-56-4.

[18] Deville, E., Battani, A., Griboulard, R., Guerlais, S., Herbin, J. P., Houzay, J. P., Muller, C., & Prinzhofer, A. (2003). The origin and processes of mud volcanism: new insights from Trinidad. *Van Rensbergen P, Hillis RR, Maltman AJ, Morley CK. (ed.) Subsurface Sediment Mobilization. Geol. Soc. London, Spec. Publs.*, 216, 475-490.

[19] Yassir, N. (2003). The role of shear stress in mobilizing deep-seated mud volcanoes: geological and geomechanical evidence from Trinidad and Taiwan. *Van Rensbergen P, Hillis RR, Maltman AJ, Morley CK. (ed.) Subsurface Sediment Mobilization. Geol. Soc. London, Spec. Publs.*, 216, 461-474.

[20] Shakirov, R., Obzhirov, A., Suess, E., Salyuk, A., & Biebow, N. (2004). Mud volcanoes and gas vents in the Okhotsk Sea area. *Geo-Marine Letters*, 24, 140-149.

[21] Stewart, S. A., & Davies, R. J. (2006). Structure and emplacement of mud volcano systems in the South Caspian Basin. *AAPG Bulletin*, 90(5), 771-786.

[22] Barber, A. J., Tjokrosapoetro, S., & Charlton, T. R. (1986). Mud volcanoes, shale diapirs, wrench faults and me´langes in accretionary complexes, eastern Indonesia. *AAPG Bull*, 70, 1729-1741.

[23] Orange, D. L. (1990). Criteria helpful in recognizing shear-zone and diapiric melenges, examples from the Hoh accretionary complex, Olympic Peninsula, Washington. *Geol. Soc. Am. Bull.*, 102, 935-951.

[24] Brown, K. M., & Orange, D. L. (1993). Structural aspects of diapiric melange emplacement: The Duck Creek diaper. *Journal of Structural Geology*, 13, 831-847.

[25] Treves, B. (1985). Mud volcanoes and shale diapirs, their implications in accretionary processes. *A review. Acta Naturalia de l'Ateneo Parmense*, 21, 31-37.

[26] Guliyiev, I. S., & Feizullayev, A. A. (1997). All About Mud Volcanoes. *Azerbaijan, Baku: Publ. House, Nafta Press.*

[27] Ivanov, M. K., Limonov, A. F., & Woodside, J. M. (1998). Extensive deep fluid flux through the sea floor on the Crimean continental margin (Black Sea). *Henriet JP, Mienert J. (ed.) Gas hydrates: relevance to world margin stability and climate change. Geol. Soc. London, Spec. Publs.*, 137, 195-213.

[28] Limonov, A. F., Woodside, J. M., Cita, M. B., & Ivanov, M. K. (1998). The Mediterranean Ridge and related mud diapirism: a background. *Mar. Geol.*, 132, 7-19.

[29] Kvenvolden, K. A., & Lorenson, T. D. (2000). The global occurrence of natural gas hydrate. *Paull C.K., Dillon W. P. (ed.) Natural Gas Hydrates: Occurrence, Distribution, and Dynamics, AGU Monograph*, 55.

[30] Tinivella, U., Lodolo, E., Camerlenghi, A., & Boehm, G. (1998). Seismic tomography study of a bottom simulating reflector off the South Shetland Islands (Antarctica). *Henriet J-P, Mienert J, (ed.) Gas hydrate: relevance to world margin stability and climate change. Geol. Soc. London, Spec. Publs.*, 147, 141-151.

[31] Tinivella, U., Accaino, F., & Della Vedova, B. (2007). Gas hydrates and active mud volcanism on the South Shetland continental margin, Antarctic Peninsula. *Geo-Mar Lett*, doi s00367-007-0093-z.

[32] Tinivella, U., Loreto, M. F., & Accaino, F. (2009). Regional versus detailed velocity analysis to quantify hydrate and free gas in marine sediments: the South Shetland margin target study. *Long D., Lovell M.A., Ress J.G., Rochelle CA. (ed.) Sediment-Hosted Gas Hydrates: New Insights on Natural and Synthetic Systems. Geol. Soc. London, Spec. Publs.* , 319, 103-119.

[33] Rakhmanov, R. R. (1987). Mud volcanoes and their importance in forecasting of subsurface petroleum potential. *Nedra (in Russian).*

[34] Lancelot, Y., & Embley, R. W. (1977). Piercement structures in deep oceans. *Bull. Amer. Assoc. Pet. Geol.*, 61, 1991-2000.

[35] Graue, K. (2000). Mud volcanoes in deep water Nigeria. Marine and Petroleum Geology. 17, 959-974.

[36] Huguen, C., Mascle, J., Chaumillon, E., Kopf, A., Woodside, J., & Zitter, T. (2004). Structural setting and tectonic control of mud volcanoes from the Central Mediterranean Ridge (Eastern Mediterranean). *Marine Geology*, 209, 245-263.

[37] Huseynov, D. A., & Guliyev, I. S. (2004). Mud volcanic natural phenomena in the South Caspian Basin: geology, fluid dynamics and environmental impact. *Environmental Geology*, 46, 1012-1023.

[38] Martinelli, G. (1998). Mudvolcanoes of Italy: proceedings of V Int. *Conference on "Gas in Marine Sediments"*, 40-42.

[39] Arhangelski, A. (1932). Some words about genesis of mud volcanoes on the Apsheron peninsula and Kerch-Taman area. *Bull. MOIP, Ser. Geol.*, 3: 269-285 (in Russian).

[40] Gubkin, I., & Feodorov, S. (1932). Mud volcanoes of the USSR in connection with oil and gas prospects. Proceedings of 27th Int. Geol. Congr. Moscow -(in Russian) , 4, 33-67.

[41] Jakubov, A. A., Ali-Zade, A. A., & Zeinalov, M.M. (1971). Mud volcanoes of the Azerbaijan, SSR-Atlas. *Baku: Publishing house of the Academy Sciences of the Azerbaijan SSR*, 245.

[42] Ginsburg, G. D., & Soloviev, V. A. (1994). Mud volcano gas hydrates in the Caspian Sea. *Bull. Geol. Soc. Denm*, 41, 95-100.

[43] Higgins, G. E., & Saunders, J. B. (1973). Mud volcanoes-their nature and origin: contribution to the geology and paleobiology of the Carribbean and adjacent areas. *Verh. Naturforsch. Geschel. Basel*, 84, 101-152.

[44] Williams, P., Pigram, C., & Dow, D. (1984). Melange production and the importance of shale diapirism in accretionary terrains. *Nature*, 309, 145-146.

[45] Mazurenko, L. L., Soloviev, V. A., Belenkaya, I., Ivanov, M. K., & Pinheiro, L. M. (2002). Mud volcano gas hydrates in the Gulf of Cadiz. *Terra Nova*, 14, 321-329.

[46] Müller, C., Theilen, F., & Milkereit, B. (2001). Large gasprospective areas indicated by bright spots. *World Oil*, 222(1).

[47] Perez-Belzuz, F., Alonso, B., & Ercilla, G. (1997). History of mud diapirism and triggering mechanisms in the Western Alboran Sea. *Tectonophysics*, 282, 399-423.

[48] Vogt, P. R., Cherkashev, G., Ginsburg, G., Ivanov, G., Milkov, A., Crane, K., Lein, A., Sunvor, E., Pimenov, N., & Egorov, A. (1997). Haakon Mosby mud volcano provides unusual example of venting. *EOS Trans Am Geophys Union*, 78(48), 556-557.

[49] Paine, W. R. (1968). Recent peat diapirs in the Netherlands: A comparison with Gulf Coast salt structures. In: Braunstein G.,. O'Brien G.D. (ed.) Diapirism and Diapirs., AAPG Mem., 8, 271-274.

[50] Neurauter, T. W., & Roberts, H. H. (1994). Three generations of mudvolcanoes on the Louisiana continental slope. *Geo-Mar. Lett.*, 14, 120-125.

[51] Planke, S., Svensen, H., Hovland, M., Banks, D. A., & Jamtveit, B. (2003). Mud and fluid migration in active mud volcanoes in Azerbaijan. *Geo-Marine Letters*, 23, 258-268.

[52] Etiope, G. (2005). Methane emission from mud volcanoes. *Martinelli G, Panahi B (ed.) Mud Volcanoes, Geodynamics and Seismicity. Springer*.

[53] Sassen, R., Losh, S. L., Cathles, L., Roberts, H. H., Whelan, J. K., Milkov, A. V., Sweet, S. T., & De Freitas, D. A. (2001). Massive vein filling gas hydrate: relation to ongoing migration from the deep subsurface in the Gulf of Mexico. *Mar. Pet. Geol.*, 18(5), 551-560.

[54] Limonov, A. F., van Weering, T. C. E., Kenyon, N. H., Ivanov, M. K., & Meisner, L. B. (1997). Seabed morphology and gas venting in the Black Sea mudvolcano area: observations with the MAK-1 deep-tow sidescan sonar and bottom profiler. *Mar. Geol.*, 137, 121-136.

[55] Foucher, J. P., & De Lange, G. (1999). Submersible observations of cold seeps on eastern Mediterranean mud volcanoes. *Proceedings of J. Conf. Abstr. EUG 10, Strasbourg, France,* B13 4(1).

[56] Reed, D. L., Silver, E. A., Tagudin, E., Shipley, H., & Volijk, P. (1990). Relations between mud volcanoes, thrust deformation, slope sedimentation, and gas hydrate, offshore north Panama. *Mar. Pet. Geol.*, 7, 44-54.

[57] Woodside, J. M., Ivanov, M. K., & Limonov, A. F. (1998). Shipboard Scientists of the Anaxiprobe Expeditions. Shallow gas and gas hydrates in the Anaximander Mountains regions, eastern Mediterranean Sea. *Henriet JP, Mienert J. (ed.) Gas Hydrates: Relevance to World Margin Stability and Climate Change. Geol. Soc. London, Spec. Publ.* , 137, 177-193.

[58] Ginsburg, G. D., Milkov, A. V., Soloviev, V. A., Egorov, A. V., Cherkashev, G. A., Crane, K., Lorenson, T. D., & Khutorskoy, . (1999). Gas hydrate accumulation at the Haakon Mosby mud volcano. *Geo-Mar. Lett.*, 19, 57-67.

[59] Dimitrov, L. I., & Woodside, J. (2003). Deep sea pockmark environments in the eastern Mediterranean. *Marine Geology*, 195, 263-276.

[60] Evans, R. J., Davies, R. J., & Stewart, S. A. (2006). Internal structure and eruptive history of a kilometrescale mud volcano system, South Caspian Sea. *Basin Research*, 19, 153-163.

[61] Hedberg, H. D. (1974). Relation of methane generation to undercompacted shales, shale diapirs and mud volcanoes. *Bull. Am. Assoc. Pet. Geol*, 58, 661-673.

[62] Ali-Zade, A. A., Shnyukov, E. F., Grigoryants, B. V., Aliyev, A. A., & Rakhmanov, R. R. (1984). Geotectonic conditions of mud volcano manifestation in the world and their role in prediction of gas and oil content in the earth's interior: proceedings of 27th International Geological Congress. 13, 377-393.

[63] Guliyiev, I. S. (1992). A review of mud volcanism. *Translation of the report by: Azerbaijan Academy of Sciences Institute of Geology*, 65.

[64] Manga, M., & Brodsky, E. (2006). Seismic triggering of Eruptions in the Far Field: Volcanoes and Geysers. *Annual Review of Earth and Planetary Science*, 34, 263-291.

[65] Svensen, H., Jamtveit, B., Planke, S., & Chevallier, L. (2006). Structure and evolution of hydrothermal vent complexes in the Karoo Basin,South Africa. *Journal of the Geological Society*, 163, 671-682.

[66] Skogseid, J., Pedersen, T., Eldholm, O., & Larsen, B. T. (2010). Tectonism and magmatism during NE Atlantic continental break-up: the Vøring Margin. *Storey B C, Alabaster T, Pankhurst, R J, Editors. Magmatism and the Causes of Continental Break-up. London, Geological Society* , 68, 305-320.

[67] Svensen, H., Planke, S., Jamtveit, B., & Pedersen, T. (2003). Seep carbonate formation controlled by hydrothermal vent complexes: a case study from the Vøring Basin, the Norwegian Sea. *Geo-Marine Letters*, 23, 351-358.

[68] Planke, S., Rassmussen, T., Rey, S. S., & Myklebust, R. (2005). Seismic characteristics and distribution of volcanic intrusions and hydrothermal vent complexes in the Vøring and Møre basins. *Dore A, Vining B, Editors. Petroleum Geology: North-West Europe and Global Perspectives. Proceedings of the 6th Geology Conference. London*, 833-844.

[69] Bell, B., & Butcher, H. (2002). On the emplacement of sill complexes: evidence from the Faroe-Shetland Basin. *Jolley D W, Bell B R, Editors. The North Atlantic Igneous Province: Stratigraphy, Tectonic, Volcanic and Magmatic Processes. London Geological Society, Special Publications*, 197, 307-329.

[70] Du, Toit. A. L. (2002). Geological Survey of Elliot and Xalanga, Tembuland. *Annual Report of the Geological Commission Cape of Good Hope for 1903*, 8, 169-205.

[71] Du, Toit. A. L. (1912). Geological Survey of Part of the Stormbergen. *Annual Report of the Geological Commission Cape of Good Hope for 1911*, 16, 112-136.

[72] Gevers, T. W. (1928). The volcanic vents of the Western Stormberg. *Transactions of the Geological Society of South Africa*, 31, 43-62.

[73] Stockley, G. M. (1947). Report on the Geology of Basutoland. Maseru, Authority of the Basutoland Government.

[74] Dingle, R. V., Siesser, W. G., & Newton, A. R. (1983). Mesozoic and Tertiary Geology of Southern Africa. Rotterdam, Balkema.

[75] Jamtveit, B., Svensen, H., Podladchikov, Y. Y., & Planke, S. (2004). Hydrothermal vent complexes associated with sill intrusions in sedimentary basins. *Breitkreuz C, Petford N, Editors. Physical Geology of High-Level Magmatic Systems. London, Geological Society of London, Special Publications*, 234, 233-241.

[76] Grapes, R. H., Reid, D. L., & Mc Pherson, J. G. (1973). Shallow dolerite intrusions and phreatic eruption in the Allan Hills region, Antarctica. New Zealand. *Journal of Geology and Geophysics*, 17, 563-577.

[77] Hanson, R. E., & Elliot, D. H. (1996). Rift-related Jurassic basaltic phreatomagmatic volcanism in the central Transantarctic Mountains: precursory stage to floodbasalt effusion. *Bulletin of Volcanology*, 58, 327-347.

[78] White, J. D. L., & Mc Clintock, M. K. (2001). Immense vent complex marks floodbasalt eruption in a wet, failed rift: Coombs Hills, Antarctica. *Geology*, 29, 935-938.

[79] Zolotukhin, V. V., & Al'mukhamedov, A. I. (1988). Traps of the Siberian Platform. In: Macdougall JD, Editors. *Continental Flood Basalts. Kluwer, Dordrecht*, 273-310.

[80] Svensen, H., Planke, S., Malthe-Sørenssen, A., Jamtveit, B., Myklebust, R., Eidem, T., & Rey, S. S. (2004). Release of methane from a volcanic basin as a mechanism for initial Eocene global warming. *Nature*, 429, 542-545.

[81] Wignall, P. B. (2001). Large igneous provinces and mass extinctions. *Earth-Science Reviews*, 53, 1-33.

[82] Courtillot, V. E., & Renne, P. R. (2003). On the ages of flood basalt events. Comptes Rendus de l'Acade´mie des Sciences. *Geoscience*, 335, 113-140.

[83] Svensen, H., Bebout, G., Kronz, A., Li, L., Planke, S., Chevallier, L., & Jamtveit, B. (2008). Nitrogen geochemistry as a tracer of fluid flow in a hydrothermal vent complex in the Karoo Basin, South Africa. *Geochimica et Cosmochimica Acta*, 72, 4929-4947.

[84] Mazzini, A., Nermoen, A., Krotkiewski, M., Podladchikov, Y., Planke, S., & Svensen, H. (2009). Strike-slip faulting as a trigger mechanism for overpressure release through piercement structures. *Implications for the Lusi mud volcano, Indonesia. Marine and Petroleum Geology*, 26, 1751-1765.

[85] Mazzini, A., Svensen, H., Akhmanov, G. G., Aloisi, G., Planke, S., Malthe-Sorenssen, A., & Istadi, B. (2007). Triggering and dynamic evolution of the LUSI mud volcano, Indonesia. *Earth and Planetary Science Letters*, 261(3-4), 375-388.

[86] Botha, B. V. J., & Theron, J. C. (1966). New evidence for the early commencement of Stormberg volcanism. *Tydskrif vir Natuurwetenskappe*, 7, 469-473.

[87] Coetzee, C. B. (1966). An ancient volcanic vent on Boschplaat 369 in the Bloemfontein district, Orange Free State. *Transactions of the Geological Society of South Africa*, 69, 127-137.

[88] Taylor, N. C. (1970). The volcanic vents of the Stormberg series. *BSc(Hons) project, Rhodes University, Grahamstown, South Africa.*

[89] Seme, U. T. (1997). Diatreme deposits near Rossouw, north Eastern Cape:sedimentology-volcanology and mode of origin. *BSc(Hons) project, Rhodes University, Grahamstown, South Africa.*

[90] Woodford, A. C., Botha, J. F., Chevallier, L., et al. (2001). Hydrogeology of the main Karoo Basin: Current Knowledge and Research Needs. *Water Research Commission Pretoria Report, 860.*

[91] Navikov, L. A., & Slobodskoy, R. M. (1979). Mechanism of formation of diatremes. *International Geology Review*, 21, 1131-1139.

[92] Lorenz, V. (1985). Maars and diatremes of phreatomagmatic origin: a review. *Transactions of the Geological Society of South Africa*, 88, 459-470.

[93] Clement, C. R., & Reid, A. M. The origin of kimberlite pipes: an interpretation based on a synthesis of geological features displayed.

[94] Webb, K. J., Smith, B. H., Paul, J. L., & Hetman, C. M. (2004). Geology of the Victor Kimberlite, Attawapiskat, Northern Ontario, Canada: cross-cutting and nested craters. *Lithos*, 76, 29-50.

[95] Surtee southern African occurrences. (1989). *Ross, J. et al. Editors. Kimberlites and Related Rocks. Geological Society of Australia, Special Publications*, 14, 632-646.

[96] Mc Clintock, M. K., Houghton, B. F., Skilling, I. P., & White, J. D. L. (2002). The volcaniclastic opening phase of Karoo flood basalt volcanism, Drakensberg Formation, South Africa. *EOS Transactions of the American Geophysical Union*, 83(47).

[97] White, J. D. L., & Ross, P. S. (2011). Maar-diatreme volcanoes: A review. *J Volcanol Geotherm Res*, 201, 1-4, 1-29.

[98] Svensen, H., Planke, S., Chevallier, L., Malthe-Sorenssen, A., Corfu, F., & Jamtveit, B. (2007). Hydrothermal venting of greenhouse gases triggering Early Jurassic global warming. *Earth Planet. Sci. Lett.*, 256, 3-4, 554-566.

[99] Svensen, H., Planke, S., Polozov, A. G., Schmidbauer, N., Corfu, F., Podladchikov, Y. Y., & Jamtveit, B. (2009). Siberian gas venting and the end-Permian environmental crisis. *Earth Planet. Sci. Lett.*, 277(3-4), 490-500.

[100] Sloan, E. D., & Koh, C. A. (2008). Clathrate Hydrates of Natural Gases, third ed. CRC Press, New York, Taylor and Francis Group, Publishers.

[101] Smelik, E. A., & King, H. E. (1997). Crystal-growth studies of natural gas clathrate hydrates using a pressurized optical cell. *American Mineralogist*, 82, 88-98.

[102] Paull, C. K., & Dillon, W. (2001). Natural gas hydrates: occurrence, distribution, and detection. *AGU Monogr*, 124, 1-315.

[103] Holder, G. D., Malone, R. D., & Lawson, W. F. (1987). Effects of gas composition and geothermal properties on the thickness and depth of natural-gas-hydrate zone. *Journal of Petroleum Technology*, 1147-1152.

[104] Wright, J. F., Dallimore, S. R., Nixon, F. M., & Duchesne, C. (2005). In situ stability of gas hydrate in reservoir sediments of the JAPEX/JNOC/GSC et al. Mallik 5L-38 gas hydrate production research well. *Paper presented at Scientific Results from the Mallik 2002 Gas Hydrate Production Research Well Program, Mackenzie Delta, Geol. Surv. Canada, Northwest Territories, Canada.*

[105] Dickens, G. R., & Quinby-Hunt, M. S. (1994). Methane hydrate stability in seawater. *Geophys. Res. Lett*, 21, 2115-2118.

[106] Subramanian, S., Kini, R.A., Dec, S.F., & Sloan, E. D. (2000). Evidence of structure II hydrate formation from methane + ethane mixtures. *Chemical Engineering Science*, 55, 1981-1999.

[107] Lu, Z., & Sultan, N. (2008). Empirical expressions for gas hydrate stability law, its volume fraction and mass-density at temperature 273.15 K to 290.15 K. *Geochemical Journal*, 42, 163-175.

[108] Sultan, N., Cochonat, P., Foucher, J. P., & Mienert, J. (2004). Effect of gas hydrates melting on seafloor slope instability. *Marine Geology*, 213, 379-401.

[109] Liu, X., & Flemings, P. B. (2007). Dynamic multiphase flow model of hydrate formation in marine sediments. *Journal of Geophysical Researcher*, 112(B03101), doi: 10.1029/2005JB004227.

[110] Henriet, J. P., & Mienert, J. (1998). Gas hydrates: relevance to world margin stability and climate change. *Geol Soc Spec Publ* , 137.

[111] Holder, G. D., & Bishnoi, P. R. (2000). Gas Hydrates: Challenges for the Future. Ann. N.Y. *Acad. Sci.*, 912.

[112] Kennett, J. P., Cannariato, K. G., Hendy, I. L., & Behl, R. J. (2007). Carbon isotopic evidence for methane hydrate stability during Quaternary Interstadials. *Science*, 288, 128-133.

[113] Brooks, J. M., Cox, H. B., Bryant, W. R., Kennicutt, M. C., Mann, R. G., & Mc Donald, T. J. (1986). Association of gas hydrates and oil seepage in the Gulf of Mexico. *Organic Geochemistry*, 10, 221-234.

[114] Severinghaus, J. P., Sowers, T., Brook, E. J., Alley, R. B., & Bender, M. L. (1998). Timing of abrupt climate change at the end of the Younger Dryas interval from thermally fractionated gases in polar ice. *Nature*, 391, 141.

[115] Reagan, M. T., & Moridis, G. J. (2007). Oceanic gas hydrate instability and dissociation under climate change scenarios. *Geophysical Research Letters*, 34(L22709), doi: 10.1029/2007GL031671.

[116] Kennett, J. P., Cannariato, K. G., Hendy, L. L., & Behl, R. J. (2002). Methane Hydrates in Quaternary Climate Change: The Clathrate Gun Hypothesis. *Spec. Publ., 54, AGU, Washington, D. C.*

[117] Haq, B. U. (1998). Natural gas hydrates: searching for the long-term climatic and slope-stability records. *Geological Society, London, Special Publications*, 137, 303-318, doi:10.1144/GSL.SP.1998.137.01.24.

[118] Shipley, T. H., Houston, M. H., Buffler, R. T., et al. (1979). Seismic reflection evidence for widespread occurrence of possible gas-hydrate horizons on continental slopes and rises. *AAPG Bull*, 63, 2204-2213.

[119] Shipley, T. H., & Didyk, B. M. (1982). Occurrence of methane hydrates offshore southern Mexico. *Watkins JS, Moore JC, et al., Editors. Init. Repts. DSDP, 66. Washington, U.S. Govt. Printing Office*, 547-556.

[120] Minshull, T. A., Singh, S. C., Westbrook, G. K., et al. (1994). Seismic velocitystructure at a gas hydrate reflector,offshore western Colombia, from full waveform inversion. *J. Geophys. Res.*, 99, 4715-4733.

[121] Sain, K., Minshull, T. A., Singh, S. C., & Hobbs, R. W. (2000). Evidence for a thick free gas layer beneath the bottom simulating reflector in the Makran accretionary prism Marine Geology. 164, 37-51.

[122] Tinivella, U., & Accaino, F. (2000). Compressional velocity structure and Poisson's ratio in marine sediments with gas hydrate and free gas by inversion of reflected and refracted seismic data (South Shetland Islands, Antarctica). *Mar Geol*, 164, 13-27.

[123] Dai, J,., Xu, H,., Snyder, F,., & Dutta, N. (2004). Detection and estimation of gas hydrates using rock physics and seismic inversion: Examples from the northern deepwater Gulf of Mexico The Leading Edge. .

[124] Chand, S., & Minshull, T. A. (2003). Seismic constraints on the effects of gas hydrate on sediment physical properties and fluid flow: a review. *Geofluids*, 3, 1-15.

[125] Zillmer, M. (2006). A method for determining gas-hydrate or free-gas saturation of porous media from seismic measurements. *Geophysics*, 71, 21-32.

[126] Singh, S. C., Minshull, T. A., & Spence, G. D. (1993). Velocity structure of a gas hydrate reflector. *Science*, 260, 204-207.

[127] Tinivella, U. (2002). The seismic response to overpressure versus gas hydrate and free gas concentration. *J Seismic Explor*, 11, 283-305.

[128] Tinivella, U. (1999). A method to estimate gas hydrate and free gas concentrations in marine sediments. *Bollettino di Geofisica Teorica ed Applicata*, 40, 19-30.

[129] Tinivella, U., & Carcione, M. (2001). Estimation of gas-hydrate concentration and free-gas saturation from log and seismic data. *The Leading Edge*, 20.

[130] Ledley, T. S., Sundquist, E. T., Schwartz, S. E., Hall, D. K., Fellows, J. D., & Killeen, T. L. (1999). Climate Change and Greenhouse Gases. EOS, Transactions of the American Geophysical Union. 80, 453-458.

[131] Nakano, S., Moritoki, M., & Ohgaki, K. (1998). High-Pressure Phase Equilibrium and Raman Microprobe Spectroscopic Studies on CO2 Hydrate System. J. Chem. Eng. Data; 43, 807-810.

[132] Lackner, K. S. (2003). A Guide to CO2 Sequestration. *Science*, 300, 1677-1678, doi: science.1079033.

[133] Chadwick, R. A., Arts, R., & Eiken, O. 4D seismic quantification of a growing CO2 plume at Sleipner, North Sea. *Geological Society, London, Petroleum Geology Conference series*, 6, 1385-1399, doi:.

[134] Conti, S., & Fontana, D. (2011). Possible Relationships between Seep Carbonates and Gas Hydrates in the Miocene of the Northern Apennines. *Journal of Geological Research*, 20727, doi:10.1155/2011/920727.

[135] Liu, X., & Flemings, P. B. (2007). Dynamic multiphase flow model of hydrate formation in marine sediments. *Journal of Geophysical Research*, 112: B03101, doi: 10.1029/2005JB004227.

[136] Schowalter, T. T. (1979). Mechanics of secondary hydrocarbon migration and entrapment. *American Association of Petroleum Geologists Bulletin*, 63, 723-760.

[137] Milkov, A. V., Claypool, G. E., Lee-J, Y., & Sassen, R. (2005). Gas hydrate systems at Hydrate Ridge offshore Oregon inferred from molecular and isotopic properties of hydrate-bound and void gases, Geochim. *Cosmochim. Acta*, 69, 1007-1026.

[138] Trehu, A. M., Flemings, P. B., Bangs, N. L., Chevallier, J., Gra`cia, E., Johnson, J. E., Liu-S, C., Liu, X., Riedel, M., & Torres, . (2004). Feeding methane vents and gas hydrate deposits at south Hydrate Ridge. *Geophys. Res. Lett.*, 31: L23310, doi: 10.1029/2004GL021286.

[139] Flemings, P. B., Liu, X., & Winter, W. (2003). Critical pressure and multiphase flow in Blake Ridge gas hydrates. *Geology*, 31, 1057-1060.

[140] Gorman, A. R., Holbrook, W. S., Hornbach, M. J., & Hackwith, K. L. (2002). Migration of methane gas through the hydrate stability zone in a low-flux hydrate province. *Geology*, 30, 327-330.

[141] Hornbach, M., Saffer, D. M., & Holbrook, W. S. (2003). Critically pressured free-gas reservoirs below gas-hydrate provinces. *Nature*, 427, 142-144.

[142] Nimblett, J., & Ruppel, C. (2003). Permeability evolution during the formation of gas hydrates in marine sediments. *J. Geophys. Res.*, 108(B9), 2420, doi: 10.1029/2001JB001650.

[143] Bouriak, S., Vanneste, M., & Saoutkine, A. (2000). Inferred gas hydrates and clay diapirs near the Storegga slide on the southern edge of the Voring Plateau, offshore Norway. *Mar Geol*, 163, 125-148.

[144] Buenz, S., & Mienert, J. (2004). Acoustic imaging of gas hydrate and free gas at the Storegga Slide. *J. Geophys. Res*, 109, B04102, doi:10.1029/2003JB002863.

[145] Ginsburg, G. D., Ivanov, V. L., & Soloviev, V. A. (1984). Natural gas hydrates of the World's Oceans. In: Oil and gas content of the World's Oceans. PGO Sevmorgeologia in Russian)., 141-158.

[146] Ginsburg, G. D., Kremlev, A. N., Grigor'ev, M. N., Larkin, G. V., Pavlenkin, A. D., & Saltykova, N. A. (1990). Filtrogenic gas hydrates in the Black Sea (twenty-first voyage of the research vessel Evpatoria). *Sov. Geol. Geophys*, 31, 8-16.

[147] Limonov, A. F., Woodside, J. M., & Ivanov, M. K. (1994). Mud volcanism in the Mediterranean and Black Seas and shallow structure of the Eratosthenes Seamount. *Initial results of the geological and geophysical investigations during the Third UNESCO-ESF "Training Through Research" Cruise of RV Gelendzhik, June-July 1993. UNESCO Rep. Mar. Sci.*, 64.

[148] Woodside, J. M., Ivanov, M. K., & Limonov, A. F. (1997). Neotectonics and fluid flow through seafloor sediments in the Eastern Mediterranean and Black Seas- Parts I and II. *IOC Tech. Ser.*, 48.

[149] Mac, Donald. R. I., Sager, W. W., & Peccini, M. B. (2003). Gas hydrate and chemosynthetic biota in mounded bathymetry at mid-slope hydrocarbon seepsNorthern Gulf of Mexico. *Marine Geology*, 198, 133-158.

[150] Judd, A., Davies, G., Wilson, J., Holmes, R., Baron, G., & Bryde, I. (1997). Contribution to atmospheric methane by natural seepages on the UK continental shelf. *Mar. Geol*, 137, 165-189.

[151] Lance, S., Henry, P., Le Pichon, X., Lallemant, S., Chamley, H., Rostek, F., & Faugeres-C, J. (1998). Submersible study of mud volcanoes seaward of the Barbados accretionary wedge: sedimentology, structure and rheology. *Mar. Geol*, 145, 255-292.

[152] Cronin, B. T., Ivanov, M. K., Limonov, A. F., Egorov, A., Akhmanov, G. G., Akhmetjanov, A. M., & Kozlova, E. (1997). Shipboard Scientific Party TTR-5, New discoveries of mud volcanoes on the Eastern Mediterranean Ridge. *J. Geol. Soc.*, 154, 173-182.

[153] Murton, B. J., & Biggs, J. (2003). Numerical modelling of mud volcanoes and their flows using constraints from the Gulf of Cadiz. *Marine Geology*, 195, 223-236.

[154] Soderberg, P., & Floden, T. (1991). Pockmark developments along a deep crustal structure in the northern Stockholm Archipelago, Baltic Sea. *Beitr. Meereskd*, 62, 79-102.

[155] Soderberg, P., & Floden, T. (1992). Gas seepages, gas eruptions and degassing structures in the seafloor along the Stromma tectonic lineament in the crystalline Stockholm Archipelago, east Sweden. *Cont. Shelf Res.*, 12, 1157-1171.

[156] Cohen, H. A., & Mc Clay, K. (1996). Sedimentation and shale tectonics of the northwestern Niger Delta front. *Mar. Pet. Geol.*, 13, 313-328.

[157] Bouma A.H., Roberts H.H. (1990). Northern Gulf of Mexico continental slope. *Geo-Mar. Lett.*, 10, 177-181.

[158] Langseth, M. G., Westbrook, G. K., & Hobart, A. (1988). Geophysical survey of mud volcano seaward of the Barbados Ridge Complex. *J. Geophys. Res.*, 93, 1049-1061.

[159] Martin, J. B., Kastner, M., Henry, P., Le Pichon, X., & Lallemant, S. (1996). Chemical and isotopic evidence for sources of fluid in a mud volcano field seaward of the Barbados accretionary wedge. *J. Geophys. Res.*, 101, 20325-20345.

[160] Heggland, R., & Nygaard, E. (1998). Shale intrusions and associated surface expressions-examples from Nigerian and Norwegian deepwater areas: proceedings Offshore Technology Conference. *Houston, TX*, 1, 111-124.

[161] Corthay, J. E. (1998). Delineation of a massive seafloor hydrocarbon seep, overpressured aquifer sands, and shallow gas reservoir, Louisiana continental slope. *Proceedings Of Offshore Technology Conference.Houston, Texas*, 37-56.

[162] Neurauter, T. W., & Bryant, W. R. (1998). Gas hydrates and their association with mud diapirs/mud volcanoes on the Louisiana continental slope. *Proceedings of Offshore Technology Conference, Houston, Texas*, 1, 599-607.

[163] Milkov, A. V., Lee-J, Y., Borowski, W. S., Torres, Xu. W., Tomaru, H., Tre´hu, A. M., Schultheiss, P., Dickens, G. R., & Claypool, G. E. (2004). Co-existence of gas hydrate, free gas, and brine within the regional gas hydrate stability zone at Hydrate Ridge (Oregon margin): Evidence from prolonged degassing of a pressurized core. *Earth Planet. Sci. Lett.*, 222, 829-843.

[164] Zatsepina, O. Y., & Buffett, B. A. (1997). Phase equilibrium of gas hydrate: Implications for the formation of hydrate in the deep sea floor. *Geophys. Res. Lett.*, 24, 1567-1570.

[165] Tomkeieff, S. I. (1983). Dictionary of Petrology. Wiley, New York.

[166] Ben-Avraham, Z., Smith, G., Reshef, M., & Jungslager, E. (2002). Gas hydrate and mud volcanoes on the southwest African continental margin off South Africa. *Geology (Boulder)*, 30, 927-930.

[167] Jungslager, E. H. A. (1999). Petroleum habitats of the South Atlantic margin. *Cameron NR, Bate RH, Clure VS. (ed.) The Oil and Gas Habitats of the South Atlantic, Spec. Publ. geol. Soc. Lond*, 153-168.

[168] Tinivella, U., Accaino, F., & Camerlenghi, A. (2002). Gas hydrate and free gas distribution from inversion of seismic data on the South Shetland margin (Antarctica). *Mar Geophys Res*, 23, 109-123.

[169] Klepeis, K., & Lowver, L. A. (1996). Tectonics of the Antarctic-Scotia plate boundary near Elephant and Clarence Islands, West Antarctica. *J. Geophys Res*, 101, 20211-20231.

[170] Loreto, M. F., Tinivella, U., Accaino, F., & Giustiniani, M. (2011). Offshore Antarctic Peninsula Gas Hydrate Reservoir Characterization by Geophysical Data Analysis. *Energies*, 4(11), 39-56, doi:10.3390/en4010039.

[171] Vargas-Cordero, I., Tinivella, U., Accaino, F., Loreto, M. F., & Fanucci, F. (2010). Thermal state and concentration of gas hydrate and free gas of Coyhaique, Chilean Margin (44°30′ S). *Marine and Petroleum Geology*, 27, 1148-1156.

[172] Roberts, H. H. (2001). Fluid expulsion on the Northern Gulf of Mexico continental slope: mud-prone to mineral-prone responses. *Paull CK, Dillon WP. (ed.) Natural gas hydrates: occurrence, distribution, and detection. AGU Monogr 124*, 145-161.

[173] Roberts, H. H., Hardage, Shedd. W. W., & Hunt, J. (2006). Seafloor reflectivity-an important seismic property for interpreting fluid/ gas expulsion geology and the presence of gas hydrate. *Leading Edge*, 25, 620-628.

[174] Milkov, A. V., & Sassen, R. (2002). Economic geology of offshore gas hydrate accumulations and provinces. *Marine and Petroleum Geology*, 19, 1-11.